10

CRM
SERIES

Centro
di Ricerca
Matematica
Ennio De Giorgi

Probability, Uncertainty and Rationality

edited by
Hykel Hosni
Franco Montagna

EDIZIONI
DELLA
NORMALE

FONDAZIONE
MONTE DEI PASCHI
DI SIENA

With the support of the Fondazione
Monte dei Paschi di Siena

© 2010 Scuola Normale Superiore Pisa

ISBN: 978-88-7642-347-5

Contents

4. Rationality   243

Introduction

There is at present wide consensus on the idea that any reasonable attempt to grasp the logical basis of rationality needs taking *uncertainty* and *sociality* very seriously. First of all, a substantial part of rational agents' reasoning is dotted with various sources of uncertainty which make their 'knowledge about the world' precarious at best. Secondly, agents tend to exercise their rationality in the public arena. Some of the most interesting aspects of rationality emerge in interactive situations: voting, bargaining, coordinated group action are just a few obvious examples. Indeed it has been extensively argued in philosophy and linguistics that sociality is a precondition for rational behaviour *tout court*.

From a logical point of view, that is, granting that concepts and tools from logic are essential in our endeavour to capture rational reasoning, uncertainty and sociality lead us naturally to seek extensions of classical logics. Just a little bit of scratching below the surface reveals that an intriguing virtuous circle is in place here. While logic lends its methods, techniques and ideas to the investigation of rationality, the practical problems which arise in modelling rational behaviour, especially in the social sciences, motivate logicians to develop more refined logical formalisms. This is why *non classical logics* play such a fundamental role in the construction of formal models of rationality, as witnessed by the ever increasing number of conferences and publications dedicated to the topic.

This volume is meant as a contribution to this line of research. It collects a selection of papers which have been presented and discussed during two events held in November 2009: *Probability, Uncertainty and Rationality*, which took place at the *Certosa di Pontignano*, Siena and *Logical Foundations of Rational Interaction*, which took place at the Centro di Ricerca Matematica 'E. De Giorgi', Pisa. Both events have been organized by the *Logical fOundations of Rational Interaction* (LORI) group[1]

---

[1] http://www.crm.sns.it/lori

with the explicit aim of bringing together established scholars as well as young researchers who share a manifold interest in the themes featuring in the title of this volume: Probability, Uncertainty and Rationality.

While most of the papers collected here can be viewed as touching on all those topics, we have grouped the contributions into four sections: *Foundations*, *Probability*, *Uncertainty* and *Rationality*. This is clearly not meant to suggest that the four topics are, or should be taken as, independent. The opposite is true. Indeed, the intended purpose of this arrangement is that of highlighting the way in which the individual papers contribute towards covering the multifaceted connection between rationality and logic.

Foundations

For over two millennia, logic, sound argumentation and rational reasoning have been one and the same. The birth of modern 'mathematical logic', while giving the subject full mathematical citizenship, caused, as a side effect, a severe restriction of the scope of logical investigations by confining them to the deductive aspect of mathematical reasoning. It is pretty much a fact that the construction of what we now call 'classical logic', came at the expense of the virtuous circle connecting logic and rationality, especially in the form of common sense reasoning. It is no coincidence that parallel to the current interest in non classical logics runs an ever increasing attention for the history of logic, especially ancient and medieval.

Granting this scenario, it is perhaps puzzling to note that some fundamental ideas which recur today in many non classical logics had been laid down by pioneers of classical logic. A clear case in point is given by, among others, Tarski and Lindenbaum whose algebraic approach to logic paved the way for the subsequent development of a number of non classical logics. Mathematical logic, so enriched, meets rational reasoning again.

A key to this 'puzzle' is offered in ORDERED ALGEBRAS AND LOGIC by George Metcalfe, Francesco Paoli and Constantine Tsinakis. The paper shows how ordered structures, such as Boolean algebras, Heyting algebras, lattice-ordered groups, and MV-algebras prove to be invaluable in putting into a unitary perspective problems which arise in algebra, model theory, proof theory, philosophical logic, computer science and linguistics. The first part of the paper is devoted to a brief yet enlightening historical reconstruction of the roots of ordered algebras and logic which culminates, in the late 1980's, with the celebrated work by Blok and Pigozzi on algebraizable logics. Focussing on the connection between residuated lattices and substructural logics, the authors move on to show the conceptual as well as practical import of this theory, both from an ordered algebra and logic perspective. Finally, the paper illustrates how the development of ordered algebras can shed new light on the classical meta-properties of decidability, interpolation, amalgamation, and completeness.

Probability

As human beings we face, on a daily basis, an overwhelming number of decisions. Some are relatively unimportant, say the shirt we choose to wear, what to have for lunch, or even where to seat in a seminar room. Others can be much more involved and their consequences may be of vital importance to us and to our peers. What is common to the vast majority of our decisions, however, is the fact that we face them in a state of partial *ignorance*. There are simply too many circumstances, past present and future, that we do not know sufficiently well, if not plainly ignore. Yet we can guess, and we often do this correctly. Not all guesses, however, are equally rational. As argued by many authors, from de Finetti and Ramsey to Savage to Cox, an agent's guessing is rational just if it conforms to the laws of probability.

Combining logic with probability is therefore a fundamental step towards reinforcing the connection between logic and rationality. This accounts for the *uncertainty* constraint that we have recalled above. Yet what happens if we consider the rationality of a plurality of interacting agents? In other words, how can we combine uncertainty and sociality under a unitary logico-mathematical framework?

This is precisely the question tackled by George Wilmers in THE SO-CIAL ENTROPY PROCESS: AXIOMATISING THE AGGREGATION OF PROBABILISTIC BELIEFS. The starting point of the investigation is the Paris-Vencovská probability logic which refines the classical idea of coherent degrees of belief through a small set of common sense principles. During the 1990's, this led to an axiomatic characterization of what is known as Maximum Entropy Inference Processes. Building on this work, Wilmers asks how the Paris-Vencovská model may be extended to a social context. That is, how can we aggregate a given number of (rational) individual belief functions into a single probability function which can be said to *represent* the collective belief of the group? By discussing a number of intuitive properties which an ideal aggregation method should satisfy, the author argues in favour of the *Social Entropy Process*. Not only the Social Entropy Process is found to satisfy the intuitive requirements, it also proves to generalize both the Paris-Vencovská model and the main pooling operators which enjoy much popularity in Statistics and Operative Reasearch.

The theory of probability has traditionally been thought of as a mathematical answer to the scarcity of our knowledge. While we can trace back logic in ancient Greece and India, it is not until the Sixteenth Century that we have records of a mathematical calculus of probability. Yet it did not take long for logicians to realize that probability was a natural way of extending logic to account for valid inference under uncertainty. Just recall the title of De Morgan's pioneering 1847 book *Formal Logic; or, the Calculus of Inference, Necessary and Probable*. It is in this spirit that de Finetti put forward his famous slogan to the effect that probability is *logic of the uncertain*.

Now, as a consequence of the foundational work which culminated in Kolmogorov's axiomatization of probability functions, the connection between probability and logic can be looked at formally. From this vantage point, David Makinson and Vincenzo Marra are able to point out that much still needs to be understood about this connection.

In CONDITIONAL PROBABILITY IN THE LIGHT OF QUALITATIVE BELIEF CHANGE, David Makinson begins by acknowledging that Kolmogorov's axiomatisation for probability seems indeed unquestionable. Moreover, it allows us to define *conditional* probability functions by means of the familiar ratio rule. Yet there are several arguments, which are reviewed in the first part of the paper, for rejecting the unqualified acceptance of the ratio rule as the appropriate definition of conditional probability. One key reason for this is the shortcoming that such a definition has with respect to what Makinson calls the *critical zone*. This can be intuitively seen as the situation in which an agent needs conditioning on a coherent, yet zero-probability sentence. Here the ratio definition is obviously inapplicable so that alternative axiomatizations for two-place probability functions need to be identified. The author reviews some classical proposals in that direction, some of which – notably Hosiasson-Lindembaum's – are not particularly well known. Then, building on intuitions and results from the AGM qualitative theory of belief revision, Makinson is able to propose a criterion for comparing the various approaches and to suggest altogether new forms of conditional probability functions.

Vincenzo Marra takes a step back and raises an intriguing question: IS THERE A PROBABILITY THEORY OF MANY-VALUED EVENTS? This question is motivated by the observation that the axiomatization of probability functions presupposes, in some specific sense, an underlying logic of events which happens to be *classical logic*. Let us now reverse the direction and take our favourite non classical logic: Can we build a theory of probability representing an agent's rational degrees of belief in events which are governed by this logic? The first part of the paper is devoted to the historically informed reconstruction of the relation between probability and classical logic. Then Marra moves on to consider the semantic extension of classical logic provided by *many-valued* logics. In his attempt to answer the main question, Marra restricts the scope of the discussion to a one-variable fragment of the many-valued logic known as Gödel logic. The final part of the paper is devoted to illustrating that, even in this relatively simple setting, the conceptual as well as formal machinery required to construct an adequate notion of many-valued probability is far from being understood.

Uncertainty

For limited beings like us the information on which we ground the vast majority of our reasoning falls short of being 'perfect' in several ways. Much epistemological work, some of it heavily motivated by the practical needs of Artificial Intelligence, has been done in order to construct an adequate taxonomy of the forms in which information can be imperfect. First of all it is bound to be *incomplete*: There are many circumstances, which are relevant to our decisions, which we simply do not know. This, as already noted, puts us in a position in which we can at best express our rational *degrees of beliefs* by assigning probabilities to the relevant events. Those events, however, are assumed to have *crisp* conditions of verification: They are either true or false (and never both!). Yet the state of affairs which are of interest in our daily reasoning can often be *vague*, meaning that their conditions of verification can lie somewhere between true and false. As a consequence, it makes certainly good sense to speak of *degrees of truth*.

The papers in the previous section show some of the most interesting aspects related to the development of logics capable of encompassing rational degrees of belief. But where does logic stand with respect to degrees of truth? Logicians have not as yet reached a wide consensus on this matter. While it is certainly the case that many-valued semantics provide the expressive power required to represent degrees of truth, the claim that the best understood many-valued logics succeed in representing rational reasoning under *vagueness* is not beyond dispute. Some perplexity arises in connection with the fact that the mathematical development of many-valued logics is often not coupled by a clear understanding of their intuitive semantics. A clear case in point is Łukasiewicz infinite valued logic, whose formal semantics has been widely appreciated for almost a century now, but whose intuitive semantics has escaped numerous attempts of clarification. Fortunately however, things seem to be changing, as pointed out by Christian Fermüller in ON GILES STYLE DIALOGUE GAMES AND HYPERSEQUENT SYSTEMS. Building on previous works by the author, the paper is centered on a natural dialogical interpretation for infinite-valued Łukasiewicz logic given via *Giles games*, which essentially involve a betting scheme based on the outcome of physical experiments. After reviewing the basic ideas of dialogical semantics, the paper moves on to show that cut-free proofs in a certain hypersequent system for Łukasiewicz logic can be understood as representing winning strategies in a Giles game for Łukasiewicz logic at an appropriate level of abstraction. This setting has two key applications. First of all, it provides an intuitively compelling semantics for Łukasiewicz logic. Secondly, it shows how (hyper)sequents could provide a useful formal tool for the analysis of rational reasoning under vagueness.

The mathematical investigation of the idea that some forms of uncertainty can be represented by giving formulas intermediate degrees of truth gave rise to what is now known as *Mathematical Fuzzy Logic* or *Fuzzy logic in a strict sense*. In his book *Metamathematics of Fuzzy Logic*, Hájek proposed the interpretation of logical conjunction as a continuous t-norm, and of implication as its residuum. A few years later, Esteva and Godo realized that *left* continuity is sufficient to guarantee the existence of an implication, and hence, to obtain a reasonable fuzzy logic. As a consequence there has been a substantial body of work on logics of left-continuous t-norms. Some of these works highlighted an interesting connection between left-continuous t-norms and substructural logics. In POSET REPRESENTATION FOR FREE RDP-ALGEBRAS, Diego Valota takes forward this line of research by focussing on the logic known as *RDP*. In the many-valued logic literature, representation results are certainly of considerable interest, not least because of their mathematical beauty. After Mundici's categorical equivalence, representation results were obtained for Gödel logic. This paper extends such results to the logic RDP.

UNCERTAINTY, INDETERMINACY AND FUZZINESS: A PROBABILISTIC APPROACH, by Martina Fedel, is an attempt to provide a uniform mathematical account of various forms of reasoning under imperfect information. The starting point of the paper is de Finetti's definition of rational degrees of belief in terms of coherent betting quotients, via the so-called Dutch Book method. Fedel then reviews two distinct generalizations of de Finetti's approach which have been discussed in the literature. The former generalizes the Dutch Book method to encompass many-valued events, so that the betting scheme can be extended to *fuzzy* events. The latter consists in relaxing the requirement that any (crisp) event should be given a single-pointed probability value, extending the betting scheme so as to allow for probability intervals. Despite the rather distant motivations which lie in the background of those two traditions, Fedel is able to combine the two ideas in a unitary framework for imprecise probabilities over fuzzy events. The key step in doing so consists in relaxing the requirement that probability should be identified with a *fair* betting quotient. However, as the author discusses in the last part of the paper, the resulting framework seems to depart in some important respects from the standard theory of imprecise probabilities.

Rationality

Most formal accounts of rationality have a *normative* or prescriptive content: They provide us with a logico-mathematical account of what an agent *should* do in order to behave rationally. Yet, when the prescriptions of the theory are compared to the actual behaviour of real agents, deviations from the theory constitute the norm rather than the exception. Granting this, it is certainly not the case that normative theories loose their interest as a consequence of negative empirical feedback. Take, for example, the case of 'democratic' elections. It is a mathematical fact, supported by centuries of unquestionable empirical evidence, that no real voting system can ever attain the ideal democratic standards of fairness and equity. This is, however, hardly an argument for rejecting democracy as a normative ideal. Let us now take the opposite side. It is just as inappropriate to neglect empirical evidence altogether: If a theory of rationality proves to be completely incapable of predicting people's behaviour, it should rightly be dismissed. There is therefore a clear tension between normative and descriptive accounts of rationality. The two papers featuring in this section, despite their differences in methods and approaches, tackle this fundamental question.

Since the pioneering work done by Hintikka, epistemic logics have been heavily criticized for their excessive idealizations concerning the logico-deductive ability of epistemic agents. Put very simply, if an agent knows a sentence, then she is required to know all the (infinitely many) logical consequences of that sentence – an obviously unreasonable requirement which points to a key aspect of the so-called *problem of logical omniscience*. After the revival of interest in epistemic logics which arose in connection with the developments of artificial multi-agent systems, what used to be just a piece of intellectual discontent turned into a daunting practical problem. One radical solution to the problem would be to drop the requirement of deductive closure altogether, but as already argued in detail by Hintikka in 1962, this would amount to the proverbial throwing away of the baby with the bath water. In TRACTABLE DEPTH-BOUNDED LOGICS AND THE PROBLEM OF LOGICAL OMNISCIENCE Marcello D'Agostino brings his previous work on tractable logics to bear on the problem of logical omniscience. The key idea consists in characterizing a class of logical deductions which are computationally feasible and then imposing deductive closure only with respect to those. The paper opens by reviewing some fundamental aspects of the problem of logical omniscience which show the inapplicability of the radical solution. Then D'Agostino recalls his *Tractable Depth-Bounded Logics*, which can be pictured intuitively as layers of logics of increasing complexity. All those layers are tractable, with the base layer, called *the Boolean Logic of depth 0* admitting of a quadratic decision procedure. The complexity grows with subsequent layers, whose infinite union coincides with classical logic. Unlike comparable works concerned with the issue of feasibility, Tractable Depth-Bounded Logics, have the remarkable feature of imposing no restriction to any particular syntactic fragment of classical logic. As a consequence, they enjoy a number of key model theoretic and proof theoretic properties, which are discussed in detail in the last part of the paper.

In RATIONAL BEHAVIOUR AT TRUST NODES Hykel Hosni and Silvia Milano discuss the empirical feedback on the normative theory of rationality in the specific case of rational decisions involving trust. Trust has long been acknowledged as a fundamental ingredient in economic interactions, yet its formalization has proved to be particularly difficult. Indeed, there is a rather straightforward argument based on backward induction which can be used to show that in extensive form games, rational agents should never trust. This is at odds with the substantial empirical evidence which shows how real people, given certain conditions, show a clear inclination to trust other parties. Starting from a simple empirical example, the authors identify the essential ingredients of beliefs about trust. It is argued that *informal institutions*, *reputation* and *cheap talk* are the key elements in determining an agent's behaviour at what the authors call *trust nodes*. The last part of the paper is devoted to the analysis of the empirical evidence on the trust game, which is found to support the view according to which informal institutions play fundamental role in determining trusting behaviour.

Acknowledgements

The programme committee for the two events from which this volume originates included, besides the editors, Mariano Giaquinta (Scuola Normale Superiore, Pisa), Stefano Marmi (Scuola Normale Superiore, Pisa), Massimo Mugnai (Scuola Normale Superiore, Pisa) and Daniele Mundici (Università di Firenze). Without their work this volume would have not been possible. Thanks also to the the referees whose thorough comments resulted, on many occasions, in significant improvements to the chapters of this book. Our gratitude also goes to the secretaries of the Centro di Ricerca Matematica 'E. De Giorgi', whose help proved invaluable in settling many practicalities. Finally, we would like to thank Giovanni Casini for his help in the preparation of some manuscripts and Luisa Ferrini for her outstanding management of the production phase.

Hykel Hosni and Franco Montagna
Pisa and Siena
28 January 2010

Workshops Participants

Matteo Bianchi (Università degli Studi di Milano), Riccardo Bruni (Università di Firenze), Olivia Caramello (Scuola Normale Superiore), Giovanni Casini (Scuola Normale Superiore), Eugenio Civardi (Scuola Normale Superiore), Pietro Codara (Università degli Studi di Milano), Marcello D'Agostino (Università di Ferrara), Martina Fedel (Università di Siena), Christian Fermüller (Technische Universität Wien), Anna Rita Ferraioli (Università di Salerno), Tommaso Flaminio (Università di Siena), Andrea Giampiccolo (Scuola Normale Superiore), Lluis Godo Lacasa (IIIA - CSIC, Barcelona), Luigi Guiso (European University Institute), Ioana Leustean (University of Bucharest), Riccardo Komesar (Università di Torino), Roberto Mana (Scuola Normale Superiore), John Mandereau (Università di Pisa), David Makinson (LSE, London), Vincenzo Marra (Università degli Studi di Milano), Silvia Milano (Scuola Normale Superiore), Luigi Luini (Università di Siena), Lorenzo Luperi Baglini (Università di Siena), Luisa Peruzzi (Università di Firenze), Gerardo Rescigno (Monte dei Paschi), Giacomo Sillari (University of Pennsylvania), Luca Spada (Università di Salerno), Umberto Straccia (ISTI-CNR, Pisa), Constantine Tsinakis (Vanderbilt University), Diego Valota (Università degli Studi di Milano), Giorgio Venturi (Scuola Normale Superiore), George Wilmers (University of Manchester).

Authors' affiliations

MARCELLO D'AGOSTINO – Dipartimento di Scienze Umane, Università di Ferrara
dgm@unife.it

MARTINA FEDEL – Dipartimento di Scienze Matematiche e Informatiche, Università di Siena
fedel@dm.unipi.it

CHRISTIAN G. FERMÜLLER – Technische Universität Wien
chrisf@logic.at

HYKEL HOSNI – Centro di Ricerca Matematica "E. De Giorgi", Scuola Normale Superiore, Pisa
hykel.hosni@sns.it

DAVID MAKINSON – Department of Philosophy, Logic and Scientific Method, London School of Economics
david.makinson@gmail.com

VINCENZO MARRA – Dipartimento di Informatica e Comunicazione, Università di Milano
vincenzo.marra@unimi.it

GEORGE METCALFE – Mathematics Institute, University of Bern
george.metcalfe@math.unibe.ch

SILVIA MILANO – Classe di Lettere, Scuola Normale Superiore, Pisa
s.milano@sns.it

FRANCESCO PAOLI – Facoltà di Scienze della Formazione, Università di Cagliari
paoli@unica.it

CONSTANTINE TSINAKIS – Department of Mathematics, Vanderbilt University
constantine.tsinakis@vanderbilt.edu

DIEGO VALOTA – Dipartimento di Scienze dell'Informazione, Università di Milano
valota@dsi.unimi.it

GEORGE WILMERS – School of Mathematics, University of Manchester
George.Wilmers@manchester.ac.uk

1

FOUNDATIONS

Ordered algebras and logic

George Metcalfe, Francesco Paoli and Constantine Tsinakis

This work is dedicated to Bjarni Jónsson on the occasion of his 90th birthday.

1 Introduction

Ordered algebras such as Boolean algebras, Heyting algebras, lattice-ordered groups, and MV-algebras have long played a decisive role in logic, both as the models of theories of first (or higher) order logic, and as algebraic semantics for the plethora of non-classical logics emerging in the twentieth century from linguistics, philosophy, mathematics, and computer science. Perhaps only in recent years, however, has the full significance of the relationship between ordered algebras and logic begun to be recognized and exploited. The first pioneering and revelatory step was taken already by Tarski and Lindenbaum back in the 1930's, who showed that despite their different origins and motivations, Boolean algebras and propositional classical logic may be viewed in a certain sense as two sides of the same coin.

A crucial element in the development of this relationship was Tarski's definition of an abstract concept of logical consequence. He was guided by the intuition that a consequence relation should specify when a single formula (the conclusion) follows from a set of formulas (the premises), satisfying only three natural constraints: (i) every formula follows from itself (*reflexivity*); (ii) whatever follows from a set of premises also follows from any larger set of premises (*monotonicity*); and (iii) whatever follows from consequences of a set of premises also follows from the set itself (*cut*).

Prototypical sources of consequence relations are axiomatic logical calculi, via the standard definition of formal proof: a formula α is a consequence of a set of formulas Γ (relative to the calculus C) when α is provable in C using assumptions in Γ. However, in Tarski's work there is also the implicit idea that behind each *class of algebras* lies hidden (at least) one consequence relation. In fact, every class $\mathcal{K}$ of algebras has an associated *equational consequence relation* $\vdash_{Eq(\mathcal{K})}$ that holds between a set of equations E and a single equation ε (in the appropriate language) if and only if ε is rendered true by every evaluation for $\mathbf{A} \in \mathcal{K}$ that makes all equations of E true.

In itself, this is not a consequence relation in Tarski's sense – it is defined on equations rather than on formulas – but is readily seen to enjoy similar properties of reflexivity, monotonicity, and cut. A few stipulations, moreover, ensure that it can be converted into a Tarskian consequence relation. First, an equation $\alpha \approx \beta$ can be viewed as a pair (α, β) of formulas. Second, a map τ from formulas to sets of equations should be specified that defines for every $\mathbf{A} \in \mathcal{K}$, a "truth set" T in the following sense: the algebra $\mathbf{A} \in \mathcal{K}$ satisfies $\tau(\alpha)$ just in case every valuation of α on $\mathbf{A}$ maps it to a member of T (intuitively: the meaning of α in $\mathbf{A}$ belongs to the set of "true values" and hence is true). This immediately yields a consequence relation in Tarski's sense, defined by

$$\Gamma \vdash_{\mathcal{K}}^{\tau} \alpha \quad \text{iff} \quad \{\tau(\beta) \mid \beta \in \Gamma\} \vdash_{Eq(\mathcal{K})} \tau(\alpha).$$

Things, of course, become interesting when this consequence relation happens to coincide with the deducibility relation of a well understood logical calculus; this is the situation precisely for the variety $\mathcal{BA}$ of Boolean algebras and classical propositional logic. What Tarski (developing an idea by Lindenbaum) showed, is that the consequence relation $\vdash_{HCL}$ extracted from the axiomatic calculus HCL for classical propositional logic is exactly $\vdash_{\mathcal{BA}}^{\tau}$ for $\tau(\alpha) = \{\alpha \approx 1\}$. In other words, Boolean algebras form an *algebraic semantics* for classical propositional logic.

Although Tarski's approach was successfully extended to non-classical logics in succeeding years, becoming a standard tool for their investigation, it took until the 1980's and the work of Blok and Pigozzi for a formal account of this general correspondence between logics and classes of algebras to appear. In particular, Blok and Pigozzi pointed out that the algebraic semantics of a given logic need not be unique; moreover, the link provided by this relation is weak in several respects. They introduced a stronger notion of *equivalent algebraic semantics* for a logic, calling a logic *algebraizable* if it has a class $\mathcal{K}$ of algebras (unique, if $\mathcal{K}$ is a quasivariety) as equivalent algebraic semantics. This account was subsequently extended in many directions, and indeed now forms the basis for the active area of research known as *abstract algebraic logic*.

One particularly satisfying development, due to Jónsson, is a generalization of the approach away from focussing only on formulas and equations, to a theory of *equivalent consequence relations*. A limitation of the Blok-Pigozzi perspective, in fact, is its exclusive reliance on the Tarskian concept of consequence relation, which only makes sense of formulas. As we have seen, a notion of consequence can be "extracted" from classes of algebras, but only via a trick consisting of reducing their equational consequence relation to a consequence relation on formulas. Jónsson sidesteps this problem by suggesting an even more abstract concept of consequence relation, defined on any *set X* whatsoever. Examples of this general concept are Tarskian consequence relations (*X* is a set of formulas), equational consequence relations (*X* is a set of equations), consequence relations from Gentzen systems (*X* is a set of sequents). According to this perspective, given a class of algebras $\mathcal{K}$, we do not need to "convert" $\vdash_{Eq(\mathcal{K})}$ into a consequence relation $\vdash_{\mathcal{K}}^{\tau}$ on formulas: $\vdash_{Eq(\mathcal{K})}$ *is already* a consequence relation on equations. In the interests of space, this development will not be further considered in the sequel, but the reader may consult [12] for additional details and [54] for an even more abstract treatment of consequence relations.

Narrowing our scope for the purposes of this article, we will focus here on the relationship between *substructural logics* and *residuated lattices*. On the one hand, substructural logics encompass many important non-classical logics such as the full Lambek calculus, linear logic, relevance logics, and fuzzy logics. On the other hand, residuated lattices, as well as providing algebraic semantics for these logics, also feature in areas interesting from the order-algebraic perspective such as the theory of lattice-ordered groups, vector lattices (or Riesz spaces), and abstract ideal theory. Intriguingly, in tackling problems from these fields, methods from both algebra and logic seem to be essential. In particular, algebraic methods have been used to address completeness problems for Gentzen systems, while these systems have themselves been used to establish decidability and amalgamation properties for classes of algebras.

The first aim of this survey article is, through Sections 2 and 3, to briefly trace the distinct historical roots of ordered algebras and logic, culminating with the theory of algebraizable logics, based on the pioneering work of Lindenbaum and Tarski and Blok and Pigozzi, that demonstrates the complementary nature of the two fields. The second aim is to explain and illustrate the usefulness of this theory for ordered algebra and logic in the context of residuated lattices and substructural logics, described in Sections 4 and 5, respectively. In particular, we will explain in Section 6, how completions on the ordered algebra side, and Gentzen systems on the logic side, are used to address properties such as decidability, interpolation and amalgamation, and completeness.

We will assume that the reader is familiar with the basic facts, definitions, and terminology from universal algebra and lattice theory. In particular, the notions *partially ordered set* or *poset* for short, *lattice*, and *algebra* are central to this paper, as are the concepts of *congruence rela-tion*, *homomorphism*, and *variety*. For an introduction to universal alge-bra, the reader may wish to consult [24, 65], or [99], while any of [7, 38], or [64] would serve as a suitable lattice theory reference. A comprehen-sive treatment of residuated lattices and propositional substructural logics is presented in [52].

2 The logic of mathematics

No survey on the relationship between algebra and logic can eschew the more general issue of an assessment of the role of logic in mathemat-ics, and, for that matter, of the role of mathematics in logic. The very expression "mathematical logic" has repeatedly been recognized as am-biguous, at least to some extent, between an investigation into the logical foundations of mathematical theories and an analysis of logical reason-ing carried out with the aid of mathematical tools. Roughly expressed, "mathematical logic" is ambiguous between "the mathematics of logic" and the "logic of mathematics."[1] In the first part of this paper, we will trace the antecedents of these two approaches by means of a short and necessarily perfunctory historical survey.

2.1 Aristotle and the Stoics

The study of logic as an independent discipline began with Aristotle (384-322 B.C.E.). In his *Organon* ($O\rho\gamma\alpha\nu o\nu$), as is well known, he analyzed the structure of arguments having syllogistic form, providing a canon of valid reasoning that would be virtually undisputed for centuries to come. Despite appearances to the contrary – after all, Aristotelian logic is what the founders of modern mathematical logic wanted to go beyond – it does not seem inappropriate to include Aristotle's contribu-tions among the attempts to carry out a logical analysis of mathematical reasoning. Let us recall, in fact, that he sharply distinguished between the formal development of his general theory of syllogism and its application to scientific discourse on the one hand, carried out in the *Prior Analyt-ics* ($A\nu\alpha\lambda\upsilon\tau\iota\kappa\acute{\alpha}$ $\Pi\rho\acute{o}\tau\epsilon\rho\alpha$) and in the *Posterior Analytics* ($A\nu\alpha\lambda\upsilon\tau\iota\kappa\acute{\alpha}$ $Y\sigma\tau\epsilon\rho\alpha$) respectively, and the analysis of the logical structure of ev-eryday reasoning on the other, reserved instead for the *Topics* and the *Sophistical Refutations* ($T o\pi\iota\kappa o\acute{\iota}$ and $\Sigma o\varphi\iota\sigma\tau\iota\kappa o\acute{\iota}$ $E\lambda\epsilon\gamma\chi o\iota$).

---

[1] This observation is not as trite as it might seem. Kreisel, for one, does not accept the former meaning: "'Mathematical logic' [...] refers to mathematical methods used in logic [...]; in analogy to 'mathematical physics' which means the mathematics of physics (and not the physics of mathe-matics)" [89].

Moreover, it should be observed that, although this may sound blasphemous to a modern ear – syllogistic logic being just a decidable barely expressive fragment of first order logic – Aristotle considered his theory of syllogism completely adequate for the formalization of *all* mathematical inference. Actually, Aristotle even tried to provide a *proof* of this claim in two interesting passages from his *Prior Analytics* (I, 25.41b36-42a32 and I, 23.40b18-41a21).

More precisely, what Aristotle attempts is the following. The definition of syllogism initially given in the *Prior Analytics* is extremely general in nature:

> A syllogism is discourse in which, certain things being stated, something other than what is stated follows of necessity from their being so. I mean by the last phrase that they produce the consequence, and by this, that no further term is required from without in order to make the consequence necessary (I, 1.24b19).

This definition would seem to subsume (although it is of course not limited to) all kinds of mathematical reasoning. Later in the book, however, Aristotle provides a different, more technical, definition of syllogism where precise constraints are introduced both as to the form of the propositions involved and regarding the structure of the inferences themselves. In the above-mentioned passages, Aristotle essentially argues that every syllogism in the first sense is also a syllogism in the second sense. His purported proof is quite obscure and full of holes, relying on a number of controversial assumptions; moreover, Aristotle is forced to constrain somehow his vague initial concept of syllogism in order to compare it with the other one, so much so that it is not at all clear whether all mathematical inferences can still be encompassed by the definition ([34,91,125]). Nonetheless, his inconclusive argument is clear evidence that Aristotle considered his theory of syllogism as describing, among other things, *the* logical structure of mathematical reasoning.

This conviction was explicitly challenged already in the Antiquity. The Stoics, who are as a rule collectively mentioned as the second most important contributors to the development of ancient logic, as well as being credited with the first noteworthy analysis of propositional connexion, drew a sharp distinction between syllogisms (whether Aristotle's categorical syllogisms or their own new brands of propositional syllogisms, hypothetical and disjunctive) and non-formally valid enthymemes, called "unmethodically valid arguments" (λόγοι ἀμεθόδως περαίνοντες). Examples of such arguments, according to the Stoics, abound in concrete mathematical proofs; therefore Aristotle's syllogistics does not provide a complete basis for mathematical reasoning ([8]).

2.2   Leibniz and Wolff

Gottfried Wilhelm Leibniz (1646-1716) is sometimes cited as a forerunner of modern mathematical logic, and even as the thinker who established the discipline long before Boole or Frege. Be this as it may, the importance of Leibniz in the development of logic and his appreciation of logical reasoning as a basis for science are acknowledged and unquestioned. Instead, the contribution of another XVII c. German philosopher, Christian Wolff (1679-1754), is less well-known. Under the influence of Leibniz, he disavowed his own early devaluative views on the role of logic in mathematics and came to embrace a revised version of Aristotle's thesis: the theory of categorical syllogism, as supplemented by the Scholastics, is sufficient to formalize the inferences contained in *all* mathematical proofs. In his *Philosophia Rationalis Sive Logica* (1728), he boldly tried to prove his claim following a two-step strategy:

1. First, he tried to show that some representative examples of mathematical proofs[2] were reducible to a finite sequence of sentences, each of which was either a definition, or an axiom, or a previously established theorem, or could be obtained from preceding sentences through the application of a (categorical, hypothetical, or disjunctive) syllogism or of a one-premise propositional inference.
2. Second, he completed his reduction process with an argument to the effect that hypothetical and disjunctive syllogisms and one-premise inferences can be reformulated as categorical syllogisms.

True to form, Wolff's attempt was far from successful. If we read his analysis carefully, we see that he interspersed his reconstructions of geometrical proofs with "intuitive judgments" (*iudicia intuitiva*), introduced by such phrases as "It is intuitively evident that..." or "Looking at the figure, we intuitively know that...", and drawn from the observation of the figure or from the constructions he had previously carried out. Of course, these "intuitive judgments" do not count as definitions or axioms; hence their presence invalidates the first step of Wolff's strategy from the start (by the way, his second step was just as gappy). Influenced by the traditional Euclidean model of geometrical proof, which provided for a sharp distinction between the logico-deductive part – the

---

[2] Interestingly enough, all the examples of proofs analyzed by Wolff belong to elementary Euclidean geometry, although he was familiar (as witnessed by his teaching syllabi and notes collected in his *Ratio Praelectionum Wolffianarum*) with the latest developments of calculus and algebra. Most probably, only Euclidean geometry met the standards of rigor he deemed necessary for his investigation.

apódeixis ($\dot{\alpha}\pi\acute{o}\delta\varepsilon\iota\xi\iota\varsigma$), or proof proper – and the constructive, synthetic part embodied by the constructions carried out on the figure – the *ékthesis* ($\check{\varepsilon}\kappa\theta\varepsilon\sigma\iota\varsigma$) and the *kataskeué* ($\kappa\alpha\tau\alpha\sigma\kappa\varepsilon\upsilon\acute{\eta}$): ([72]) – Wolff was only concerned with a logical reconstruction of the former, while he regarded the latter as somehow foreign to the body of the proof. In sum: categorical syllogistics may or may not have been sufficient for a formalization of the narrow strictly deductive fragment of a standard Euclidean proof, but of course it was not enough for a formalization of the proof as a whole[3].

2.3 Bolzano

Bernard Bolzano (1781-1848) brought reflection on the logical foundations of mathematics to unprecedented levels of awareness and depth. A mathematician by trade, he sought from the very beginning of his career (e.g., in his *Beyträge zu einer begründeteren Darstellung der Mathematik*, 1810) to establish on a firmer ground the foundations of the mathematical disciplines. In accordance with the time-honored tradition of "doctrine of method" leading from the French XVII c. theorists (Pascal, Descartes, Arnauld) to his more recent antecedents Lambert, Crusius, and Kant, Bolzano believed that logic is instrumental for mathematics in that it serves as a preliminary methodological framework for stating the rules that properly found each mathematical discipline.

The second volume of his monumental work *Wissenschaftslehre* (1837) contains a detailed and not yet fully appreciated development of a powerful logical system, including an analysis of logical consequence, viewed by some as an anticipation of Tarski. In particular, he considers the two relations of *derivability* (*Ableitbarkeit*) and *consecution* (*Abfolge*) among propositions. The former relation is, somewhat surprisingly, a multiple-conclusion one: the propositions $q_1, ..., q_m$ are said to be derivable from the propositions $p_1, ..., p_n$ with respect to the component concepts $a_1, ..., a_r$ iff the following two conditions are satisfied:

1. There exists a sequence of concepts $b_1, ..., b_r$ such that, denoting by $p_i(a_1/b_1, ..., a_r/b_r)$ the result of uniformly substituting in p_i every occurrence of a_j by an occurrence of b_j, all the propositions

$$p_1(a_1/b_1, ..., a_r/b_r), ..., p_n(a_1/b_1, ..., a_r/b_r)$$

are true (*compatibility clause*).

---

[3] According to some Kantian interpreters, Kant's claim to the effect that mathematics is synthetic a priori does not contradict Wolff's thesis that mathematics is based on syllogistic logic: Kant is referring to mathematical proof as a whole, Wolff to the *apódeixis* ([50,77]).

2. Every such sequence of concepts $b_1, ..., b_r$ also makes all the conclusions true, *i.e.*, the propositions

$$q_1(a_1/b_1, ..., a_r/b_r), ..., q_m(a_1/b_1, ..., a_r/b_r)$$

are true.

This notion is meant to formalize several features Bolzano requires of "good" mathematical proofs: the compatibility clause reflects Bolzano's desire, inherited from Arnauld and Kant, to assign direct proofs a higher status than ex absurdo proofs, while the fact that the derivability relation is relativized to a sequence of components – possibly a *proper* subsequence of the sequence of all non-logical concepts contained in the propositions at issue – is an attempt to describe an *enthymematic* consequence relation broad enough to encompass forms of reasoning currently employed in standard mathematical practice (*cf.* the Stoics' "unmethodically conclusive arguments" mentioned above).

The relation of consecution is even more interesting, since it goes some way towards investigating a stronger *causal* concept of consequence. To illustrate the difference between the two notions, consider the following propositions:

(1) In Rome the temperature is higher in August than in January.
(2) In Rome the mercury columns of thermometers are higher in August than in January.

(1) and (2) are derivable from each other (w.r.t. the component concepts "Rome", "August", "January"), but it is only (1) that causally implies (2), not vice versa. According to Bolzano, only the best mathematical proofs are made up by causal inferences, while in other cases there is at most a "transfer of evidence" from the premises down to the conclusions.

Consecution is not formally defined by Bolzano, but only characterized by means of a list of properties. Among such properties there is a remarkable one: if p and q are mathematical propositions and p causally implies q, p cannot be "more complex" than q. In one of the most extraordinary passages of his *Wissenschaftslehre* (Sections 216-221), Bolzano investigates in detail the formal structure of proof trees explicitly taken as mirroring the structure of actual mathematical proofs, and viewed as bottom-up proof processes of "going from a consequence up to its reasons": nodes in the trees correspond to mathematical propositions, and arcs to instances of the consecution relation. Taking advantage of the above-mentioned postulate of non-decreasing complexity, he tries to show that such proof trees must be analytic and therefore finite. His proof turns out to be conclusive only under additional assumptions; however, it remains one of the earliest and most limpid examples of formal analysis of the logical structure of mathematical proofs.

2.4   From Frege to Hilbert

Gottlob Frege (1848-1925) articulated a view of the relation between logic and mathematics that was at once clear-cut and utterly controversial: mathematics *is* logic. Frege, indeed, defended a strong reductionist thesis nowadays known as *logicism*: mathematical concepts can be defined in terms of purely logical concepts, and mathematical principles can be derived from the laws of logic alone. As part of the job needed to prove this claim, he had of course to define numbers in logical terms. Frege did not start from scratch: he built upon the work of Dedekind, Cantor, and Weierstrass, who had managed around 1870 to reduce real numbers to rational numbers (Cantor, *e.g.*, defined real numbers as certain equivalence classes of Cauchy sequences of rationals), and given similar reductions of rational numbers to integers and of integers to natural numbers. What remained then to complete the process was a reduction of natural numbers to logic.

To do this, Frege needed an adequate logical framework. Indeed, one of his greatest accomplishments was the introduction of a logical system that closely resembles an axiom system for second order quantifier logic in the modern sense, appearing first in his early work *Begriffsschrift* (1879) and then, in a more mature formulation, in his two-volume *Grundgesetze der Arithmetik* (1893-1903). Within this framework, he was able to provide a powerful argument for the logicist claim, proceeding via a (supposedly) logical definition of the concept of natural number.

It is well-known that Frege's attempt was not crowned with success. His logic relied on an unrestricted comprehension principle asserting the existence of a set (or a concept, as Frege would have put it) for every open formula with one free variable. This existential claim, apart from casting a shadow on the purely logical nature of his system, was shown to be inconsistent by Russell. The ultimate failure of Frege's program, however, should not obscure its merits and partial achievements. Frege, in fact, succeeded in deriving the Dedekind-Peano axioms for arithmetic in a consistent subsystem of his logic ([20, 73]). Moreover, as a by-product of his foundational work, he laid the groundwork for modern mathematical logic as an investigation into the logical foundations of mathematical theories.

In *Principia Mathematica* (1910-1913), coauthored by Bertrand Russell (1872-1970) and Alfred N. Whitehead (1861-1947), Frege's seminal

A1. $\alpha \to (\beta \to \alpha)$

A2. $(\alpha \to (\beta \to \gamma)) \to ((\alpha \to \beta) \to (\alpha \to \gamma))$

A3. $\alpha \wedge \beta \to \alpha$

A4. $\alpha \wedge \beta \to \beta$

A5. $(\alpha \to \beta) \to ((\alpha \to \gamma) \to (\alpha \to \beta \wedge \gamma))$

A6. $\alpha \to \alpha \vee \beta$

A7. $\beta \to \alpha \vee \beta$

A8. $(\alpha \to \gamma) \to ((\beta \to \gamma) \to (\alpha \vee \beta \to \gamma))$

A9. $(\alpha \to \beta) \to ((\alpha \to \neg \beta) \to \neg \alpha)$

A10. $\alpha \to (\neg \alpha \to \beta)$

A11. $\neg \neg \alpha \to \alpha$

A12. $(\alpha \to \alpha) \to 1$

A13. $1 \to (\alpha \to \alpha)$

A14. $0 \to \neg 1$

A15. $\neg 1 \to 0$

R1. $\dfrac{\alpha \quad \alpha \to \beta}{\beta}$ (Modus ponens)

Table 1. Hilbert-style calculus HCL for classical propositional logic.

work on the codification of logical principles in a formal calculus was improved, both notationally and conceptually. It is Russell and White-head's, not Frege's, formalization of logic that constituted the backdrop against which most logical research in the first thirty years of the last century – including Gödel's completeness and incompleteness theorems – was carried out. Building on this monumental piece of work, by the end of the 1920's David Hilbert (1862-1943) had essentially formulated the modern concepts of logical language, axiomatic calculus, and formal proof. Today, as a matter of fact, axiomatic logical calculi are antono-mastically named *Hilbert-style calculi* in his honor. Later, he also cod-ified in his two-volume *Grundlagen der Mathematik* [76], written with Paul Bernays, what is now considered to be the standard presentation of first order classical logic. Its propositional fragment HCL is reproduced, with a few inessential variants, in Table 1.

A common feature of the logicians we have just mentioned was their insistence on *syntactic* aspects of logic. Actually, it should be said that, prior to Frege, the distinction between syntax and semantics did not make any sense at all, in logic or any other mathematical theory: there was no difference, say, between a provable and a valid sentence in syllogistic logic, or between a consistent axiom system for geometry and a system admitting models. The revolution of the axiomatic method changed this state of affairs once and for all. Following Hilbert, a set of axioms for a mathematical theory was no longer a body of self-evident truths about an intended domain of objects, and proofs were no longer viewed as means to transfer evidence from axioms to theorems. Rather, axioms were thought of as arbitrarily designated sentences which implicitly define their own objects, and proofs as means to ensure that the relations enunciated by theorems hold in every possible domain of entities in which the axioms also hold. In logic, the availability of a rigorous concept of logical calculus made it possible to reduce all the claims concerning truth or validity, which otherwise would have to be checked in an "outer" domain of objects, to provability claims that it was possible to verify *within* the calculus itself.

3 The mathematics of logic

Uncovering the logical structure of mathematical proofs and mathematical theories is not the same as trying to formalize reasoning – on *any* subject matter – by mathematical means. Although the mainstream tradition in early XX c. mathematical logic, leading from Frege and Russell to Hilbert and Gödel, can be categorized under the former heading, outstanding contributions to the shaping of contemporary logic were also made by a second important stream, generally referred to as "algebra of logic", which eventually converged with the first stream into a unique research domain. George Boole (1815-1864) was a pioneer of this approach, while Stanley Jevons (1835-1882), Charles Sanders Peirce (1839-1914) and Ernst Schröder (1841-1902) followed in his footsteps. Two leading figures in the foundational research of the pre-World War Two period, Leopold Löwenheim (1878-1957) and Thoralf Skolem (1887-1963) can also be seen as belonging, at least to some extent, to this tradition.

3.1 The early tradition in the algebra of logic

According to many interpreters, Boole's *Mathematical Analysis of Logic*, published in 1847, and the expanded version of the treatise appearing in 1854 under the title *An Investigation of the Laws of Thought,* mark the official birth of modern mathematical logic. In these milestone volumes, Boole admittedly sought no less than to "investigate the fundamental laws of those operations of the mind by which reasoning is performed, to give expression to them in the symbolical language of a calculus, and upon this foundation to establish the science of Logic and construct its method."

Despite this somewhat bombastic statement of purpose, Boole did not depart from tradition as radically as it might seem, because one of his main concerns – as for his contemporary Augustus De Morgan (1806-1871) – was providing an algebraic treatment of Aristotelian syllogistic logic, which occupies about one third of his *Mathematical Analysis of Logic*. Boole started with the assumption that ordinary logic is concerned with assertions that can be considered assertions about *classes* of objects. He then translated the latter into equations in the language of classes. His approach contained numerous errors, partly due to his insistence that the algebra of logic should behave like ordinary algebra, but offered significant new perspectives.

Although the name "Boolean algebra" might suggest that the inventor of this concept was Boole, there is by now widespread agreement among the scholarly community that he was not. Not that it is always easy to clearly understand what Boole had in mind when working on his calculus of classes: as Hailperin puts it ([70, page 61]),

> If we look carefully at what Boole actually did [...], we find him carrying out operations, procedures, and processes of an algebraic character, often inadequately justified by present-day standards and, at times, making no sense to a modern mathematician. [...] Boole considered this acceptable so long as the end result could be given a meaning.

The reason why Boole's calculus cannot be considered as a first incarnation of a Boolean algebra of sets is the fact that the two operations of *combination* of two classes x and y (written xy in Boole's notation) and of *aggregation* of x and y (written $x + y$) do correspond to intersection and union, respectively, but the latter only makes sense when x and y are disjoint classes. The algebras to be found in Boole's work bear therefore some resemblance to partial algebras in the modern sense, except for the fact that often Boole happily disregarded his disjointness condition throughout his calculations, seemingly finding such a procedure unobjectionable provided the final result did not violate the condition itself.

Boole preceded an array of researchers who tried to develop further his idea of turning logical reasoning into an algebraic calculus. The already mentioned Stanley Jevons, Charles Sanders Peirce, and Ernst Schröder took their cue from Boole's investigations, yet all of them suggested their own improvements and modifications to the work of their predecessor.

Jevons, in particular, was dissatisfied with Boole's choice of primitive set-theoretical operations. He did not like the fact that aggregation was a partial operation, undefined on pairs of classes with nonempty intersection. In his *Pure Logic* (1864), he suggested a variant of Boole's calculus which he proudly advertised as based only on "processes of self-evident meaning and force." He viewed + as a total operation making sense for any pair of classes, and essentially corresponding to set-theoretic union. He also showed that all the expressions he used remained interpretable throughout the intermediate steps of his calculations, thereby overcoming one of the main drawbacks of Boole's work ([69]).

The main contribution to logic of the versatile Peirce – a philosopher, mathematician, and authority on several pure and applied sciences – is usually attributed to the foundation of the algebra of relations, for which Schröder also made significant developments. However, in the 12,000 pages of his published work – rising to an astounding 90,000 if we take into account his unpublished manuscripts – much more can be found. He investigated the laws of propositional logic, discovering that all the usual propositional connectives were definable in term of the single connective NAND; he introduced quantifiers, although, unlike Frege, he did not go so far as to suggest an axiomatic calculus for quantified logic. He conceived of complex and fascinating graphs by which he could represent logical syntax in two or even three dimensions. For all these achievements, however, his impact on logic would not be even remotely comparable to that exerted by Frege ([63]).

Let us finally mention that, contrary to the dominant paradigm of Hilbertian formalism described in Subsection 2.4, adherents to the algebra of logic tradition inherited from Boole a markedly anti-formalistic stance on logic and mathematics: mathematical language was seen as a system of *interpreted* symbols, and semantical notions like validity or satisfiability were accorded priority over their syntactical counterparts. This view had some unfortunate consequences. Skolem, for example, came very close to proving the completeness theorem for first order logic, but refrained from giving it an explicit formulation because he viewed consistency of a system as equivalent *by definition* to satisfiability [25].

3.2 Lindenbaum and Tarski

Roughly at the same time as Hilbert and Bernays wrote what is now considered as the standard presentation of first order classical logic, Emil Leon Post (1897-1954), Jan Łukasiewicz (1878-1956), and Clarence Irving Lewis (1883-1964), among others, introduced the first Hilbert-style calculi for some propositional (many-valued or modal) non-classical logics. If we also count Heyting's calculus for intuitionistic logic, we can see how already in the 1930's classical logic had quite a number of competitors, each one of which tried to capture a different concept of *logical consequence*. But what should count, abstractly speaking, as a concept of logical consequence? In answering this question, the Polish logicians Adolf Lindenbaum (1904-1941) and Alfred Tarski (1901-1983) initiated a confluence of the Fregean and algebra of logic traditions into one unique stream. Lindenbaum and Tarski showed how it is possible to associate in a canonical way, at least at the propositional level, logical calculi (and their attendant consequence relations) with classes of algebras.

To give some idea of their accomplishments, we will subordinate historical accuracy to the needs of a more systematic treatment. As a first step, we will define the general concept of consequence relation along the lines of Tarski's 1936 paper ([132]). A (propositional) *language* over a countably infinite set X, whose members are referred to as *variables*, is a nonempty set $\mathcal{L}$ (disjoint from X), whose members are called *connectives*, such that a nonnegative integer n is assigned to each member c of $\mathcal{L}$. This integer is called the *arity* of c. The set Fm of $\mathcal{L}$ *-formulas* over X is defined as follows:

- *Inductive beginning*: Every member p of X is a formula.
- *Inductive step*: If c is a connective of arity n and $\alpha_1, ..., \alpha_n$ are formulas, then so is $c(\alpha_1, ..., \alpha_n)$.

We will confine ourselves to considering cases where $\mathcal{L}$ is finite; we also observe that most connectives used in logic have arities 0 through 2. In this last case – *i.e.*, for *binary* connectives – the customary infix notation will be employed.

On the algebraic side, the same concept of language can be adopted to specify the symbols denoting the primitive operations of an algebra (or class of algebras). In this case, we sometimes use the phrase "similarity type" or simply "type" in place of "language". If $\mathcal{L} = \{c_1, ..., c_n\}$ is a language over X, then by the inductive definition of formula

$$\mathbf{Fm} = \langle Fm, c_1, ..., c_n \rangle$$

is an algebra of type $\mathcal{L}$, called the *formula algebra* of $\mathcal{L}$. In the following, given a formula $\alpha(p_1, ..., p_n)$ containing at most the indicated variables, an algebra $\mathbf{A}$ of type $\mathcal{L}$ and elements $a_1, ..., a_n \in A$, we will denote by $\alpha^{\mathbf{A}}(a_1, ..., a_n)$ (or $\alpha^{\mathbf{A}}(\vec{a})$) when the length of the string is either clear from the context or inessential) the result of the application to α of the unique homomorphism $h : \mathbf{Fm} \to \mathbf{A}$ such that $h(p_i) = a_i$ for all $i \leq n$. This notation will be sometimes extended to sets of formulas in the obvious way. An *equation* of type $\mathcal{L}$ is a pair (α, β) of $\mathcal{L}$-formulas, written as $\alpha \approx \beta$. Endomorphisms on $\mathbf{Fm}$ are called *substitutions*, and whenever there is some substitution σ such that $\alpha = \sigma(\beta)$, we say that α is a *substitution instance* of β. The dependence of **Fm** on $\mathcal{L}$ will be tacitly acknowledged in what follows.

We now have all we need to define consequence relations.

A *consequence relation* over the formula algebra **Fm** is a relation $\vdash\, \subseteq \wp(Fm) \times Fm$ with the following properties:

1. $\alpha \vdash \alpha$ *(reflexivity)*.
2. If $\Gamma \vdash \alpha$ and $\Gamma \subseteq \Delta$, then $\Delta \vdash \alpha$ *(monotonicity)*.
3. If $\Gamma \vdash \alpha$ and $\Delta \vdash \gamma$ for every $\gamma \in \Gamma$, then $\Delta \vdash \alpha$ *(cut)*.

A (propositional) *logic* is a pair $L = (\mathbf{Fm}, \vdash_L)$, where **Fm** is the formula algebra of some given type $\mathcal{L}$ and $\vdash_L$ is a *substitution-invariant* consequence relation over **Fm**; in other words, if $\Gamma \vdash_L \alpha$ and σ is a substitution on **Fm**, then $\sigma(\Gamma) \vdash_L \sigma(\alpha)$ (where $\sigma(\Gamma) = \{\sigma(\gamma) \mid \gamma \in \Gamma\}$). Informally speaking, we have a logic whenever we can specify a logical language and a concept of consequence among formulas of that language according to which: (i) every formula follows from itself; (ii) whatever follows from a set of premises also follows from any larger set of premises; (iii) whatever follows from consequences of a set of premises also follows from the set itself; (iv) whether a conclusion follows or not from a set of premises only depends on the *logical form* of the premises and the conclusion themselves. A formula α is a *theorem* of the logic $L = (\mathbf{Fm}, \vdash_L)$ if $\emptyset \vdash_L \alpha$.

Our next step will consist of rigorously defining Hilbert-style calculi. An *inference rule* over **Fm** is a pair $\mathsf{R} = (\Gamma, \alpha)$, where Γ is a finite (possibly empty) subset of Fm and $\alpha \in Fm$. If Γ is empty, the rule is an *axiom*; otherwise, it is a *proper rule*. A *Hilbert-style calculus* (over **Fm**) is a set of inference rules over **Fm** that contains at least one axiom and at least one proper rule. For axioms, we will henceforth omit outer brackets and the empty set symbol; also, proper rules $(\{\alpha_1, ..., \alpha_n\}, \alpha)$ will be written in the fractional form

$$\frac{\alpha_1 \quad \cdots \quad \alpha_n}{\alpha}$$

If $\Delta \cup \{\beta\} \subseteq Fm$ and HL is a Hilbert-style calculus over **Fm**, then a *derivation* of β from Δ in HL is a finite sequence $\beta_1, ..., \beta_n$ of formulas in Fm such that $\beta_n = \beta$ and for each β_i $(i \leq n)$:

1. either β_i is a member of Δ; or
2. β_i is a substitution instance of an axiom of HL; or else
3. there are a substitution σ and an inference rule $(\Gamma, \alpha) \in$ HL such that $\beta_i = \sigma(\alpha)$ and, for every $\gamma \in \Gamma$, $\sigma(\gamma) \in \{\beta_1, ..., \beta_{i-1}\}$.

From any Hilbert-style calculus HL over **Fm** we can extract a logic $(\mathbf{Fm}, \vdash_{HL})$ by specifying that $\Gamma \vdash_{HL} \alpha$ whenever there is a derivation of α from Γ in HL. Such logics are called *deductive systems* and have the property that whenever $\Gamma \vdash_{HL} \alpha$, there is always some *finite* $\Delta \subseteq \Gamma$ such that $\Delta \vdash_{HL} \alpha$ (this much is clear from the very definition of derivation: after we prune Γ of all that is not necessary to derive α, we are left with a finite set). This compactness property, on the other hand, need not be available in general. Logics having the property are called *finitary*; hence, we may rephrase the above by saying that deductive systems are finitary logics.

An example of the preceding rather abstract discussion is the Hilbert-style calculus HCL for classical propositional logic illustrated in Table 1 of Subsection 2.4. Its language $\mathcal{L}$ contains the connectives $\neg, \wedge, \vee, \rightarrow$, 0, and 1. Classical propositional logic can be now identified with the deductive system $CL = (\mathbf{Fm}, \vdash_{HCL})$.

Developing an idea by Lindenbaum, Tarski showed in 1935 in what sense Boolean algebras can be considered the algebraic counterpart of CL ([131]). Actually, Tarski pointed out a rather weak kind of correspondence between Boolean algebras and classical logic: he showed that the former are an *algebraic semantics* for the latter, a notion that we now proceed to explain in full generality.

Let $L = (\mathbf{Fm}, \vdash_L)$ be a logic in the language $\mathcal{L}$, and let $\tau = \{\gamma_i(p) \approx \delta_i(p)\}_{i \in I}$ be a set of equations in a single variable of $\mathcal{L}$. To avoid overloading notation, it will be convenient to think of τ as a function which maps formulas in Fm to sets of equations of the same type. Therefore we let $\tau(\alpha)$ stand for the set

$$\{\gamma_i(p/\alpha) \approx \delta_i(p/\alpha)\}_{i \in I}.$$

Now let $\mathcal{K}$ be a class of algebras also of the same type. We say that $\mathcal{K}$ is an algebraic semantics for L if, for some such τ, the following condition holds for all $\Gamma \cup \{\alpha\} \in Fm$:

$$\Gamma \vdash_L \alpha \quad \text{iff} \quad \text{for every } \mathbf{A} \in \mathcal{K} \text{ and every } \vec{a} \in A^n,$$
$$\text{if } \tau(\gamma)^{\mathbf{A}}(\vec{a}) \text{ for all } \gamma \in \Gamma, \text{ then } \tau(\alpha)^{\mathbf{A}}(\vec{a}),$$

a condition which will sometimes be rewritten as

$$\Gamma \vdash_L \alpha \quad \text{iff} \quad \{\tau(\gamma) \mid \gamma \in \Gamma\} \vdash_{Eq(\mathcal{K})} \tau(\alpha).$$

In particular, if $\mathcal{L}$ contains a nullary connective 1, it is sometimes possible to choose τ to be the singleton $\{p \approx 1\}$. Informally speaking, what we do in such cases is the following: given an algebra $\mathbf{A} \in \mathcal{K}$, we interpret its elements as "meanings of propositions" or "truth values" and, in particular, the element $1^{\mathbf{A}}$ as "true"; homomorphisms $h : \mathbf{Fm} \to \mathbf{A}$, where $\mathbf{A} \in \mathcal{K}$, are interpreted as "assignments of meanings" to elements of Fm. Moreover, we want α to follow from the set of premises Γ just in case, whenever we assign the meaning "true" to all the premises in some algebra in $\mathcal{K}$, the conclusion is also assigned the meaning "true".

Let us now apply this definition to our classical logic example. We just remind the reader that a Boolean algebra is usually defined as an algebra $\mathbf{A} = \langle A, \wedge, \vee, \neg, 1, 0 \rangle$ such that $\langle A, \wedge, \vee, 1, 0 \rangle$ is a bounded distributive lattice and, for every $a \in A$, $a \wedge \neg a = 0$ and $a \vee \neg a = 1$. To apply the definition of algebraic semantics, Boolean algebras must be algebras of the same type as the formula algebra of CL, whence it is expedient to include in the type the derived operation symbol $\to$ (defined via $p \to q = \neg p \vee q$). Once this is done, setting $\tau = \{p \approx 1\}$ it is possible to show that:

Theorem 3.1. *The class $\mathcal{BA}$ of Boolean algebras is an algebraic semantics for CL.*

Proof. (Sketch). The left-to-right implication can be established by induction on the length of a derivation of α from Γ in HCL: we show that axioms A1 to A15 are always evaluated at $1^{\mathbf{A}}$ for every $\mathbf{A} \in \mathcal{BA}$ and that the proper inference rule R1 preserves this property.

The converse implication is trickier. We show the contrapositive: we suppose that $\Gamma \nvdash_{\text{HCL}} \alpha$ and prove that there exist a Boolean algebra $\mathbf{A}$ and a sequence of elements $\vec{a}$ such that $\gamma(\vec{a}) = 1^{\mathbf{A}}$ for all $\gamma \in \Gamma$, yet $\alpha(\vec{a}) \neq 1^{\mathbf{A}}$.

Let T be the smallest set of $\mathcal{L}$-formulas that includes Γ and is closed under the consequence relation of CL (that is, for every $\beta \in Fm$, $T \vdash_{\text{HCL}} \beta$ implies that $\beta \in T$). Define a binary relation Θ_T on Fm by stipulating that

$$(\beta, \gamma) \in \Theta_T \quad \text{iff} \quad \beta \to \gamma, \gamma \to \beta \in T.$$

The whole business of finishing our completeness proof amounts to establishing the following two assertions:

1. Θ_T is a congruence on $\mathbf{Fm}$, and the coset $\left[1^{\mathbf{Fm}}\right]_{\Theta_T}$ is just T;
2. the quotient $\mathbf{Fm}/\Theta_T$ is a Boolean algebra.

Proofs of (1) and (2) make heavy use of syntactic lemmas established for HCL. Once this laborious task has been carried out, in order to construct our falsifying model, it suffices to take $\mathbf{A} = \mathbf{Fm}/\Theta_T$ (we are justified in so doing by (2) above) and evaluate each variable p in $\Gamma \cup \{\alpha\}$ as its own congruence class modulo Θ_T: then $\gamma^{\mathbf{A}}(\overrightarrow{[p]_{\Theta_T}}) = 1^{\mathbf{A}}$ for all $\gamma \in \Gamma$ (since $\Gamma \subseteq T$) yet $\alpha^{\mathbf{A}}(\overrightarrow{[p]_{\Theta_T}}) \neq 1^{\mathbf{A}}$ (since $\alpha \notin T$).   □

3.3 Blok and Pigozzi: algebraizable logics

Algebraic logic rapidly developed after World War Two, once again to the credit of Polish logicians. Although Tarski had permanently settled in the States before that time, establishing in Berkeley what would later become the leading research group in algebraic logic worldwide, his compatriots Jerzy Łoś, Roman Suszko, Helena Rasiowa, and Roman Sikorski kept the flag of Polish algebraic logic flying, developing in detail throughout the 1950's and 1960's the theory of logical matrices initiated twenty years earlier by Łukasiewicz and Tarski himself. A major breakthrough in the discipline came about in 1989, when Wim Blok and Don Pigozzi (one of Tarski's students) published their monograph on algebraizable logics [13], considered a milestone in the area of abstract algebraic logic. This subsection summarizes some of the main results of their text.

The concept of algebraic semantics described above is too weak in at least two respects. First, in the relationship between a given logic and a candidate algebraic semantics there is room for much promiscuity. In fact:

- There are logics with no algebraic semantics. For example, let HI be the Hilbert-style calculus whose sole axiom is $\alpha \rightarrow \alpha$ and whose sole inference rule is modus ponens. Then the deductive system $(\mathbf{Fm}, \vdash_{\mathrm{HI}})$ has no algebraic semantics [14].
- The same logic can have more than one algebraic semantics. As we have recalled, every Boolean algebra is a complemented distributive lattice. A Boolean algebra $\mathbf{B}$ is, in particular, a bounded distributive lattice such that for every $a, b \in B$, $a \rightarrow b$ (defined in this particular case as $\neg a \vee b$) is (w.r.t. the induced order of the underlying lattice) the top element in the set

$$\{x \in B \mid a \wedge x \leq b\}.$$

A *Heyting algebra* can be defined as an algebra $\mathbf{H} = \langle B, \wedge, \vee, \rightarrow, 1, 0 \rangle$ satisfying exactly the above conditions. Moreover, the class of Heyting algebras can be equationally defined and so forms a variety $\mathcal{HA}$. Any Boolean algebra therefore forms a Heyting algebra by letting $\neg x = x \rightarrow 0$; more precisely, an equational basis for Boolean algebras relative to Heyting algebras is given by the single equation

$x \approx (x \to 0) \to 0$, expressing the fact that negation is an *involution*. Now, by Glivenko's Theorem ([61]), CL admits not only $\mathcal{BA}$ as an algebraic semantics, but also $\mathcal{HA}$, by choosing $\tau = \{\neg\neg p \approx 1\}$ ([14]).

- There can be different logics with the same algebraic semantics. Since Heyting algebras are an algebraic semantics for intuitionistic logic IL, the previous example shows that both CL and IL have $\mathcal{HA}$ as an algebraic semantics.

Second, the relation between a logic L and its algebraic semantics $\mathcal{K}$ is asymmetric. In fact, the property of belonging to the set of "true values" of an algebra $\mathbf{A} \in \mathcal{K}$ must be definable by means of the set of equations τ, whence the class $\mathcal{K}$ has the expressive resources to indicate when a given *formula* is valid in L. On the other hand, the logic L need not have the expressive resources to indicate when a given *equation* holds in $\mathcal{K}$. For example, there is no way in CL to express, by means of a condition involving a set of formulas, when it is the case that an equation $\alpha \approx \beta$ holds in its algebraic semantics $\mathcal{HA}$. This means that there is a sense of "faithful representation" according to which the relation $\vdash_{Eq(\mathcal{HA})}$ faithfully represents the consequence relation of CL, but not conversely. This makes a sharp contrast to the situation we have with the other algebraic semantics of CL we examined, namely $\mathcal{BA}$: we gather from the proof of Theorem 3.1 that an equation $\alpha \approx \beta$ holds in $\mathcal{BA}$ just in case $\vdash_{CL} \alpha \to \beta$ and $\vdash_{CL} \beta \to \alpha$. The notion of *algebraizability* ([13]) aims at making precise this stronger relation between a logic and a class of algebras which holds between CL and $\mathcal{BA}$, but not between CL and its "unofficial" semantics $\mathcal{HA}$.

Before stating the formal definition of this concept, let us establish some further notational conventions: given an equation $\alpha \approx \beta$ and a set of formulas in two variables $\rho = \{\alpha_j(p,q)\}_{j \in J}$, we will abbreviate by $\rho(\alpha, \beta)$ the set $\{\alpha_j(p/\alpha, q/\beta)\}_{j \in J}$, and ρ will be regarded as a function mapping equations to sets of formulas. If Γ, Δ are sets of formulas, by $\Gamma \vdash_L \Delta$ we will mean $\Gamma \vdash_L \alpha$ for every $\alpha \in \Delta$; similarly, if E, E' are sets of equations, by $E \vdash_{Eq(\mathcal{K})} E'$ we will mean $E \vdash_{Eq(\mathcal{K})} \varepsilon$ for every $\varepsilon \in E'$.

A logic $L = (\mathbf{Fm}, \vdash_L)$ is said to be *algebraizable* with *equivalent algebraic semantics* $\mathcal{K}$ (where $\mathcal{K}$ is a class of algebras of the same type as $\mathbf{Fm}$) iff there exist a map τ from formulas to sets of equations, and a map ρ from equations to sets of formulas such that the following conditions hold for any $\alpha, \beta \in Fm$:

AL1: $\Gamma \vdash_L \alpha$ iff $\tau(\Gamma) \vdash_{Eq(\mathcal{K})} \tau(\alpha)$;

AL2: $E \vdash_{Eq(\mathcal{K})} \alpha \approx \beta$ iff $\rho(E) \vdash_L \rho(\alpha, \beta)$;

AL3: $\alpha \dashv\vdash_L \rho(\tau(\alpha))$;

AL4: $\alpha \approx \beta \dashv\vdash_{Eq(\mathcal{K})} \tau(\rho(\alpha, \beta))$.

The sets $\tau(p)$ and $\rho(p, q)$ are respectively called a *system of defining equations* and a *system of equivalence formulas* for L and $\mathcal{K}$. A logic L is algebraizable (tout court) iff, for some $\mathcal{K}$, it is algebraizable with equivalent algebraic semantics $\mathcal{K}$.

Put differently, a logic L is algebraizable with equivalent algebraic semantics $\mathcal{K}$ when:

- the relation $\vdash_L$ is faithfully interpretable via the map τ into the relation $\vdash_{Eq(\mathcal{K})}$ (AL1);
- the relation $\vdash_{Eq(\mathcal{K})}$ is faithfully interpretable via the map ρ into the relation $\vdash_L$ (AL2);
- the two maps ρ and τ are mutually inverse, meaning that if we apply them in succession we end up with a formula, respectively an equation, which is equivalent to the one we started with according to $\vdash_L$, respectively $\vdash_{Eq(\mathcal{K})}$ (AL3 and AL4).

This definition can be drastically simplified, in that one can show that a logic L is algebraizable with equivalent algebraic semantics $\mathcal{K}$ iff it satisfies either AL1 and AL4, or else AL2 and AL3.

As an example, we can strengthen Theorem 3.1 by showing that

Theorem 3.2. CL *is algebraizable with equivalent algebraic semantics* $\mathcal{BA}$.

Proof. Let

$$\tau(p) = \{p \approx 1\};$$
$$\rho(p, q) = \{p \to q, q \to p\}.$$

By the previous observation, we need only check that τ and ρ satisfy conditions AL1 and AL4 above. However, AL1 is just Theorem 3.1. As for AL4,

$$\alpha \approx \beta \dashv\vdash_{Eq(\mathcal{BA})} \tau(\rho(\alpha, \beta)) \text{ iff } \alpha \approx \beta \dashv\vdash_{Eq(\mathcal{BA})} \tau(\alpha \to \beta, \beta \to \alpha)$$
$$\text{iff } \alpha \approx \beta \dashv\vdash_{Eq(\mathcal{BA})} \{\alpha \to \beta \approx 1, \beta \to \alpha \approx 1\}.$$

However, given any $\mathbf{A} \in \mathcal{BA}$ and any $\overrightarrow{a} \in A^n$, $\alpha^{\mathbf{A}}(\overrightarrow{a}) = \beta^{\mathbf{A}}(\overrightarrow{a})$ just in case $\alpha \to \beta^{\mathbf{A}}(\overrightarrow{a}) = 1^{\mathbf{A}}$ and $\beta \to \alpha^{\mathbf{A}}(\overrightarrow{a}) = 1^{\mathbf{A}}$, which proves our conclusion. $\square$

Every equivalent algebraic semantics for L is, in particular, an algebraic semantics for L in virtue of AL1. However, as we have seen, the converse need not hold. The concept of equivalent algebraic semantics is therefore a genuine strengthening of Tarski's definition.

If L is algebraizable with equivalent algebraic semantics $\mathcal{K}$, then $\mathcal{K}$ might not be the unique equivalent algebraic semantics for L. However, in case L is finitary, any two equivalent algebraic semantics for L generate the same quasivariety. Clearly, this quasivariety is in turn an equivalent algebraic semantics for the same logic, whence we are justified in talking about *the* equivalent quasivariety semantics for L. On the other hand, it is not uncommon to find different algebraizable logics with the same equivalent algebraic semantics (see, *e.g.*, [113]); however, if L and L' are algebraizable with equivalent quasivariety semantics $\mathcal{K}$ and with the *same set* of defining equations $\tau(p)$, then L and L' must coincide.

In [13], one finds several elegant equivalent characterizations of algebraizability. Some of them use concepts and tools from the *theory of logical matrices*, which was, as we have recalled, one of the early developments in abstract algebraic logic (see [142]). We just mention a syntactic characterization which, unlike the others, does not presuppose any technical prerequisites:

Theorem 3.3. *A logic* $L = (\mathbf{Fm}, \vdash_L)$ *is algebraizable iff there exist a set* $\rho(p, q)$ *of formulas in two variables and a set of equations* $\tau(p)$ *in a single variable such that, for any* $\alpha, \beta, \gamma \in Fm$, *the following conditions hold:*

1. $\vdash_L \rho(\alpha, \alpha)$;
2. $\rho(\alpha, \beta) \vdash_L \rho(\beta, \alpha)$;
3. $\rho(\alpha, \beta), \rho(\beta, \gamma) \vdash_L \rho(\alpha, \gamma)$;
4. *For every n-ary connective* c^n *and for every* $\overrightarrow{\alpha}, \overrightarrow{\beta} \in (Fm)^n$,
   $\rho(\alpha_1, \beta_1), ..., \rho(\alpha_n, \beta_n) \vdash_L \rho(c^n(\overrightarrow{\alpha}), c^n(\overrightarrow{\beta}))$;
5. $\alpha \dashv\vdash_L \rho(\tau(\alpha))$.

In this case $\rho(p, q)$ *and* $\tau(p)$ *are, respectively, a set of equivalence formulas and a set of defining equations for* L.

4 The algebras of logic

The focus of this section is *residuated lattices*, algebraic counterparts of the propositional substructural logics discussed in the next section. The defining properties that describe the class $\mathcal{RL}$ of residuated lattices are few and easy to grasp, and concrete examples are readily constructed that illustrate their key features. However, the theory is also sufficiently robust that the class $\mathcal{RL}$ encompasses a large portion of the ordered algebras arising in logic. Notably, the rich algebraic theory of residuated lattices has produced powerful tools for the comparative study of substructural logics. Moreover, the bridge provided by algebraic logic yields significant benefits to algebra. In fact, one can argue convincingly that an in-depth study of residuated lattices is impossible at this time without the use of logical (in particular, proof-theoretic) techniques.

Our primary aim in this section is to present basic facts from the theory of residuated lattices, including the description of their congruence relations, and provide a brief historical account of the development of the concept of residuated structure in algebra.

4.1 Preliminaries

A subset F of a poset $\mathbf{P}$ is said to be an *order-filter* of $\mathbf{P}$ if whenever $y \in P$, $x \in F$, and $x \leq y$, then $y \in F$. Note that the empty set $\emptyset$ is an order-filter. For an element $a \in P$, we write $\uparrow a = \{x \in P \mid a \leq x\}$ for the *principal* order-filter generated by a; more generally, for $A \subseteq P$, $\uparrow A = \{x \in P \mid a \leq x,$ for some a $\in$ A$\}$ denotes the *order-filter generated by A. Order-ideals* are defined dually.

We denote the *least* element of a poset $\mathbf{P}$, if it exists, by $\perp_{\mathbf{P}}$. Similarly, $\top_{\mathbf{P}}$ denotes the *greatest* element. Obviously, least elements and greatest elements, when they exist, are unique. Let $X \subseteq P$ be any subset (possibly empty). We use $\bigvee_{\mathbf{P}} X$ and $\bigwedge_{\mathbf{P}} X$, respectively, to denote the *join* (or least upper bound) and *meet* (or greatest lower bound) of X in $\mathbf{P}$ whenever they exist. We use the terms *isotone* and *order-preserving* synonymously to describe a map $\varphi \colon P \to Q$ between posets $\mathbf{P}$ and $\mathbf{Q}$ with the property that for all $x, y \in P$, if $x \leq y$ then $\varphi(x) \leq \varphi(y)$. If for all elements $x, y \in P$, $x \leq y$ implies $\varphi(x) \geq \varphi(y)$, then φ will be called *anti-isotone* or *order-reversing*. The poset subscripts appearing in some of the notation of this paragraph will henceforth be omitted whenever there is no danger of confusion.

4.2 Residuated maps and residuated lattices

Let $\mathbf{P}$ and $\mathbf{Q}$ be posets. A map $\varphi : P \to Q$ is called *residuated* provided there exists a map $\varphi_* : Q \to P$ such that $\varphi(x) \leq y \iff x \leq \varphi_*(y)$, for all $x \in P$ and $y \in Q$. We refer to φ_* as the *residual* of φ.

We have the following simple but useful result (see, *e.g.*, [58]):

Lemma 4.1.

1. *If $\varphi_* : Q \to P$ is residuated with residual φ_*, then φ preserves all existing joins in $\mathbf{P}$ and φ_* preserves any existing meets in $\mathbf{Q}$.*

2. *Conversely, if* **P** *is a complete lattice and* $\varphi : \mathbf{P} \to \mathbf{Q}$ *preserves all joins, then it is residuated.*

A binary operation $\cdot$ on a partially ordered set $\mathbf{P} = \langle P, \leq \rangle$ is said to be *residuated* if there exist binary operations $\backslash$ and $/$ on P such that for all $x, y, z \in P$,

$$x \cdot y \leq z \text{ iff } x \leq z/y \text{ iff } y \leq x \backslash z.$$

Note that $\cdot$ is residuated if and only if, for all $a \in P$, the maps $x \mapsto ax$ $(x \in P)$ and $x \mapsto xa$ $(x \in P)$ are residuated in the sense of the preceding definition. Their residuals are the maps $y \mapsto a \backslash y$ $(y \in P)$ and $y \mapsto y/a$ $(y \in P)$, respectively. The operations $\backslash$ and $/$ are referred to as the *right residual* and the *left residual* of $\cdot$, respectively. Observe also that $\cdot$ is residuated if and only if it is order-preserving in each argument and, for all $x, y \in P$, the sets $\{z \mid x \cdot z \leq y\}$ and $\{z \mid z \cdot x \leq y\}$ both contain greatest elements, $x \backslash y$ and y/x, respectively. In particular, note that $\cdot$ and $\leq$ uniquely determine $\backslash$ and $/$.

It is suggestive to think of the residuals as generalized division operations. The expression y/x is read as "y over x" while $x \backslash y$ is read as "x under y." In either case, y is considered the *numerator* and x the *denominator*. We tend to favor $/$ in our calculations, but any statement about residuated structures has a "mirror image" obtained by replacing $x \cdot y$ by $y \cdot x$ and interchanging x/y with $y \backslash x$. It follows directly from the preceding definition that a statement is equivalent to its mirror image, and we often state results in only one form. As usual, we write xy for $x \cdot y$ and adopt the convention that, in the absence of parenthesis, $\cdot$ is performed first, followed by $\backslash$ and $/$, and finally $\vee$ and $\wedge$. We also define $x^1 = x$ and $x^{n+1} = x^n \cdot x$.

As a consequence of Lemma 4.1, multiplication preserves all existing joins in each argument, and $\backslash$ and $/$ preserve all existing meets in the "numerator." Moreover, it is easy to check that they convert all existing joins in the "denominator" to meets. More specifically, we have the following result:

Proposition 4.2. *Let* $\cdot$ *be a residuated map on a poset* **P** *with residuals* $\backslash$ *and* $/$.

1. *The operation* $\cdot$ *preserves all existing joins in each argument; i.e., if* $\bigvee X$ *and* $\bigvee Y$ *exist for* $X, Y \subseteq P$, *then* $\bigvee_{x \in X, y \in Y} xy$ *exists and*

$$\left(\bigvee X \right) \left(\bigvee Y \right) = \bigvee_{x \in X, \, y \in Y} xy.$$

2. *The residuals preserve all existing meets in the numerator, and convert existing joins to meets in the denominator, i.e., if $\bigvee X$ and $\bigwedge Y$ exist for $X, Y \subseteq P$, then for any $z \in P$, $\bigwedge_{x \in X} z/x$ and $\bigwedge_{y \in Y} y/z$ exist and*

$$z \Big/ \Big(\bigvee X \Big) = \bigwedge_{x \in X} z/x \quad \text{and} \quad \Big(\bigwedge Y \Big) \Big/ z = \bigwedge_{y \in Y} y/z.$$

A *residuated lattice*, or a *residuated lattice-ordered monoid*, is an algebra

$$\mathbf{L} = \langle L, \wedge, \vee, \cdot, \backslash, /, 1 \rangle$$

such that $\langle L, \wedge, \vee \rangle$ is a lattice; $\langle L, \cdot, 1 \rangle$ is a monoid; and $\cdot$ is residuated, in the underlying partial order, with residuals $\backslash$ and $/$.

An *FL-algebra* $\mathbf{L} = \langle L, \wedge, \vee, \cdot, \backslash, /, 1, 0 \rangle$ is an algebra such that: (i) $\langle L, \wedge, \vee, \cdot, \backslash, /, 1 \rangle$ is a residuated lattice, and (ii) 0 is a distinguished element (nullary operation) of L.

We use the symbols $\mathcal{RL}$ and $\mathcal{FL}$ to denote the class of all residuated lattices and FL-algebras respectively. The following lemma collects basic properties of residuated lattices, most of which by now can be ascribed to the "folklore" of the subject.

Proposition 4.3. *The following equations (and their mirror images) hold in any residuated lattice (in particular, in any FL-algebra).*

1. $x(y/z) \leq xy/z$
2. $x/y \leq xz/yz$
3. $(x/y)(y/z) \leq x/z$
4. $x/yz \approx (x/z)/y$
5. $x\backslash(y/z) \approx (x\backslash y)/z$
6. $(1/x)(1/y) \leq 1/yx$
7. $(x/x)x \approx x$
8. $(x/x)^2 \approx x/x$

If a residuated lattice $\mathbf{L}$ has a bottom element $\bot$, then $\bot\backslash\bot$ is its top element $\top$. Moreover, for all $x \in P$, we have

$$x\bot = \bot = \bot x \quad \text{and} \quad \bot\backslash x = \top = x\backslash\top.$$

Sometimes it is useful to have $\bot$ and $\top$ in the signature. In particular, a *bounded FL-algebra* is an algebra $\mathbf{L} = \langle L, \wedge, \vee, \cdot, \backslash, /, 1, 0, \bot, \top \rangle$ such that: (i) $\langle L, \wedge, \vee, \cdot, \backslash, /, 1, 0 \rangle$ is an FL-algebra, and (ii) $\bot$ and $\top$ are, respectively, the bottom and top elements of L.

The next result, whose proof is left to the reader, provides a straightforward way to verify that $\mathcal{RL}$ and $\mathcal{FL}$ are equational classes.

Lemma 4.4. *An algebra* $\mathbf{L} = \langle L, \wedge, \vee, \cdot, \backslash, /, 1 \rangle$ *is a residuated lattice if and only if* $\langle L, \cdot, 1, \rangle$ *is a monoid,* $\langle L, \wedge, \vee \rangle$ *is a lattice, and for all* $a, b \in L$,

1. *the maps* $x \mapsto ax$ *and* $x \mapsto xa$ *preserve finite joins*;
2. *the maps* $x \mapsto a\backslash x$ *and* $x \mapsto x/a$ *are isotone*;
3. $a(a\backslash b) \leq b \leq a\backslash ab$; *and*
4. $(b/a)a \leq b \leq ba/a$.

Hence, we have the following equational characterization of $\mathcal{RL}$ and $\mathcal{FL}$:

Proposition 4.5. *The classes* $\mathcal{RL}$ *and* $\mathcal{FL}$ *are finitely based equational classes. Their defining equations consist of the defining equations for lattices and monoids together with the six equations given below:*

1. $x(y \vee z) \approx xy \vee xz$ $\qquad$ $(y \vee z)x \approx yx \vee zx$
2. $x\backslash y \leq x\backslash(y \vee z)$ $\qquad$ $y/x \leq (y \vee z)/x$
3. $x(x\backslash y) \leq y \leq x\backslash xy$
4. $(y/x)x \leq y \leq yx/x$

Two varieties of particular interest are the variety $\mathcal{CRL}$ of commutative residuated lattices and the variety $\mathcal{CFL}$ of commutative FL-algebras. These varieties satisfy the equation $xy \approx yx$, and hence the equation $x\backslash y \approx y/x$. In what follows, we use the symbol $\rightarrow$ to denote both the operations $\backslash$ and $/$. While we always think of these varieties as subvarieties of $\mathcal{RL}$ and $\mathcal{FL}$, respectively, we slightly abuse notation by listing only one occurrence of the operation $\rightarrow$ in describing their members.

We also mention that it is convenient sometimes to add an extra nullary operation 0 to the type of $\mathcal{RL}$, and think of $\mathcal{RL}$ as the subvariety of $\mathcal{FL}$ axiomatized, relative to $\mathcal{FL}$, by the equation $1 \approx 0$.

Example 4.6. The variety of Boolean algebras can be identified with the subvariety of $\mathcal{CFL}$, which we may again harmlessly call $\mathcal{BA}$, satisfying the additional equations $xy \approx x \wedge y$, $(x \rightarrow y) \rightarrow y \approx x \vee y$, and $x \wedge 0 \approx 0$. More specifically, every Boolean algebra $\mathbf{B} = \langle B, \wedge, \vee, \neg, 1, 0 \rangle$ satisfies the equations above with respect to $\wedge, \vee, 0$, and Boolean implication $x \rightarrow y = \neg x \vee y$. Conversely, if a (commutative) residuated lattice $\mathbf{L}$ satisfies these equations and we define $\neg x = x \rightarrow 0$, then $\langle L, \wedge, \vee, \neg, 1, 0 \rangle$ is a Boolean algebra. In stricter mathematical terms, the variety of Boolean algebras is *term-equivalent* to the subvariety of $\mathcal{CFL}$ satisfying the equations $xy \approx x \wedge y$, $(x \rightarrow y) \rightarrow y \approx x \vee y$, and $x \wedge 0 \approx 0$.

Likewise, the variety of Heyting algebras is term-equivalent to the sub-variety of $\mathcal{CFL}$, which we again call $\mathcal{HA}$, satisfying the additional equations $xy \approx x \wedge y$ and $x \wedge 0 \approx 0$.

Example 4.7. Ring theory constitutes a historically remarkable source for residuated structures. (Refer to Subsection 4.6 for further details.) Let **R** be a ring with unit and let I(**R**) denote the lattice of two-sided ideals of **R**. Then I(**R**) $= \langle I(R), \cap, \vee, \cdot, \backslash, /, R, \{0\} \rangle$ is a (not necessarily commutative) FL-algebra, where, for $I, J \in I(R)$,

$$IJ = \left\{ \sum_{k=1}^{n} a_k b_k \mid a_k \in I; b_k \in J; n \geq 1 \right\};$$

$$I \backslash J = \{x \in R \mid Ix \subseteq J\}; \text{ and}$$

$$J/I = \{x \in R \mid xI \subseteq J\}.$$

For a related interesting example, consider an integral domain **R** and its field of quotients **K**. Let L(**K**) denote the lattice of **R**-submodules of **K**. Then L(**K**) $= \langle L(K), \cap, \vee, \cdot, \backslash, /, R, \{0\} \rangle$ is an FL-algebra, where, for $I, J \in L(K)$,

$$IJ = \left\{ \sum_{k=1}^{n} a_k b_k \mid a_k \in I; b_k \in J; n \geq 1 \right\}.$$

Example 4.8. Lattice-ordered groups play a fundamental role in the study of algebras of logic. (Refer to Subsection 4.6 for a short account of their role in mathematics.) A *lattice-ordered group*, ℓ-*group* for short, is an algebra $\mathbf{G} = \langle G, \wedge, \vee, \cdot, ^{-1}, 1 \rangle$ such that (i) $\langle G, \wedge, \vee \rangle$ is a lattice; (ii) $\langle G, \cdot, ^{-1}, 1 \rangle$ is a group; and (iii) addition is order-preserving in each argument (equivalently, it satisfies the equation $x(y \vee z)w \approx (xyw) \vee (xzw)$). The variety of ℓ-groups is term-equivalent to the subvariety $\mathcal{LG}$ of $\mathcal{RL}$ defined by the additional equation $(1/x)x \approx 1$.

More specifically, if $\mathbf{G} = \langle G, \wedge, \vee, \cdot, ^{-1}, 1 \rangle$ is an ℓ-group and we define $x/y = xy^{-1}$ and $y \backslash x = y^{-1}x$, then $\mathbf{G} = \langle G, \wedge, \vee, \cdot, \backslash, /, 1 \rangle$ is a residuated lattice satisfying the equation $(1/x)x \approx 1$. Conversely, if a residuated lattice $\mathbf{L} = \langle L, \wedge, \vee, \cdot, \backslash, /, 1 \rangle$ satisfies the last equation and we define $x^{-1} = 1/x$, then $\mathbf{L} = \langle L, \wedge, \vee, \cdot, ^{-1}, 1 \rangle$ becomes an ℓ-group. Moreover, this correspondence is bijective.

Example 4.9. MV-algebras (refer to the discussion in Subsection 5.2) are the algebraic counterparts of the infinite-valued Łukasiewicz propositional logic. An *MV-algebra* is traditionally defined as an algebra $\mathbf{M} = \langle M, \oplus, \neg, 0 \rangle$ of type $\langle 2, 1, 0 \rangle$ that satisfies the following equations:

(MV1) $x \oplus (y \oplus z) \approx (x \oplus y) \oplus z$

(MV2) $x \oplus y \approx y \oplus x$

(MV3) $x \oplus 0 \approx x$

(MV4) $\neg\neg x \approx x$

(MV5) $x \oplus \neg 0 \approx \neg 0$

(MV6) $\neg(\neg x \oplus y) \oplus y \approx \neg(\neg y \oplus x) \oplus x$

The variety of MV-algebras is term-equivalent to the subvariety, $\mathcal{MV}$, of $\mathcal{CFL}$ satisfying the extra equations $x \vee y \approx (x \to y) \to y$ and $x \wedge 0 \approx 0$.

In more detail, if $\mathbf{M} = \langle M, \oplus, \neg, 0 \rangle$ is an MV-algebra and we define, for $x, y \in M$, $xy = \neg(\neg x \oplus \neg y)$, $x \wedge y = x(\neg x \oplus y)$, $x \vee y = x \oplus (\neg xy)$, $x \to y = \neg(x \neg y)$, and $1 = \neg 0$, then $\mathbf{M} = \langle M, \wedge, \vee, \cdot, \to, 1, 0 \rangle$ is in $\mathcal{MV}$. Conversely, if $\mathbf{M} \in \mathcal{MV}$ and we define, for all $x, y \in M$, $\neg x = x \to 0$ and $x \oplus y = \neg(\neg x \neg y)$, then $\mathbf{M} = \langle M, \oplus, \neg, 0 \rangle$ is an MV-algebra. We refer to [31] (Section 4.2) or [141] (Subsection 3.4.5) for the details of the proof.

Example 4.10. Many varieties of ordered algebras arising in logic – including Boolean algebras, Abelian ℓ-groups, and MV-algebras, but not Heyting algebras and ℓ-groups – are *semilinear*, that is, generated by their totally ordered members. An equational basis for the variety of semilinear residuated lattices $\mathcal{SemRL}$ relative to $\mathcal{RL}$ consists of the equation (λ_z and ρ_w are defined in the next subsection)

$$\lambda_z(x/(x \vee y)) \vee \rho_w(y/(x \vee y)) \approx 1.$$

The proof of this result makes heavy use of the material of Subsection 4.5 (see [19] and [82]). A simplified equational basis given in [71] for the variety of commutative semilinear residuated lattices $\mathcal{CSemRL}$ relative to $\mathcal{CRL}$ consists of the equations

$$[(x \to y) \vee (y \to x)] \wedge 1 \approx 1 \quad \text{and} \quad 1 \wedge (x \vee y) \approx (1 \wedge x) \vee (1 \vee y).$$

Semilinear varieties play a fundamental role in fuzzy logics. In particular, the varieties $\mathcal{UL}$ of commutative semilinear bounded FL-algebras and $\mathcal{MTL}$ of commutative semilinear bounded FL-algebras satisfying the equations $1 \approx \top$ and $0 \approx \bot$ form algebraic semantics for *uninorm logic* ([101]) and *monoidal t-norm logic* ([48]) respectively (see Subsection 5.2).

4.3 The class $\mathcal{RL}$ is an ideal variety

The main result of this subsection is Theorem 4.17 below, first established in [19]. It shows that the congruences of members of $\mathcal{RL}$ are determined by their convex normal subalgebras (to be defined below). In particular, $\mathcal{RL}$, and hence $\mathcal{FL}$, is a *1-regular variety*, that is, each congruence relation of an algebra in $\mathcal{RL}$ is determined by its equivalence class of 1. A more economical proof of 1-regularity for $\mathcal{RL}$ can be given by observing that this property is a *Mal'cev property*, meaning that one can establish if a variety has the property by checking whether it satisfies certain quasi-equations involving finitely many terms (two, in this special case). However, a concrete description of these equivalence classes is essential for developing the structure theory of residuated lattices and its applications to substructural logics.

If **L** is a residuated lattice, the set $L^- = \{a \in L \mid a \leq 1\}$ is called the *negative cone* of **L**. Note that the negative cone is a submonoid of $\langle L, \cdot, 1 \rangle$. As such, we will denote it by $\mathbf{L}^-$.

Let $\mathbf{L} \in \mathcal{RL}$. For each $a \in L$, define $\rho_a(x) = (ax/a) \wedge 1$ and $\lambda_a(x) = (a \backslash xa) \wedge 1$. We refer to ρ_a and λ_a respectively as *right and left conjugation* by a. An *iterated conjugation* map is a finite composition of right and left conjugation maps.

A subset $X \subseteq L$ is called *(order-)convex* if for any $x, y \in X$ and $a \in L$, $x \leq a \leq y$ implies $a \in X$; X is called *normal* if it is closed with respect to all iterated conjugations.

Let **L** be a residuated lattice. For $a, b \in L$ define $[a, b]_r = (ab/ba) \wedge 1$ and $[a, b]_l = (ba \backslash ab) \wedge 1$. We call $[a, b]_r$ and $[a, b]_l$ respectively the *right and left commutators* of a with b.

We will say that a subset X is *closed with respect to commutators* if for any $a \in L$ and $x \in X$, both the commutators $[a, x]_r$ and $[x, a]_l$ lie in X. Normality and "closure with respect to commutators" are identical properties for certain "nice" subsets as we show in the next two lemmas.

Lemma 4.11. *Let* **H** *be a convex subalgebra of* **L**. *Then* **H** *is normal if and only if it is closed with respect to commutators.*

Proof. Suppose that **H** is normal. Then $1 \geq [a, h]_r = (ah/ha) \wedge 1 = ((ah/a)/h) \wedge 1 \geq (((ah/a) \wedge 1)/h) \wedge 1 = (\rho_a(h)/h) \wedge 1 \in H$ so that $[a, h]_r \in H$ by convexity. The proof that $[h, a]_l \in H$ is analogous.

Conversely, suppose that H is closed with respect to commutators. We have $[a, h]_r h \wedge 1 \in H$ and $[a, h]_r h \wedge 1 = ((ah/ha) \wedge 1)h \wedge 1 \leq (ah/ha)h \wedge 1 = ((ah/a)/h)h \wedge 1 \leq (ah/a) \wedge 1 = \rho_a(h) \leq 1$ so $\rho_a(h) \in H$ by convexity. The proof that $\lambda_a(h) \in H$ is analogous. $\qquad\square$

The same result holds for convex submonoids of the negative cone of $\mathbf{L}$:

Lemma 4.12. *If* $\mathbf{S}$ *is a convex submonoid of* $\mathbf{L}^-$, *then* $\mathbf{S}$ *is normal if and only if* S *is closed with respect to commutators.*

Proof. Let $s \in S$ and $a \in L$ and suppose that $\mathbf{S}$ is normal. Then $1 \geq [a, s]_r = (as/sa) \wedge 1 = ((as/a)/s) \wedge 1 \geq (as/a) \wedge 1 = \rho_a(s) \in S$ where the last inequality above follows since $s \leq 1$. Similarly, $[s, a]_l \in S$. Conversely, if S is closed with respect to commutators, then $[a, s]_r s \in S$. But $[a, s]_r s = (((as/a)/s) \wedge 1)s \leq ((as/a)/s)s \wedge s \leq (as/a) \wedge s \leq (as/a) \wedge 1 = \rho_a(s) \leq 1$ and by convexity we have $\rho_a(s) \in S$. Similarly, $\lambda_a(s) \in S$. $\qquad\square$

We often find it useful to convert one of the division operations into its dual. The following two equations, which are referred to as *switching equations* and can be verified by straightforward calculation, provide a means to do so in any residuated lattice:

$$z/y \leq py\backslash z \ , \text{ where } p = [z/y, y]_r \ , \text{ and}$$
$$x\backslash z \leq z/xq \ , \text{ where } q = [x, x\backslash z]_l.$$

Note that the above equations still hold if the "$\wedge 1$" factor is omitted from the commutators.

Lemma 4.13. *Let* $\mathbf{L}$ *be a residuated lattice and* $\Theta \in Con(\mathbf{L})$. *Then the following are equivalent:*

1. $a \Theta b$
2. $[(a/b) \wedge 1] \Theta 1$ *and* $[(b/a) \wedge 1] \Theta 1$
3. $[(a\backslash b) \wedge 1] \Theta 1$ *and* $[(b\backslash a) \wedge 1] \Theta 1$

Proof. Suppose that $a \Theta b$. Then $(a/a) \Theta (b/a)$ so that

$$1 = [(a/a) \wedge 1] \Theta [(b/a) \wedge 1]$$

and the other relations in (2) and (3) follow similarly. Conversely, suppose that both $[(a/b) \wedge 1] \Theta 1$ and $[(b/a) \wedge 1] \Theta 1$. Setting $r = [(a/b) \wedge 1]b$ and $s = [(b/a) \wedge 1]a$, we have $r \Theta b$ and $s \Theta a$. Moreover, $r \leq (a/b)b \leq a$ and $s \leq (b/a)a \leq b$ so that $r = (a \wedge r) \Theta (a \wedge b)$ and $s = (b \wedge s) \Theta (b \wedge a)$ whence $b \Theta r \Theta (a \wedge b) \Theta s \Theta a$; we have shown (2) $\rightarrow$ (1). (3) $\rightarrow$ (1) is proved in an analogous manner. $\qquad\square$

Lemma 4.14. *Let Θ be a congruence relation on a residuated lattice* **L.** *Then $[1]_\Theta = \{a \in A \mid a \, \Theta \, 1\}$ is a convex normal subalgebra of* **L.**

Proof. Since 1 is idempotent with respect to all the binary operations of **L**, it immediately follows that $[1]_\Theta$ forms a subalgebra of **L**. Convexity can be checked directly, or is a consequence of the well-known fact that the equivalence classes of any lattice congruence are convex. Finally, let $a \in [1]_\Theta$ and $c \in L$. Then

$$\lambda_c(a) = (c\backslash ac) \wedge 1 \, \Theta \, (c\backslash 1c) \wedge 1 = (c\backslash c) \wedge 1 = 1$$

so that $\lambda_c(a) \in [1]_\Theta$. Similarly, $\rho_c(a) \in [1]_\Theta$. □

Lemma 4.15. *Suppose that* **H** *is a convex normal subalgebra of* **L.** *For any $a, b \in L$,*

$$(a/b) \wedge 1 \in H \; \Leftrightarrow \; (b\backslash a) \wedge 1 \in H.$$

Proof. Suppose that $(a/b) \wedge 1 \in H$. Since **H** is normal, we have

$$h = b\backslash(((a/b) \wedge 1)b) \wedge 1 \in H.$$

But $h \leq [b\backslash(a/b)b] \wedge 1 \leq (b\backslash a) \wedge 1 \leq 1 \in H$ so $(b\backslash a) \wedge 1 \in H$. The reverse implication is proved similarly. □

Next we characterize the congruence corresponding to a given convex normal subalgebra.

Lemma 4.16. *Let* **H** *be a convex normal subalgebra of a residuated lattice* **L.** *Then*

$$\begin{aligned}
\Theta_{\mathbf{H}} &= \{(a, b) \mid \exists h \in H, ha \leq b \text{ and } hb \leq a\} \\
&= \{(a, b) \mid (a/b) \wedge 1 \in H \text{ and } (b/a) \wedge 1 \in H\} \\
&= \{(a, b) \mid (a\backslash b) \wedge 1 \in H \text{ and } (b\backslash a) \wedge 1 \in H\}
\end{aligned}$$

is a congruence on **L.**

Proof. First we show that the three sets defined above are indeed equal. That the second and third sets are identical follows from Lemma 4.15. If (a, b) is a member of the second set, then letting $h = (a/b) \wedge (b/a) \wedge 1$, we have $h \in H$, $ha \leq (b/a)a \leq b$ and $hb \leq (a/b)b \leq b$, so that (a, b) is a member of the first set. Conversely, if (a, b) is a member of the first set, then for some $h \in H$ we have $ha \leq b$ or $h \leq b/a$, and hence $h \wedge 1 \leq (b/a) \wedge 1 \leq 1$. By convexity, we get $(b/a) \wedge 1 \in H$. Similarly, $(a/b) \wedge 1 \in H$.

It is a simple matter to verify that $\Theta_{\mathbf{H}}$ is an equivalence relation. To prove that it is a congruence relation, we must establish its compatibility with respect to multiplication, meet, join, right division, and left division. We just verify compatibility for multiplication and right division:

Θ is compatible with multiplication
Suppose that $a \; \Theta \; b$ and $c \in L$. Then

$$(a/b) \wedge 1 \leq (ac/bc) \wedge 1 \leq 1$$

so $(ac/bc) \wedge 1 \in H$. Similarly, $(bc/ac) \wedge 1 \in H$ so $(ac) \; \Theta \; (bc)$. Next, using the normality of $\mathbf{H}$,

$$\rho_c((a/b) \wedge 1) = (c[(a/b) \wedge 1]/c) \wedge 1 \in H.$$

But $\rho_c((a/b) \wedge 1) \leq [c(a/b)/c] \wedge 1 \leq [ca/b/c] \wedge 1 = (ca/cb) \wedge 1 \leq 1 \in H$ so $(ca/cb) \wedge 1 \in H$. Similarly, $(cb/ca) \wedge 1 \in H$ so $(ca) \; \Theta \; (cb)$.

Θ is compatible with right division
Suppose that $a \; \Theta \; b$ and $c \in L$. Then

$$(a/b) \wedge 1 \leq [(a/c)/(b/c)] \wedge 1 \leq 1$$

so $[(a/c)/(b/c)] \wedge 1 \in H$. Similarly, $[(b/c)/(a/c)] \wedge 1 \in H$ so $(a/c) \; \Theta \; (b/c)$. Next,

$$(b/a) \wedge 1 \leq [(c/b)\backslash(c/a)] \wedge 1 \leq 1 \in H$$

so $[(c/b)\backslash(c/a)] \wedge 1 \in H$. Hence it follows by Lemma 4.15 that $[(c/b)/(c/a)] \wedge 1 \in H$. Similarly, $[(c/a)/(c/b)] \wedge 1 \in H$ so $(c/a) \; \Theta \; (c/b)$.

$\square$

Theorem 4.17. *The lattice $CN(\mathbf{L})$ of convex normal subalgebras of a residuated lattice $\mathbf{L}$ is isomorphic to its congruence lattice $Con(\mathbf{L})$. The isomorphism is given by the mutually inverse maps $\mathbf{H} \mapsto \Theta_{\mathbf{H}}$ and $\Theta \mapsto [1]_{\Theta}$.*

Proof. We have shown both that $\Theta_{\mathbf{H}}$ is a congruence and that $[1]_{\Theta}$ is a member of $CN(\mathbf{L})$, and it is clear that the maps $\mathbf{H} \mapsto \Theta_{\mathbf{H}}$ and $\Theta \mapsto [1]_{\Theta}$ are isotone. It remains only to show that these two maps are mutually inverse, since it will then follow that they are lattice homomorphisms.

Given $\Theta \in \mathrm{Con}(\mathbf{L})$, set $H = [1]_\Theta$; we must show that $\Theta = \Theta_\mathbf{H}$. But this is easy; using Lemma 4.13,

$$a \ \Theta \ b \ \Leftrightarrow \ [((a/b) \wedge 1) \ \Theta \ 1 \text{ and } ((b/a) \wedge 1) \ \Theta \ 1] \ \Leftrightarrow$$

$$[((a/b) \wedge 1) \in H \text{ and } ((b/a) \wedge 1) \in H] \Leftrightarrow a \ \Theta_\mathbf{H} \ b.$$

Conversely, for any $\mathbf{H} \in \mathcal{C}N(\mathbf{L})$ we must show that $H = [1]_{\Theta_\mathbf{H}}$. But

$$h \in H \ \rightarrow \ [(h/1) \wedge 1 \in H \text{ and } (1/h) \wedge 1 \in H]$$

so $h \in [1]_{\Theta_\mathbf{H}}$. If $a \in [1]_{\Theta_\mathbf{H}}$, then $(a, 1) \in \Theta_\mathbf{H}$ and we use the first description of $\Theta_\mathbf{H}$ in Lemma 4.16 to conclude that there exists an element $h \in H$ such that $ha \leq 1$ and $h = h1 \leq a$. Now it follows from the convexity of $\mathbf{H}$ that $h \leq a \leq h\backslash 1$ implies $a \in H$. □

We remark that in the event that $\mathbf{L}$ is commutative, then every convex subalgebra of $\mathbf{L}$ is normal. Thus the preceding theorem implies the following result of [71]:

Corollary 4.18. *The lattice $\mathcal{C}(\mathbf{L})$ of convex subalgebras of a commutative residuated lattice $\mathbf{L}$ is isomorphic to its congruence lattice $\mathrm{Con}(\mathbf{L})$. The isomorphism is given by the mutually inverse maps $\mathbf{H} \mapsto \Theta_\mathbf{H}$ and $\Theta \mapsto [1]_\Theta$.*

4.4 Convex normal submonoids and deductive filters

In the previous subsection we saw that the congruences of a residuated lattice $\mathbf{L}$ correspond to its convex normal subalgebras. Here we show that these subalgebras in turn correspond to both the convex normal submonoids of $\mathbf{L}^-$ and the deductive filters of $\mathbf{L}$ (defined below). The original references for the first correspondence are [18, 19]. Deductive filters (under the name "filters") and their correspondence with congruences of residuated lattices were introduced in [16]. See also [82, 141], and [52].

The next lemma shows that a convex normal subalgebra is completely determined by its negative cone:

Lemma 4.19. *Let $\mathbf{S}$ be a convex normal submonoid of $\mathbf{L}^-$. Then $H_S := \{a \mid s \leq a \leq s\backslash 1, \text{ for some } s \in S\}$ is the universe of a convex normal subalgebra $\mathbf{H}_S$ of $\mathbf{L}$, and $S = H_S^-$. Conversely, if $\mathbf{H}$ is any convex normal subalgebra of $\mathbf{L}$ then, setting $S_H = H^-$, $\mathbf{S}_H$ is a convex normal submonoid of $\mathbf{L}^-$ and H can be recovered from S_H as described above. Moreover, the mutually inverse maps $\mathbf{H} \mapsto \mathbf{S}_H$ and $\mathbf{S} \mapsto \mathbf{H}_S$ establish a lattice isomorphism between $\mathcal{C}N(\mathbf{L})$ and $\mathcal{C}NM(\mathbf{L}^-)$.*

Proof. Given a convex, normal subalgebra **H** of **L**, the assertions about S_H are easy to verify. Thus we turn our attention to the other direction. Let **S** be a convex normal submonoid of $\mathbf{L}^-$ and define H_S as above. It is easy to show that H_S is convex and normal. Moreover, it is immediate that $H_S^- = S$. It remains to prove closure under the binary operations. We just verify closure under left and right division. To this end, let $a, b \in H_S$. Then there are $s, t \in S$ so that $s \leq a \leq s\backslash 1$ and $t \leq b \leq t\backslash 1$.

Closure under left division

We have $a\backslash b \leq s\backslash(t\backslash 1) = (ts)\backslash 1$, but to find a lower bound for $a\backslash b$ is a little trickier. First notice that $t \leq b$ and $sa \leq 1$ imply that $tsa \leq b$. From this we derive $ats(ats\backslash tsa) \leq tsa \leq b$ and $ts(ats\backslash tsa) \leq a\backslash b$. Setting $p = (ats)\backslash(tsa)$ and $q = ts(p \wedge 1)$, we know that $p \wedge 1 = [ts, a]_l \in S$ and so $q \in S$. But now $q \leq tsp \leq a\backslash b$ and we have found the desired lower bound. Finally, setting $r = qts$, it follows that $r \leq a\backslash b \leq r\backslash 1$.

Closure under right division

Observe that $s \leq a$ and $tb \leq 1$ imply the inequalities $stb \leq a$ and $st \leq a/b$, but to find an upper bound requires extra work: $a/b \leq (s\backslash 1)/t \leq pt\backslash(s\backslash 1) = spt\backslash 1$, where $p = [(s\backslash 1)/t, t]_r$ as given by the switching equation. But $p \in S$ by the comments following Lemma 4.11 and we have found an appropriate upper bound. Finally, we can set $r = (st)(spt)$ and it follows that $r \leq a/b \leq r\backslash 1$.

We have shown that the maps between the two lattices are well-defined and mutually inverse. Since they are clearly isotone, the theorem is proved.   □

A subset F of a residuated lattice is a *deductive filter* provided:

(DF1) $\uparrow\{1\} \subseteq F$,

(DF2)   if $x, x\backslash y \in F$, then $y \in F$,

(DF3)   if $x, y/x \in F$, then $y \in F$,

(DF4)   if $x, y \in F$, then $x \wedge y \in F$,

(DF5)   if $x \in F$ and $y \in L$, then $y\backslash(xy) \in F$ and $(yx)/y \in F$.

An alternative description of a deductive filter is provided by the following result of [141].

Lemma 4.20. *A subset F of a residuated lattice* **L** *is a deductive filter if and only if it is a non-empty order-filter of* **L** *closed under multiplication and conjugation.*

Let $CN(\mathbf{L})$, $CNM(\mathbf{L}^-)$, and $\mathcal{DF}(\mathbf{L})$ denote respectively the lattices under set-inclusion of convex normal subalgebras of **L**, convex normal submonoids of $\mathbf{L}^-$, and deductive filters of **L**. We have the following result (see [141] or [52]):

Proposition 4.21. *In a residuated lattice* **L**, *the lattice* $CNM(\mathbf{L}^-)$ *of convex normal submonoids of* $\mathbf{L}^-$ *is isomorphic to the lattice* $\mathcal{DF}(\mathbf{L})$ *of deductive filters of* **L**. *The isomorphism is given by the mutually inverse maps* $M \mapsto \uparrow M$ *and* $F \mapsto F^-$, *for* $M \in CNM(\mathbf{L}^-)$ *and* $F \in \mathcal{DF}(\mathbf{L})$.

4.5 Convex normal subalgebra generation

The original references for the results of this subsection are [18] and [19]. (See also [82].) They provide intrinsic descriptions of convex normal submonoids, convex normal subalgebras, and deductive filters. The local deduction theorem for the logic corresponding to the variety of commutative residuated lattices, Theorem 6.3 (1), and the parametrized local deduction theorem in [53] are the logical counterparts of and follow easily from Corollary 4.28 (1) and Proposition 4.24, respectively.

Lemma 4.22. *For all* $a_1, a_2, \ldots, a_n, b \in L$, *if* $a = \prod a_j$, *then*

$$\prod \rho_b(a_j) \le \rho_b(a) \quad and \quad \prod \lambda_b(a_j) \le \lambda_b(a).$$

Proof. We prove only the case $n = 2$; the proof can be completed by the obvious induction.

$$\rho_b(a_1)\rho_b(a_2) = [(ba_1/b) \wedge 1][(ba_2/b) \wedge 1] \le [(ba_1/b)(ba_2/b)] \wedge 1$$

$$\le [((ba_1/b)ba_2)/b] \wedge 1 \le (ba_1a_2/b) \wedge 1 = \rho_b(a_1a_2).$$

In the last two inequalities, we use Lemma 4.3 (5) and (4) respectively. The proof for λ_b is analogous. □

The next result provides an element-wise description of a convex normal submonoid of the negative cone generated by a subset.

Proposition 4.23. *Let* **L** *be a residuated lattice and* $S \subseteq L^-$. *An element* $x \le 1$ *belongs to the convex normal submonoid of* $\mathbf{L}^-$ *generated by* S *iff there exist iterated conjugates* $\gamma_1(s_1), \ldots, \gamma_n(s_n)$ *of elements of* S *such that* $\gamma_1(s_1) \ldots \gamma_n(s_n) \le x$.

Proof. Let $cnm(S)$ denote the set described in the statement of the proposition. It is clear that $1 \in M(S)$, that $cnm(S)$ is convex and closed under multiplication, and that any convex normal submonoid of $\mathbf{L}^-$ containing S must contain $cnm(S)$. Moreover, since $S \subseteq L^-$, $S \subseteq cnm(S)$. It only remains to show that $cnm(S)$ is normal. But this follows from Lemma 4.22 and the convexity of $cnm(S)$: if $x \in cnm(S)$, then for some iterated conjugates $\gamma_1(s_1), \ldots, \gamma_n(s_n)$ of elements of S, $\gamma_1(s_1) \ldots \gamma_n(s_n) \le x \le 1$. By Lemma 4.22, for all $a \in L$, $\rho_a(\gamma_1(s_1)) \ldots \rho_a(\gamma_n(s_n)) \le \rho_a(x) \le 1$, and similarly for $\lambda_a(x)$.   □

For any subsets $S \subseteq L^-$ and $T \subseteq L$, we write $cnm(S)$ for the convex normal submonoid of $\mathbf{L}^-$ generated by S, $cn(T)$ for the convex normal subalgebra of $\mathbf{L}$ generated by T, and $df(T)$ for the deductive filter of $\mathbf{L}$ generated by T.

We have as a direct consequence of Lemma 4.19 and Proposition 4.23:

Proposition 4.24. *Let* $\mathbf{L}$ *be a residuated lattice and* $S \subseteq L^-$. *Then* $x \in df(S)$ *iff there exist iterated conjugates* $\gamma_1(s_1), \ldots, \gamma_n(s_n)$ *of elements of* S *such that* $\gamma_1(s_1) \ldots \gamma_n(s_n) \le x$.

Likewise, Propositions 4.21 and 4.23 easily yield:

Proposition 4.25. *Let* $\mathbf{L}$ *be a residuated lattice and* $S \subseteq L^-$. *Then* $x \in cn(S)$ *iff there exist iterated conjugates* $\gamma_1(s_1), \ldots, \gamma_n(s_n)$ *of elements of* S *such that* $\gamma_1(s_1) \ldots \gamma_n(s_n) \le x \le (\gamma_1(s_1) \ldots \gamma_n(s_n)) \backslash 1$.

The natural question arises as to whether there are analogous descriptions of $cn(S)$ and $df(S)$ for arbitrary subsets S of L. Let us write $cnm(a)$ for $cnm(\{a\})$ (for $a \in L^-$), $df(a)$ for $df(\{a\})$, and $cn(a)$ for $cn(\{a\})$.

Lemma 4.26. *For any* $a \in L$, $cn(a) = cn(a')$ *and* $df(a) = df(a')$, *where* $a' = a \wedge (1/a) \wedge 1$.

Proof. Clearly $a' \in cn(a)$. On the other hand,

$$a' \le a \le (1/a) \backslash 1 \le a' \backslash 1,$$

so $a \in cn(a')$, and likewise for $df(a)$.   □

Thus we have the following corollary:

Corollary 4.27. *Let $S \subseteq L$ and set $S^* = \{s \wedge (1/s) \wedge 1 \mid s \in S\}$. Then:*

1. $x \in cn(S)$ *iff there exist iterated conjugates* $\gamma_1(s_1), \ldots, \gamma_n(s_n)$ *of elements of S^* such that*

$$\gamma_1(s_1) \ldots \gamma_n(s_n) \leq x \leq x \leq (\gamma_1(s_1) \ldots \gamma_n(s_n)) \backslash 1.$$

2. $x \in df(S)$ *iff there exist iterated conjugates* $\gamma_1(s_1), \ldots, \gamma_n(s_n)$ *of elements of S^* such that*

$$\gamma_1(s_1) \ldots \gamma_n(s_n) \leq x.$$

We close this subsection by noting that the above intrinsic descriptions become substantially simpler whenever **L** is a commutative residuated lattice. In this case, the convex normal submonoids of $\mathbf{L}^-$ are its convex submonoids, the convex normal subalgebras of **L** are the convex subalgebras, and the deductive filters of **L** are the non-empty order-filters closed under multiplication. Thus, we immediately get the following result from [71]:

Corollary 4.28. *Let **L** be a commutative residuated lattice, and let $S \cup \{a\} \subseteq L^-$. Then:*

1. $x \in cnm(a)$ *iff there exists a natural number n such that $a^n \leq x \leq 1$.*
2. $x \in cnm(S)$ *iff there exist $s_1, \ldots, s_k \in S$ and natural numbers $n_1, \ldots, n_k$ such that $s_1^{n_1} \ldots s_k^{n_k} \leq x \leq 1$.*
3. $x \in cn(a)$ *iff there exists a natural number n such that $a^n \leq x \leq a \backslash 1$.*
4. $x \in cn(S)$ *iff there exist $s_1, \ldots, s_k \in S$ and natural numbers $n_1, \ldots, n_k$ such that $s_1^{n_1} \ldots s_k^{n_k} \leq x \leq (s_1^{n_1} \ldots s_k^{n_k}) \backslash 1$.*
5. $x \in df(a)$ *iff there exists a natural number n such that $a^n \leq x$.*
6. $x \in df(S)$ *iff there exist $s_1, \ldots, s_k \in S$ and natural numbers $n_1, \ldots, n_k$ such that $s_1^{n_1} \ldots s_k^{n_k} \leq x$.*

4.6 Historical remarks

In this section we attempt to summarily reconstruct the development of the concept of residuated structure in algebra. Instead of organizing this historical survey in strict chronological order, we prefer – for the sake of greater readability – to follow five separate thematic threads, each of which has in our opinion decisively contributed to shaping the contemporary notion of residuated structure.

A word of caution is in order here: although residuated maps are almost ubiquitous in mathematics, we circumscribe our survey to examples of residuation which bear a tighter connection to the main theme of this paper. Therefore, we will mostly confine ourselves to examining the historical development of residuated lattices, disregarding, *e.g.*, the plentiful and certainly important examples of residuated pairs $(\varphi, \varphi_*) : \mathbf{P} \to \mathbf{Q}$, where $\mathbf{P}$ and $\mathbf{Q}$ are *different* posets. The content of the next subsection is the only exception to this policy, motivated by the great historical relevance of Galois theory as the first significant appearance of residuation in mathematics.

Galois theory

After the Italian Renaissance mathematicians Scipione del Ferro, Tartaglia, Cardano, and Ferrari had shown that cubic and quartic equations were solvable by radicals by means of a general formula, algebraists spent subsequent centuries striving to achieve a similar result for polynomial equations with rational coefficients of degree 5 or higher. These efforts came to an abrupt end in 1824, when Niels Abel (patching an earlier incorrect proof by another Italian, Paolo Ruffini) showed that such equations have no general solution by radicals. Since, however, it was well-known that some *particular* equations of degree greater or equal than 5 could indeed be solved, the question remained open as to whether a general criterion was available to determine which polynomial equations were solvable by radicals and which ones were not.

In 1832, the French mathematician Évariste Galois (just before meeting his death in a tragical duel) found the right approach to settle the issue once and for all. He associated to each polynomial equation ε a permutation group (now called *Galois group* in honor of its inventor) consisting of those permutations of the set of all roots of ε having the property that every algebraic equation satisfied by the roots themselves is still satisfied after the roots have been permuted. As a simple example, let ε be the equation

$$x^2 - 4x + 1 \approx 0,$$

whose roots are $r_1 = 2 + \sqrt{3}$, $r_2 = 2 - \sqrt{3}$. It can be shown that *any* algebraic equation with rational coefficients in the variables x and y which is satisfied by $x = r_1$ and $y = r_2$ (for example, $xy \approx 1$ or $x + y \approx 4$) is also satisfied by $y = r_1$ and $x = r_2$. It follows that both permutations of the two-element set $\{r_1, r_2\}$ – the identity permutation and the permutation which exchanges r_1 with r_2 – belong to the Galois group of ε, which is therefore isomorphic to the cyclic group of order 2. More generally, Galois established that a polynomial equation is solvable by radicals if and only if its Galois group $\mathbf{G}$ is *solvable* – namely, if there exist subgroups $\mathbf{G}_0, \ldots, \mathbf{G}_n$ of $\mathbf{G}$ such that

$$\{1\} = \mathbf{G}_0 \subset \mathbf{G}_1 \subset \ldots \mathbf{G}_{n-1} \subset \mathbf{G}_n = \mathbf{G},$$

and, moreover, for all $i \leq n$, $\mathbf{G}_{i-1}$ is normal in $\mathbf{G}_i$ and $\mathbf{G}_i/\mathbf{G}_{i-1}$ is Abelian.

In the above example, permutations which respect algebraic equations satisfied by r_1 and r_2 can be seen as automorphisms of the quotient field $\mathbb{Q}(r_1, r_2)/\mathbb{Q}$, where $\mathbb{Q}(r_1, r_2)$ is nothing but the field one obtains from the field of rationals by adjoining the two roots of the given equation. This approach can be taken up in general, and it is indeed this abstract perspective that underlies present-day Galois theory (see, *e.g.*, [83]), where Galois groups are seen as field automorphisms of a field extension $\mathbf{L}/\mathbf{F}$ of a given base field $\mathbf{F}$. Let us now stipulate that:

- $S(\mathbf{L}, \mathbf{F})$ is the set of all subfields of $\mathbf{L}$ that contain $\mathbf{F}$;
- for $\mathbf{M} \in S(\mathbf{L}, \mathbf{F})$, $\mathbf{Gal}_{\mathbf{M}}(\mathbf{L})$ is the group of all field automorphisms φ of $\mathbf{L}$ such that $\varphi|_M = id$;
- $Sg(\mathbf{Gal}_{\mathbf{F}}(\mathbf{L}))$ is the set of all subgroups of such a group.

Then the maps

$$f(\mathbf{M}) = \mathbf{Gal}_{\mathbf{M}}(\mathbf{L})$$
$$f_*(\mathbf{H}) = \{a \in L \mid \varphi(a) = a \text{ for all } \varphi \in H\}$$

induce a residuated pair (f, f_*) between $\mathbf{P} = \langle S(\mathbf{L}, \mathbf{F}), \subseteq \rangle$ and the order dual $\mathbf{Q}^{\partial}$ of $\mathbf{Q} = \langle Sg(\mathbf{Gal}_{\mathbf{F}}(\mathbf{L})), \subseteq \rangle$ (*i.e.*, the poset obtained from $\mathbf{Q}$ by reversing its order)[4]. Consequently, Galois theory provides us with a first mathematically significant instance of residuation.

Ideal theory of rings

There is a another respect in which polynomial equations constitute a historically remarkable source for residuated structures. Around the middle of the XIX century, it was observed by Ernst Kummer that unique factorization into primes, true of ordinary integers in virtue of the fundamental theorem of arithmetic, fails instead for algebraic integers – namely, for roots of monic polynomials with integer coefficients. Let $\mathbb{Z}\left[\sqrt{-n}\right]$

---

[4] This example of residuated pair is actually so important that, in general, a residuated pair (f, f_*) between posets $\mathbf{P}$ and $\mathbf{Q}^{\partial}$ is often referred to as a *Galois correspondence* between $\mathbf{P}$ and $\mathbf{Q}$.

denote the quadratic integer ring of all complex numbers of the form $a + b\sqrt{-n}$, with $a, b \in \mathbb{Z}$ and $n \in \mathbb{N}$. It turns out that unique factorization fails in $\mathbb{Z}[\sqrt{-n}]$ for several instances of n, although it holds in some special cases – *e.g.*, for *Gaussian integers* ($n = 1$). Kummer tried to recover a weakened form of this fundamental property with his theory of *ideal numbers*, but the adoption of a modern abstract viewpoint on the issue, that would eventually lead to the birth of contemporary ring theory and to the ideal theory of rings, must be credited to Richard Dedekind's 1871 X Supplement to the second edition of Dirichlet's *Zahlentheorie*. There, Dedekind introduced the concepts of ring and ring ideal in what essentially is their modern usage (the term *ring*, however, was coined by Hilbert only much later) and proved that every ideal of the ring of algebraic integers is uniquely representable (up to permutation of factors) as a product of prime ideals. Unique factorization, therefore, is recovered at the level of ideals (see, *e.g.*, [45]).

The ideal theory of commutative rings was intensively investigated early in the XX century by Lasker and Macaulay, who generalized the results by Dedekind to polynomial rings, and in the 1920's by Noether and Krull. In particular, Emmy Noether proved the celebrated theorem according to which, in any commutative ring whose lattice of ideals satisfies the ascending chain condition, ideals decompose into intersections of finitely many primary ideals.

This theorem is typical of Noether's general approach. The system $I(\mathbf{R})$ of ideals of a commutative ring $\mathbf{R}$ is viewed as an instance of a lattice[5] endowed with an additional operation of multiplication (see Example 4.7). The same viewpoint was taken and further developed by Morgan Ward and his student R.P. Dilworth in a series of papers during the late 1930's (see, *e.g.*, [39–41, 138–140]), whose focus is on another binary operation on ideals: the *residual* $J \rightarrow I$ of I with respect to J. Recall from Example 4.7 above that

$$J \rightarrow I = \{x \in R \mid xJ \subseteq I\}.$$

As Ward and Dilworth observed, $J \rightarrow I$ has the property that, for any ideal K of the ring $\mathbf{R}$, $JK \subseteq I$ iff $K \subseteq J \rightarrow I$. Ward and Dilworth introduced and investigated in detail, under the name of *residuated lattices*, some lattice-ordered structures with a multiplication which is abstracted from ideal multiplication, and with a residuation which is in turn

---

[5] Lattice theory underwent a remarkable development as a spin-off from Noether's abstract perspective, partly motivated by her desire to release ring theory from the concrete setting of polynomial rings.

abstracted from ideal residuation. They were thus in a position to extend to a purely lattice-theoretic setting some of the results obtained by Noether and Krull for lattices of ideals of commutative rings, including the above-mentioned Noether decomposition theorem.

Ward and Dilworth's papers did not have that much immediate impact, but began a line of research that would crop up again every so often in the following decades. Thus, the notion of residuated lattice re-emerged in the different contexts of the semantics for fuzzy logics (*e.g.*, [67]) and substructural logics (*e.g.*, [110]), and in the setting of studies with a more pronounced universal-algebraic flavor (*e.g.*, [82]), with the latter two streams eventually converging into a single one (see, *e.g.*, [52]).

Interestingly enough, neither Hájek's nor Ono's, nor our official definition of residuated lattice, given in Subsection 4.2 and essentially due to Blount and Tsinakis ([19]), exactly overlaps with the original definition given by Ward and Dilworth. Differences concern both the similarity type and, less superficially, the properties characterizing the respective algebras. The Hájek-Ono residuated lattices are invariably *bounded* as lattices and *integral* as partially ordered monoids, meaning that the top element of the lattice is the neutral element of multiplication. Moreover, multiplication is *commutative*. Residuated lattices as defined here are not necessarily bounded and, even if they are, they need not be integral; multiplication is not required to be commutative. What about the original Ward-Dilworth residuated lattices? Put in a very rough way, they lie somewhere in between the preceding concepts: in fact, Ward and Dilworth do not assume the existence of a top or bottom in the underlying lattice, but if there is a top, then it must be the neutral element of multiplication, which is supposed to be a commutative operation. It is only fair to observe that Dilworth also introduces in [40] a noncommutative variant of his notion of residuated lattice, abstracted from the residuated lattice of two-sided ideals of a noncommutative ring, but to the best of our knowledge this generalization was not taken up again until the noncommutative concept of residuated lattice was introduced and had already become established.

Boolean and Heyting algebras

One of the first *classes* of residuated lattices that received considerable attention in its own right was of course the class of Boolean algebras. As we have seen in Subsection 3.1, it is not historically correct to consider Boole as their inventor. A closer approximation to the modern understanding of Boolean algebras can instead be found in the writings of Ernst Schröder, especially his *Operationskreis des Logikkalkuls* (1854). There, he introduces something vaguely resembling an equational axiomatization of Boolean algebras, replacing Boole's partial aggregation operation by plain lattice join, and comes very close to realizing that Boolean algebras make clear-cut instances of residuated algebras: he observes, in fact, that the equation $x \wedge b = a$ is solvable iff $a \wedge \neg b = 0$ and possesses, when this condition is met, a smallest solution $x = a$ and a largest solution $x = a \vee \neg b$. What he apparently misses, however, is the fact that $a \vee \neg b$ is also a largest solution for the *inequation* $x \wedge b \leq a$, or, in other words, that such an operation residuates meet in Boolean algebras [47].

For a proper treatment of Boolean algebras as *algebras* one must wait until the turn of the century, when Edward V. Huntington provided the first equational basis for the variety, followed in the subsequent thirty years by more and more economical axiomatizations with different choices of primitive operations, due, *e.g.*, to Bernstein and Sheffer. Most of the fundamental results of the structure theory for Boolean algebras were however established in the 1930's by Marshall Stone ([126]), *in primis*:

- the term-equivalence between Boolean rings (*i.e.*, idempotent rings with identity) and Boolean algebras;
- the representation of Boolean algebras as algebras of sets, which can be seen as a corollary of a more comprehensive duality between Boolean algebras and a class of topological spaces (Boolean spaces, *i.e.*, totally disconnected compact Hausdorff spaces.)

Both results are extremely powerful, the former because a class of algebras that had been introduced for the purpose of systematizing logical reasoning turns out to be nothing but a subclass of a class of structures whose centrality in "standard" mathematics can hardly be denied – and a very natural subclass, at that. Given a Boolean algebra $\mathbf{B} = \langle B, \wedge, \vee, \neg, 1, 0 \rangle$, Stone associates to it an algebra $\mathbf{R_B} = \langle B, \cdot, +, -, 0, 1 \rangle$, where for every $a, b \in B$

$$a \cdot b = a \wedge b$$
$$a + b = (a \wedge \neg b) \vee (b \wedge \neg a)$$
$$-a = a.$$

Conversely, given a Boolean ring $\mathbf{R} = \langle R, \cdot, +, -, 0, 1 \rangle$, he constructs an algebra $\mathbf{B_R} = \langle R, \wedge, \vee, \neg, 1, 0 \rangle$, where for every $a, b \in R$

$$a \wedge b = a \cdot b$$
$$a \vee b = a + b + a \cdot b$$
$$\neg a = a + 1.$$

He then shows that: (i) $\mathbf{R_B}$ is a Boolean ring; (ii) $\mathbf{B_R}$ is a Boolean algebra; (iii) $\mathbf{R_{B_R}} = \mathbf{R}$; (iv) $\mathbf{B_{R_B}} = \mathbf{B}$.

The latter result quoted above fleshes out a natural intuition about Boolean algebras as essentially algebras of sets. Given a Boolean algebra $\mathbf{B} = \langle B, \wedge, \vee, \neg, 1, 0 \rangle$, a filter F of the lattice reduct of $\mathbf{B}$ is called an *ultrafilter* of $\mathbf{B}$ iff, for every $b \in B$, exactly one element in the set $\{b, \neg b\}$ belongs to F. Ultrafilters coincide with maximal filters of $\mathbf{B}$, *i.e.*, filters that are not properly included in any proper filter of $\mathbf{B}$. Letting

$$U(\mathbf{B}) = \{F \subseteq B \mid F \text{ is an ultrafilter of } \mathbf{B}\},$$

Stone showed that the algebra of sets

$$\langle \wp(U(\mathbf{B})), \cap, \cup, -, U(\mathbf{B}), \emptyset \rangle$$

is a Boolean algebra and that $\mathbf{B}$ can be embedded into it via the map

$$f(a) = \{F \in U(\mathbf{B}) \mid a \in F\}.$$

Around the same time, due to the independent work of various researchers (in particular, Glivenko [126] and Heyting [74]), the algebraic counterpart of Brouwer's intuitionistic logic was found in *Heyting algebras*. As we have seen in Example 4.6, both Heyting algebras and Boolean algebras are term-equivalent to varieties of FL-algebras.

ℓ-groups

The name of Richard Dedekind is recurring quite often in these historical notes; in fact, he can legitimately be said to have invented, anticipated, or first investigated at the abstract level, many of the most fundamental notions in contemporary abstract algebra. We have already mentioned his contributions to ring theory in his X Supplement to the second edition of Dirichlet's *Zahlentheorie* (1871). In his Göttingen classnotes on algebra, written between 1858 and 1868 but only published more than a century later (see [120]), Dedekind investigates the abstract concept of group without confining himself to their concrete representations as groups of permutations. In a later paper (*Über Zerlegung von Zahlen durch ihre größten gemeinsamen Teiler*, 1897), he essentially introduces and investigates the notion of lattice, as well as combining the two concepts into what is now known as a *lattice-ordered group* (or ℓ-group) (*cf.* [47]). Dedekind's approach to lattices was algebraic rather than order-theoretic: he views lattices as sets equipped with two binary operations each satisfying associativity, commutativity, and idempotency, and linked together by the absorption law, rather than as posets where every finite set has a meet and a join. Moreover, since he was drawn to lattice theory mainly by his number-theoretic interests, the privileged examples of meet and join he has in mind are not the set-theoretic operations of intersection and union but the arithmetical operations of greatest common divisor and least common multiple ([84]).

As we have mentioned, Dedekind explicitly focused on ℓ-groups. The definition of ℓ-group given in Example 4.8 is, at least in the commutative case, implicit in Dedekind's paper. In a nutshell, he considers an Abelian group endowed with an additional semilattice operation, postulating that the group binary operation distributes over such a join. In other words, the resulting algebra $\mathbf{A} = \langle A, \cdot, \vee, ^{-1}, 1 \rangle$ must satisfy the equation

$$x(y \vee z) \approx xy \vee xz.$$

Then he observes that the defined term operation $a \wedge b = ab(a \vee b)^{-1}$ turns the term reduct $\langle A, \wedge, \vee \rangle$ into a *distributive* lattice and the whole structure into a commutative ℓ-group. The same proof carries over to the noncommutative case if we only rephrase more carefully the above definition of meet as $a \wedge b = a(a \vee b)^{-1}b$, but Dedekind does not go as far as to observe this fact.

In the late 1920's and early 1930's, functional analysts like Riesz, Freudenthal, and Kantorovich developed conspicuous bits of ℓ-group theory in the context of their investigations into vector lattices. These different streams converged at last in Birkhoff's first edition (1940) of his *Lattice Theory* [11], containing a chapter on ℓ-groups where the concept is defined in full clarity and precision and Birkhoff's original contributions to the subject, as well as the main results that had been proved in other diverse research areas, are collected and systematized.

To the best of our knowledge, the first author who expressly noticed that in any ℓ-group multiplication is residuated by the two division operations was Jeremiah Certaine in his PhD thesis [27]. It was only much later, however ([18, 19]), that this observation was expanded to an explicit proof of the fact that the variety of ℓ-groups is term-equivalent to a variety of residuated lattices, as explained in Example 4.8.

Birkhoff's problem

Boolean algebras and ℓ-groups contributed towards the historical development of the theory of residuated structures also in several ways other than those reported on so far. In the already mentioned 1940 edition of his *Lattice Theory* [11, Problem 108], Garrett Birkhoff challenged his readers by suggesting the following project:

Develop a common abstraction which includes Boolean algebras (rings) and lattice ordered groups as special cases.

Over the subsequent decades, several mathematicians tried their hands at Birkhoff's intriguing problem. A minimal requirement that has to be met by a class of structures to be considered an answer to Birkhoff's problem is including both Boolean algebras (or Boolean rings) and ℓ-groups as instances and it is clear that such a desideratum can be satisfied by concepts that are very different from one another. True to form, the list of suggestions advanced in response to Birkhoff's challenge includes such disparate items as classes of partial algebras ([33, 121, 122, 143]) or classes of structures with multi-valued operations ([104]).

Following the lead of the Indian mathematician K.L.N. Swamy ([127–129]), a number of authors observed that both Boolean algebras and ℓ-groups are residuated structures and formulated their common generalizations accordingly. Indeed, Swamy's dually residuated lattice-ordered semigroups, Rama Rao's direct products of Boolean rings and ℓ-groups ([118]), and Casari's lattice-ordered pregroups ([26]) all form varieties that are term-equivalent to some subvariety of $\mathcal{RL}$ or $\mathcal{FL}$.

5   Structural proof theory

Attempting to identify a precise border between ordered algebra and logic would be unwise. Nevertheless, we may safely say that while algebra focuses primarily on structures and their properties, logic (narrowly conceived) concerns itself more with syntax and deduction. Yet despite these differences in perspective, traditional Hilbert-style presentations of propositional logics as axiom systems typically enjoy a close relationship with classes of algebras, formalized, as we have seen, via the Lindenbaum-Tarski-inspired Blok-Pigozzi method of algebraization. In particular, theorems of classical or intuitionistic logic may be translated into equations holding in all Boolean or Heyting algebras and vice versa. To the algebraist, this may suggest that propositional logic is little more than "algebra in disguise." There is something to this point of view, though a logician may quickly respond, first, that some logics are not algebraizable and, second, that the case for an algebraic approach to first order logic is not so compelling. More pertinently for our present concerns, syntactic presentations offer an alternative perspective that pure semantics cannot provide. In particular, syntactic objects such as formulas, equations, and proofs, may be investigated themselves as first-class citizens using methods such as induction on formula complexity or height of a proof. This idea was first taken seriously by Hilbert who established the field of *proof theory* with the aim of proving the consistency of arithmetic and other parts of mathematics using only so-called "finitistic" methods (see, *e.g.*, [75]).

The original goal of Hilbert's program was famously dashed in 1931 by Kurt Gödel's incompleteness theorems, but partially resurrected by Gerhard Gentzen in the mid-to-late 1930's ([56, 57]). Gentzen was able to show that the consistency of arithmetic is provable over the base theory of primitive recursive arithmetic extended with quantifier-free transfinite induction up to the ordinal ε_0. This was the first result of the area subsequently known as *ordinal analysis*. Our interest here lies, however, with the tools that Gentzen used to prove this result. A limitation of the early period of proof theory was the reliance on a rather rigid interpretation of the axiomatic method, that is, axiomatizations typically consisting of many axiom schemata and just a few rules, notably, modus ponens. The axiomatic approach is flexible but does not seem to reflect the way that mathematicians, or humans in general, construct and reason about proofs, and suffers from a lack of control over proofs as mathematical objects. These issues were addressed by Gentzen in [56] via the introduction of two new proof formalisms: *natural deduction* and the *sequent calculus*. In particular, he defined sequent calculi, LK and LJ, for first order classical logic and first order intuitionistic logic, respectively, giving birth to an area known now as *structural proof theory*. Since our interest lies here with ordered algebras, we will focus only on the propositional parts of Gentzen's systems, which, as we will see, correspond directly to Boolean algebras and Heyting algebras. *Substructural logics*, which themselves correspond to classes of the residuated lattices introduced in the previous section, are then obtained, very roughly speaking, by removing certain rules from these systems.

5.1 Gentzen's LJ and LK

Hilbert-style axiom systems, and to a certain extent Gentzen's own natural deduction systems, are hindered by the fact that they deal directly with formulas. So-called *Gentzen systems* gain flexibility by considering more complicated structures. In particular, Gentzen introduced the notion of a *sequent*, an ordered pair of finite sequences of formulas, written:

$$\alpha_1, \ldots, \alpha_n \Rightarrow \beta_1, \ldots, \beta_m.$$

Intuitively, we might think of the disjunction of the formulas $\beta_1, \ldots, \beta_m$ "following from" the conjunction of the formulas $\alpha_1, \ldots, \alpha_n$, although since sequents are purely syntactic objects, any meaning ascribed to them follows only from the role that they play in the given proof system.

Sequent rules are typically written schematically using Γ and Δ to stand for arbitrary sequences of formulas, comma for concatenation, and an empty space for the empty sequence, and consist of *instances* with a finite set of premises and a single conclusion, rules with no premises being called *initial sequents*. A *sequent calculus* L is simply a set of sequent rules, and a *derivation* in such a system of a sequent S from a set of sequents X is a finite tree of sequents with root S such that each sequent is either a leaf and a member of X, or S is the conclusion and its children (if any) are the premises of an instance of a rule of the system. When such a tree exists, we say that "S is derivable from X in L" and write $X \vdash_L S$.

Figure 5.1 displays an inessential variant of Gentzen's sequent calculus (propositional) LK in the same language as HCL that consists of simple initial sequents of the form $\alpha \Rightarrow \alpha$, a *cut rule* corresponding, like modus ponens, to the transitivity of deduction, and two distinguished collections of rules. The first collection contains rules that introduce occurrences of connectives on the left and right of the sequent arrow. Such *logical rules* may be thought of as defining (operationally) the meaning of the connectives, and roughly correspond to the elimination and introduction rules of Gentzen's natural deduction system. The second collection of *structural rules*, which also come in left/right pairs, simply manipulate the structure of sequents: *exchange rules*, *weakening rules*, and *contraction rules* allow formulas to be permuted, added, and combined, respectively.

Example 5.1. Let us take a look at a derivation in LK of Peirce's law, noting that α and β can be any formulas:

$$
\cfrac{
  \cfrac{
    \cfrac{
      \cfrac{\cfrac{}{\alpha \Rightarrow \alpha}\ (\text{ID})}{\alpha \Rightarrow \beta, \alpha}\ (\text{WR})
    }{\Rightarrow \alpha \to \beta, \alpha}\ (\Rightarrow\to)
    \quad
    \cfrac{}{\alpha \Rightarrow \alpha}\ (\text{ID})
  }{
    \cfrac{(\alpha \to \beta) \to \alpha \Rightarrow \alpha, \alpha}{(\alpha \to \beta) \to \alpha \Rightarrow \alpha}\ (\text{CR})
  }\ (\to\Rightarrow)
}{\Rightarrow ((\alpha \to \beta) \to \alpha) \to \alpha}\ (\Rightarrow\to)
$$

One of the most remarkable features of Gentzen's framework is that it also accommodates a calculus LJ for intuitionistic logic, obtained from LK simply by restricting sequents $\Gamma \Rightarrow \Delta$ so that Δ is allowed to contain at most one formula. In particular, LJ has no right exchange or right contraction rules, and right weakening is confined to premises with empty succedents (in a sense, LJ is the first example of a *substructural* sequent calculus). Hence, for instance, the derivation of Peirce's law in Example 5.1 is (rightly) blocked.

Let us show now that LK really is a sequent calculus corresponding to the Hilbert-style presentation HCL of classical propositional logic, and therefore also Boolean algebras (the same proof works also for LJ with

Initial sequents

$$\overline{\alpha \Rightarrow \alpha} \ \ (\text{ID})$$

Cut rule

$$\frac{\Gamma_2 \Rightarrow \alpha, \Delta_2 \quad \Gamma_1, \alpha, \Gamma_3 \Rightarrow \Delta_1}{\Gamma_1, \Gamma_2, \Gamma_3 \Rightarrow \Delta_1, \Delta_2} \ \ (\text{CUT})$$

Left structural rules

$$\frac{\Gamma_1, \alpha, \beta, \Gamma_2 \Rightarrow \Delta}{\Gamma_1, \beta, \alpha, \Gamma_2 \Rightarrow \Delta} \ \ (\text{EL})$$

$$\frac{\Gamma_1, \Gamma_2 \Rightarrow \Delta}{\Gamma_1, \alpha, \Gamma_2 \Rightarrow \Delta} \ \ (\text{WL})$$

$$\frac{\Gamma_1, \alpha, \alpha, \Gamma_2 \Rightarrow \Delta}{\Gamma_1, \alpha, \Gamma_2 \Rightarrow \Delta} \ \ (\text{CL})$$

Right structural rules

$$\frac{\Gamma \Rightarrow \Delta_1, \alpha, \beta, \Delta_2}{\Gamma \Rightarrow \Delta_1, \beta, \alpha, \Delta_2} \ \ (\text{ER})$$

$$\frac{\Gamma \Rightarrow \Delta_1, \Delta_2}{\Gamma \Rightarrow \Delta_1, \alpha, \Delta_2} \ \ (\text{WR})$$

$$\frac{\Gamma \Rightarrow \Delta_1, \alpha, \alpha, \Delta_2}{\Gamma \Rightarrow \Delta_1, \alpha, \Delta_2} \ \ (\text{CR})$$

Left logical rules

$$\frac{\Gamma_1, \Gamma_2 \Rightarrow \Delta}{\Gamma_1, 1, \Gamma_2 \Rightarrow \Delta} \ \ (1 \Rightarrow)$$

$$\overline{0 \Rightarrow} \ \ (0 \Rightarrow)$$

$$\frac{\Gamma \Rightarrow \alpha, \Delta}{\neg\alpha, \Gamma \Rightarrow \Delta} \ \ (\neg \Rightarrow)$$

$$\frac{\Gamma_1, \alpha, \Gamma_2 \Rightarrow \Delta}{\Gamma_1, \alpha \wedge \beta, \Gamma_2 \Rightarrow \Delta} \ \ (\wedge \Rightarrow)_1$$

$$\frac{\Gamma_1, \beta, \Gamma_2 \Rightarrow \Delta}{\Gamma_1, \alpha \wedge \beta, \Gamma_2 \Rightarrow \Delta} \ \ (\wedge \Rightarrow)_2$$

$$\frac{\Gamma_1, \alpha, \Gamma_2 \Rightarrow \Delta \quad \Gamma_1, \beta, \Gamma_2 \Rightarrow \Delta}{\Gamma_1, \alpha \vee \beta, \Gamma_2 \Rightarrow \Delta} \ \ (\vee \Rightarrow)$$

$$\frac{\Gamma_2 \Rightarrow \alpha, \Delta_2 \quad \Gamma_1, \beta, \Gamma_3 \Rightarrow \Delta_1}{\Gamma_1, \Gamma_2, \alpha \to \beta, \Gamma_3 \Rightarrow \Delta_1, \Delta_2} \ \ (\to \Rightarrow)$$

Right logical rules

$$\overline{\Rightarrow 1} \ \ (\Rightarrow 1)$$

$$\frac{\Gamma \Rightarrow \Delta_1, \Delta_2}{\Gamma \Rightarrow \Delta_1, 0, \Delta_2} \ \ (\Rightarrow 0)$$

$$\frac{\Gamma, \alpha \Rightarrow \Delta}{\Gamma \Rightarrow \Delta, \neg\alpha} \ \ (\Rightarrow \neg)$$

$$\frac{\Gamma \Rightarrow \Delta_1, \alpha, \Delta_2}{\Gamma \Rightarrow \Delta_1, \alpha \vee \beta, \Delta_2} \ \ (\Rightarrow \vee)_1$$

$$\frac{\Gamma \Rightarrow \Delta_1, \beta, \Delta_2}{\Gamma \Rightarrow \Delta_1, \alpha \vee \beta, \Delta_2} \ \ (\Rightarrow \vee)_2$$

$$\frac{\Gamma \Rightarrow \Delta_1, \alpha, \Delta_2 \quad \Gamma \Rightarrow \Delta_1, \beta, \Delta_2}{\Gamma \Rightarrow \Delta_1, \alpha \wedge \beta, \Delta_2} \ \ (\Rightarrow \wedge)$$

$$\frac{\alpha, \Gamma \Rightarrow \beta, \Delta}{\Gamma \Rightarrow \alpha \to \beta, \Delta} \ \ (\Rightarrow \to)$$

Figure 5.1. The Sequent Calculus LK.

respect to axiomatizations of intuitionistic logic and Heyting algebras). We let $\square(\alpha_1, \ldots, \alpha_n)$ stand for $\alpha_1 \square \ldots \square \alpha_n$ for $\square \in \{\wedge, \vee\}$ where $\wedge()$ is 1 and $\vee()$ is 0, and define:

$$\tau(\alpha) = \{\Rightarrow \alpha\};$$
$$\rho(\Gamma \Rightarrow \Delta) = \wedge\Gamma \to \vee\Delta.$$

Theorem 5.2. $X \vdash_{LK} S$ *if and only if* $\{\rho(S') \mid S' \in X\} \vdash_{HCL} \rho(S)$.

Proof. The left-to-right direction is established by an induction on the height of a derivation in LK (straightforward, but requiring many tedious derivations in HCL). For the right-to-left direction, it is easily checked that for any axiom α of HCL, $\tau(\alpha)$ is derivable in LK. Moreover, if $\tau(\alpha)$ and $\tau(\alpha \to \beta)$ are derivable in LK, then so is $\tau(\beta)$, using (CUT) twice with the derivable sequent $\alpha, \alpha \to \beta \Rightarrow \beta$. Note also that for any sequent S: $S \vdash_{LK} \tau(\rho(S))$ and $\tau(\rho(S)) \vdash_{LK} S$. Hence if $\{\rho(S') \mid S' \in X\} \vdash_{HCL} \rho(S)$, then $\{\tau(\rho(S')) \mid S' \in X\} \vdash_{LK} \tau(\rho(S))$ and, as required, $X \vdash_{LK} S$. $\qquad\square$

However, it is worth asking at this point what advantages, if any, LJ and LK hold over Hilbert-style systems. Proof search in the latter is hindered by the need to guess formulas α and $\alpha \to \beta$ as premises when applying modus ponens. The same situation seems to occur for these sequent calculi: we have to guess which formula α to use when applying (CUT). Certainly finding derivations would be much simpler if we could do without this rule. Then we could just apply rules where formulas in the premises are subformulas of formulas in the conclusion. In fact, this is the case, as established by Gentzen for (first order) LJ and LK in his famous Hauptsatz. Indeed, Gentzen showed not only that (CUT) is not needed for deriving sequents from empty sets of assumptions, but also that there exists a *cut elimination* algorithm that transforms such derivations into cut-free derivations.

Let us consider briefly the ideas behind Gentzen's proof. Intuitively, the idea is to push applications of the cut rule upwards in derivations until they reach initial sequents and disappear. For example, suppose that we have a derivation in LJ ending

$$\frac{\vdots \qquad\qquad \vdots}{\dfrac{\Gamma_2 \Rightarrow \alpha \quad \Gamma_1, \alpha, \Gamma_3 \Rightarrow \Delta}{\Gamma_1, \Gamma_2, \Gamma_3 \Rightarrow \Delta}} \text{(CUT)}$$

The cut-formula α occurs on the right in one premise, and on the left in the other. A natural strategy for eliminating this application of (CUT) is to look at the derivations of these premises. If one of the premises is an instance of (ID), then it must be $\alpha \Rightarrow \alpha$ and the other premise must be exactly the conclusion, derived with one fewer applications of (CUT). Otherwise, we have two possibilities. The first is that one of the premises ends with an application of a rule where α is not the decomposed formula, *e.g.*,

$$
\cfrac{
\cfrac{
\cfrac{\vdots \quad\quad \vdots}{\Gamma_2'' \Rightarrow \beta_1 \quad \Gamma_2', \beta_2, \Gamma_2''' \Rightarrow \alpha}{\Gamma_2', \Gamma_2'', \beta_1 \to \beta_2, \Gamma_2''' \Rightarrow \alpha} \ (\to\Rightarrow) \quad\quad \cfrac{\vdots}{\Gamma_1, \alpha, \Gamma_3 \Rightarrow \Delta}
}{\Gamma_1, \Gamma_2', \Gamma_2'', \beta_1 \to \beta_2, \Gamma_2''', \Gamma_3 \Rightarrow \Delta}
}{} \ (\text{CUT})
$$

In this case, we can "push the cut upwards" in the derivation to get:

$$
\cfrac{
\cfrac{\vdots}{\Gamma_2'' \Rightarrow \beta_1} \quad\quad \cfrac{\cfrac{\vdots}{\Gamma_2', \beta_2, \Gamma_2''' \Rightarrow \alpha} \quad \cfrac{\vdots}{\Gamma_1, \alpha, \Gamma_3 \Rightarrow \Delta}}{\Gamma_1, \Gamma_2', \beta_2, \Gamma_2''', \Gamma_3 \Rightarrow \Delta} \ (\text{CUT})
}{\Gamma_1, \Gamma_2', \Gamma_2'', \beta_1 \to \beta_2, \Gamma_2''', \Gamma_3 \Rightarrow \Delta} \ (\to\Rightarrow)
$$

That is, we have a derivation where the left premise in the new application of (CUT) has a shorter derivation than the application in the original derivation.

The second possibility is that the last application of a rule in both premises involves α as the decomposed formula, *e.g.*

$$
\cfrac{
\cfrac{\cfrac{\vdots}{\alpha_1, \Gamma_2 \Rightarrow \alpha_2}}{\Gamma_2 \Rightarrow \alpha_1 \to \alpha_2} \ (\Rightarrow\to) \quad\quad \cfrac{\cfrac{\vdots}{\Gamma_1'' \Rightarrow \alpha_1} \quad \cfrac{\vdots}{\Gamma_1', \alpha_2, \Gamma_3 \Rightarrow \Delta}}{\Gamma_1', \Gamma_1'', \alpha_1 \to \alpha_2, \Gamma_3 \Rightarrow \Delta} \ (\to\Rightarrow)
}{\Gamma_1', \Gamma_1'', \Gamma_2, \Gamma_3 \Rightarrow \Delta} \ (\text{CUT})
$$

Here we rearrange our derivation in a different way: we replace the application of (CUT) with applications of (CUT) with cut-formulas α_1 and α_2:

$$
\cfrac{
\cfrac{\vdots}{\Gamma_1'' \Rightarrow \alpha_1} \quad\quad \cfrac{\cfrac{\vdots}{\alpha_1, \Gamma_2 \Rightarrow \alpha_2} \quad \cfrac{\vdots}{\Gamma_1', \alpha_2, \Gamma_3 \Rightarrow \Delta}}{\Gamma_1', \alpha_1, \Gamma_2, \Gamma_3 \Rightarrow \Delta} \ (\text{CUT})
}{\Gamma_1', \Gamma_1'', \Gamma_2, \Gamma_3 \Rightarrow \Delta} \ (\text{CUT})
$$

We now have two applications of (CUT) but with cut-formulas of a smaller complexity than the original application.

This procedure, formalized using a double induction on cut-formula complexity and the combined height of derivations of the premises, eliminates applications of (CUT) for many sequent calculi. However, it encounters a problem with rules that contract formulas in one or more of the premises. Consider the following situation:

$$\frac{\begin{array}{cc} \vdots & \dfrac{\vdots}{\Gamma_1, \alpha, \alpha, \Gamma_3 \Rightarrow \Delta} \\[6pt] \overline{\Gamma_2 \Rightarrow \alpha} & \overline{\Gamma_1, \alpha, \Gamma_3 \Rightarrow \Delta} \end{array} \ (\text{CL})}{\Gamma_1, \Gamma_2, \Gamma_3 \Rightarrow \Delta} \ (\text{CUT})$$

In this case we need to perform several cuts simultaneously, *e.g.*, making use of Gentzen's "mix" rule for LK,

$$\frac{\Gamma \Rightarrow \alpha, \Delta \quad \Gamma' \Rightarrow \Delta'}{\Gamma, \Gamma'_\alpha \Rightarrow \Delta', \Delta_\alpha}$$

where Γ' has at least one occurrence of α, and Γ'_α and Δ_α are obtained by removing all occurrences of α from Γ' and Δ, respectively.

Theorem 5.3 (Gentzen 1935). *Cut-elimination holds for* LK *and* LJ

This result has many important applications. As an immediate consequence, for example, both LJ and LK are *consistent*: there cannot be any cut-free derivation of an arbitrary variable p for instance. Similarly, since any cut-free derivation of $\Rightarrow \alpha \vee \beta$ in LJ must necessarily involve a derivation of $\Rightarrow \alpha$ or $\Rightarrow \beta$, intuitionistic logic has the so-called *disjunction property*. Cut-elimination also facilitates easy proofs of decidability for the derivability of sequents for LK and LJ, and hence for checking validity in propositional classical or intuitionistic logic. Call two sequents *equivalent* if one can be derived from the other using the exchange and contraction rules. Then easily there are a finite number of non-equivalent sequents that can occur in a cut-free derivation of a sequent in LJ or LK, and hence, checking that equivalent sequents do not occur, the search for such a cut-free derivation must terminate. It follows that propositional classical logic and, more interestingly, propositional intuitionistic logic, are decidable. Algebraically of course this means that the equational theories of Boolean algebras and Heyting algebras are decidable, raising the question as to whether other classes of algebras can be proved decidable using similar methods. It is worth noting also that these calculi can also be used to prove complexity results for the respective logics and classes of algebras.

We remark finally that many variants of LJ and LK have appeared in the literature. In particular, the Finnish logician Oiva Ketonen [85] suggested a new version of LK where $(\wedge \Rightarrow)_1$, $(\wedge \Rightarrow)_2$, $(\Rightarrow \vee)_1$, and $(\Rightarrow \vee)_2$ are replaced by:

$$\frac{\Gamma_1, \alpha, \beta, \Gamma_2 \Rightarrow \Delta}{\Gamma_1, \alpha \wedge \beta, \Gamma_2 \Rightarrow \Delta} \ (\wedge \Rightarrow)' \qquad \frac{\Gamma \Rightarrow \Delta_1, \alpha, \beta, \Delta_2}{\Gamma \Rightarrow \Delta_1, \alpha \vee \beta, \Delta_2} \ (\Rightarrow \vee)'$$

Also, Haskell B. Curry [36] later considered variants obtained by replacing $(\vee \Rightarrow)$ and $(\Rightarrow \wedge)$ with:

$$\frac{\Gamma_1, \alpha \Rightarrow \Delta_1 \quad \Gamma_2, \beta \Rightarrow \Delta_2}{\Gamma_1, \Gamma_2, \alpha \vee \beta \Rightarrow \Delta_1, \Delta_2} \ (\vee \Rightarrow)' \quad \frac{\Gamma_1 \Rightarrow \alpha, \Delta_1 \quad \Gamma_2 \Rightarrow \beta, \Delta_2}{\Gamma_1, \Gamma_2 \Rightarrow \alpha \wedge \beta, \Delta_1, \Delta_2} \ (\Rightarrow \wedge)'$$

It is an easy exercise to see that these rules are interderivable with the previous rules given for $\wedge$ and $\vee$, making crucial use of the structural rules of weakening, exchange, and contraction. In the absence of such rules, the connectives $\wedge$ and $\vee$ split into two. That is, the original rules define what are often called the *additive* or *lattice* connectives $\wedge$ and $\vee$, whereas Ketonen and Curry's rules define the so-called *multiplicative* or *group* connectives, renamed $\cdot$ and $+$. Moreover, in the absence of weakening rules, as we will see below, the constants 1 and 0 also split, as in the absence of exchange rules, does the implication connective $\rightarrow$.

5.2 Substructural logics

The expression "substructural logic" was suggested by Kosta Došen and Peter Schroeder-Heister at a conference in Tübingen in 1990 to describe a family of logics emerging (post-Gentzen) with a wide range of motivations from linguistics, algebra, set theory, philosophy, and computer science. Roughly speaking, the term "substructural" refers to the fact that these logics, which all live in a certain sense "below the surface" of classical logic, fail to admit one or more classically sound structural rules. Most convincingly, logics defined by sequent calculi obtained by removing weakening, contraction, or exchange rules from LJ or LK may be deemed substructural, although even in these cases, further logical rules may be added to capture connectives that (as remarked above) split when structural rules are removed. Other classically sound structural rules, such as "weaker" versions of weakening or contraction, may also be added, giving a family of logics characterized by cut-free sequent calculi. Nevertheless, there remain important classes of logics (*e.g.*, relevant and fuzzy logics) typically accepted as substructural that do not fit comfortably into this framework, requiring more flexible formalisms such as hypersequents, display calculi, etc. More perplexing still, there are closely related logics (and classes of algebras) lacking structural rules for which no reasonable cut-free calculus is known. Are these also substructural?

A practical answer to this question, suggested by the authors of [52], is to define substructural logics by appeal to their algebraic semantics. That is, since most substructural logics correspond in some way to classes of *residuated lattices* (or slight variants thereof), this family could be identified with logics having these classes of algebras as equivalent algebraic

Axioms

$$\overline{\alpha \Rightarrow \alpha}\ (\text{ID})$$

Cut rule

$$\frac{\Gamma_1 \Rightarrow \alpha \quad \Gamma_2, \alpha, \Gamma_3 \Rightarrow \Delta}{\Gamma_1, \Gamma_2, \Gamma_3 \Rightarrow \Delta}\ (\text{CUT})$$

Left logical rules

$$\frac{\Gamma_1, \Gamma_2 \Rightarrow \Delta}{\Gamma_1, 1, \Gamma_2 \Rightarrow \Delta}\ (1 \Rightarrow)$$

$$\frac{}{0 \Rightarrow}\ (0 \Rightarrow)$$

$$\frac{\Gamma_2 \Rightarrow \alpha \quad \Gamma_1, \beta, \Gamma_3 \Rightarrow \Delta}{\Gamma_1, \beta/\alpha, \Gamma_2, \Gamma_3 \Rightarrow \Delta}\ (/ \Rightarrow)$$

$$\frac{\Gamma_2 \Rightarrow \alpha \quad \Gamma_1, \beta, \Gamma_3 \Rightarrow \Delta}{\Gamma_1, \Gamma_2, \alpha\backslash\beta, \Gamma_3 \Rightarrow \Delta}\ (\backslash \Rightarrow)$$

$$\frac{\Gamma_1, \alpha, \beta, \Gamma_2 \Rightarrow \Delta}{\Gamma_1, \alpha \cdot \beta, \Gamma_2 \Rightarrow \Delta}\ (\cdot \Rightarrow)$$

$$\frac{\Gamma_1, \alpha, \Gamma_2 \Rightarrow \Delta}{\Gamma_1, \alpha \wedge \beta, \Gamma_2 \Rightarrow \Delta}\ (\wedge \Rightarrow)_1$$

$$\frac{\Gamma_1, \beta, \Gamma_2 \Rightarrow \Delta}{\Gamma_1, \alpha \wedge \beta, \Gamma_2 \Rightarrow \Delta}\ (\wedge \Rightarrow)_2$$

$$\frac{\Gamma_1, \alpha, \Gamma_2 \Rightarrow \Delta \quad \Gamma_1, \beta, \Gamma_2 \Rightarrow \Delta}{\Gamma_1, \alpha \vee \beta, \Gamma_2 \Rightarrow \Delta}\ (\vee \Rightarrow)$$

Right logical rules

$$\frac{}{\Rightarrow 1}\ (\Rightarrow 1)$$

$$\frac{\Gamma \Rightarrow}{\Gamma \Rightarrow 0}\ (\Rightarrow 0)$$

$$\frac{\Gamma, \alpha \Rightarrow \beta}{\Gamma \Rightarrow \beta/\alpha}\ (\Rightarrow /)$$

$$\frac{\alpha, \Gamma \Rightarrow \beta}{\Gamma \Rightarrow \alpha\backslash\beta}\ (\Rightarrow \backslash)$$

$$\frac{\Gamma_1 \Rightarrow \alpha \quad \Gamma_2 \Rightarrow \beta}{\Gamma_1, \Gamma_2 \Rightarrow \alpha \cdot \beta}\ (\Rightarrow \cdot)$$

$$\frac{\Gamma \Rightarrow \alpha}{\Gamma \Rightarrow \alpha \vee \beta}\ (\Rightarrow \vee)_1$$

$$\frac{\Gamma \Rightarrow \beta}{\Gamma \Rightarrow \alpha \vee \beta}\ (\Rightarrow \vee)_2$$

$$\frac{\Gamma \Rightarrow \alpha \quad \Gamma \Rightarrow \beta}{\Gamma \Rightarrow \alpha \wedge \beta}\ (\Rightarrow \wedge)$$

Figure 5.2. The Full Lambek Calculus FL

semantics. Such a definition offers uniformity and clarity, although it may be objected that there exist both classes of algebras which have no corresponding logic and substructural logics which lack a corresponding class of residuated lattices. Here we deliberately refuse to say exactly what a substructural logic is, believing rather that an understanding of the richness of this family is best gained by a (necessarily brief, see [42,112,119], and [52] for a wealth of further material) historical survey of the most important candidates.

The Lambek calculus and residuated lattices

Chronologically, the first substructural logic occurred in the field of linguistics. In a 1958 paper [90], Joachim Lambek made use of a substructural sequent calculus (which became known, naturally enough, as the

Lambek calculus) to represent transformations on syntactic types of a formal grammar. Lambek's approach built on earlier work on *categorial grammar* in the 1930's by the Polish logician Kazimierz Ajdukiewicz, who aimed to develop an analysis of natural language by assigning syntactic types to linguistic expressions that describe their syntactic roles (*e.g.*, verb, noun phrase, verb phrase, sentence). A naive approach to this task would consist of listing a number of lexical atoms (*e.g.*, Joan, smiles, charmingly) and a number of mutually unrelated types (*e.g.*, NP = noun phrase; V = verb; Adv = adverb; VP = verb phrase; S = sentence), and then tagging each lexical atom with the appropriate type:

Joan: NP; smiles: V; charmingly: Adv.

However, Ajdukiewicz understood that the stock of basic grammatical categories can be substantially reduced by the use of *type-forming operators* \ and /, where an expression v has type $\alpha \backslash \beta$ (respectively, β / α) if whenever the expression v' has type α, the expression $v'v$ (respectively, vv') has type β. Indeed, the whole apparatus of categorial grammar can then be constructed out of just two basic types, n (noun) and s (sentence).

For example, in English, John works is a sentence, but works John is not. The intransitive verb works has type $n \backslash s$: when applied to the right of an expression of type n, it yields an expression of type s. On the other hand, the adjective poor has type n / n: when applied to the left of an expression of type n, it yields another expression of type n (a complex noun phrase). We may write these transformations, respectively, as $n, n \backslash s \Rightarrow s$ and $n / n, n \Rightarrow n$. More generally, if α, β are types, the following transformations are permissible in Ajdukiewicz's categorial grammar:

$$\alpha, \alpha \backslash \beta \Rightarrow \beta \quad \text{and} \quad \beta / \alpha, \alpha \Rightarrow \beta.$$

Lambek extended the deductive power of categorial grammar by setting up a sequent calculus for permissible transformations on types, introducing a new type-forming operation $\cdot$ such that v has type $\alpha \cdot \beta$ whenever $v = v'v''$ with v' of type α and v'' of type β, and admitting, in addition to modus ponens, patterns of *hypothetical reasoning* corresponding to right introduction rules for the implications. Adding rules for the lattice connectives $\wedge$ and $\vee$, and the constants 1 and 0, gives the *Full Lambek Calculus* FL, displayed in Figure 5.2.

FL has come to play a distinguished role in the field of substructural logics. Just as classical logic is a candidate for the top element of the lattice of such logics, so FL is a candidate for the bottom element. That is, most other substructural logics may be obtained as extensions of FL (although, non-associative substructural logics have also been investigated, not least by Lambek himself.) In particular, Hiroakira Ono and colleagues have popularized the usage of FL_X where $X \subseteq \{e, c, w\}$ to denote the extension of FL with the appropriate grouping of exchange (e), contraction (c), and weakening rules (w), and $InFL_X$ to denote the corresponding multiple-conclusion sequent calculus. In particular, FL_{ewc} and $InFL_{ewc}$ correspond to LJ and LK with split connectives.

Not surprisingly, FL corresponds to the class of FL-algebras. Moreover, a sequent calculus RL for the class of residuated lattices is obtained by removing the rules for 0, as the following result makes precise. Let $\Box(\alpha_1, \ldots, \alpha_n)$ stand for $\alpha_1 \Box \ldots \Box \alpha_n$ for $\Box \in \{\cdot, +\}$ where $\cdot()$ is 1 and $+()$ is 0, and define:

$$\tau(\alpha \approx \beta) = \{\alpha \Rightarrow \beta, \beta \Rightarrow \alpha\};$$

$$\rho(\Gamma \Rightarrow \Delta) = \{\cdot\Gamma \leq +\Delta\}.$$

Theorem 5.4. $X \vdash_{RL} S$ *if and only if* $\{\rho(S') \mid S' \in X\} \vdash_{Eq(\mathcal{RL})} \rho(S)$.

Proof. The left-to-right direction is proved by induction on the height of a derivation in RL. For the right-to-left direction, consider

$$\Sigma \cup \{(\alpha, \beta)\} \subseteq Fm^2$$

and define

$$\Sigma^{\Rightarrow} = \{\alpha' \Rightarrow \beta' \mid (\alpha', \beta') \in \Sigma\} \quad \text{and} \quad \Sigma^{\leq} = \{\alpha' \leq \beta' \mid (\alpha', \beta') \in \Sigma\}.$$

We will prove that $\Sigma^{\leq} \vdash_{Eq(\mathcal{RL})} \alpha \leq \beta$ implies $\Sigma^{\Rightarrow} \vdash_{RL} \alpha \Rightarrow \beta$; the result then follows swiftly from the fact that for any sequent S, $S \vdash_{RL} \tau(\rho(S))$ and $\tau(\rho(S)) \vdash_{RL} S$.

Define the following binary relation on Fm:

$$\alpha\Theta_\Sigma\beta \quad \text{iff} \quad \Sigma^{\Rightarrow} \vdash_{RL} \alpha \Rightarrow \beta \text{ and } \Sigma^{\Rightarrow} \vdash_{RL} \beta \Rightarrow \alpha.$$

Θ_Σ is a congruence on **Fm**. Clearly it is reflexive and symmetric, and for transitivity, if $\alpha\Theta_\Sigma\beta$ and $\beta\Theta_\Sigma\gamma$, then $\Sigma^{\Rightarrow} \vdash_{RL} \{\alpha \Rightarrow \beta, \beta \Rightarrow \gamma\}$, so by (CUT), $\Sigma^{\Rightarrow} \vdash_{RL} \alpha \Rightarrow \gamma$; similarly, $\Sigma^{\Rightarrow} \vdash_{RL} \gamma \Rightarrow \alpha$, so $\alpha\Theta_\Sigma\gamma$ as required. Suppose, moreover, that $\alpha_1\Theta_\Sigma\alpha_2$ and $\beta_1\Theta_\Sigma\beta_2$. Then $\Sigma^{\Rightarrow} \vdash_{RL} \{\alpha_1 \Rightarrow \alpha_2, \alpha_2 \Rightarrow \alpha_1, \beta_1 \Rightarrow \beta_2, \beta_2 \Rightarrow \beta_1\}$ and we can construct, *e.g.*, derivations for $\Sigma^{\Rightarrow} \vdash_{RL} \{\alpha_1\backslash\beta_1 \Rightarrow \alpha_2\backslash\beta_2, \alpha_1 \wedge \beta_1 \Rightarrow \alpha_2 \wedge \beta_2\}$ ending:

$$\cfrac{\cfrac{\alpha_2 \Rightarrow \alpha_1 \quad \beta_1 \Rightarrow \beta_2}{\alpha_2, \alpha_1\backslash\beta_1 \Rightarrow \beta_2} (\backslash\Rightarrow)}{\alpha_1\backslash\beta_1 \Rightarrow \alpha_2\backslash\beta_2} (\Rightarrow\backslash)$$

$$\cfrac{\cfrac{\alpha_1 \Rightarrow \alpha_2}{\alpha_1 \wedge \beta_1 \Rightarrow \alpha_2} (\wedge\Rightarrow)_1 \quad \cfrac{\beta_1 \Rightarrow \beta_2}{\alpha_1 \wedge \beta_1 \Rightarrow \beta_2} (\wedge\Rightarrow)_2}{\alpha_1 \wedge \beta_1 \Rightarrow \alpha_2 \wedge \beta_2} (\Rightarrow\wedge)$$

Hence (using symmetry), Θ_Σ is compatible with $\backslash$ and $\wedge$, and similarly, with the other operations.

It follows easily that the quotient algebra $\mathbf{Fm}/\Theta_\Sigma$ with equivalence classes $[\alpha]$ for $\alpha \in Fm$ as elements is a residuated lattice. So if $\Sigma^{\leq}\vdash_{Eq(\mathcal{RL})}\alpha \leq \beta$, we can define the canonical evaluation $e(x) = [x]$ and prove by induction that $e(\gamma) = [\gamma]$ for all $\gamma \in Fm_{\mathrm{RL}}$. Since $\Sigma^{\Rightarrow} \vdash_{\mathrm{RL}} \alpha' \Rightarrow \beta'$ for all $(\alpha', \beta') \in \Sigma$, we have $e(\alpha') = [\alpha'] \leq [\beta'] = e(\beta')$ for all $(\alpha', \beta') \in \Sigma$. Hence $[\alpha] = e(\alpha) \leq e(\beta) = [\beta]$ and, as required, $\Sigma^{\Rightarrow} \vdash_{\mathrm{RL}} \alpha \Rightarrow \beta$. □

This equivalence extends easily to FL_X-algebras and the calculi FL_X for $X \subseteq \{e, c, w\}$, and also to classes of bounded FL-algebras with respect to calculi FL_X^B, obtained by adding to FL_X the rules:

$$\frac{}{\Gamma_1, \perp, \Gamma_2 \Rightarrow \Delta}\ (\perp \Rightarrow) \qquad \frac{}{\Gamma \Rightarrow \top}\ (\Rightarrow \top)$$

Relevance logics

An important source of substructural logics lacking weakening rules is the philosophy of logic and the long-standing debate over *entailment*. As is well-known, contemporary modal logic originated with C.I. Lewis' dissatisfaction and critical attitude towards Russell's classical propositional calculus, notably its – supposedly counterintuitive and repugnant to common sense – "paradoxes of material implication" such as the laws of *a fortiori* and *ex absurdo quodlibet*:

$$\alpha \rightarrow (\beta \rightarrow \alpha) \quad \text{and} \quad \neg\alpha \rightarrow (\alpha \rightarrow \beta).$$

In a series of writings culminating in his *Symbolic Logic*, coauthored with C.H. Langford ([92]), Lewis introduced calculi of *strict implication* to describe a tighter notion of implication intended to be true whenever it is impossible that the antecedent holds while the consequent does not. These calculi avoid Russell's material paradoxes, yet derive "paradoxes of strict implication" such as:

$$\alpha \rightarrow (\beta \vee \neg\beta) \quad \text{and} \quad (\alpha \wedge \neg\alpha) \rightarrow \beta.$$

The Lewis-Langford analysis was therefore deemed inadequate by many commentators since it fails to take into account the *relevance* connection between the antecedent and consequent of a logical implication.

On the other hand, Lewis showed how to derive his paradoxes using just a few seemingly unobjectionable modes of inference ([92, Chapter 8]), *e.g.*, for *ex absurdo quodlibet*:

1. $\alpha \wedge \neg\alpha$ assumption;
2. α 1, simplification: from $\alpha \wedge \beta$ derive α;
3. $\neg\alpha$ 1, simplification: from $\alpha \wedge \beta$ derive β;
4. $\alpha \vee \beta$ 2, addition: from α derive $\alpha \vee \beta$;
5. β 3,4, disjunctive syllogism: from $\neg\alpha, \alpha \vee \beta$ derive β.

Hence the relevant logician must show what is wrong with this reasoning. Several possible replies were devised in the 1930's, 1940's, and 1950's. Connexive logicians (*e.g.*, [105]) worked out a notion of entailment on the basis of two principles: that α entails β just in case α is inconsistent with $\neg\beta$, and that α entails β only if α is consistent with β. Such a concept validates some classically falsifiable principles and falsifies some classical tautologies and valid inference rules – for example, the simplification and addition moves in Lewis' independent proof are not permissible. Analytic logicians (*e.g.*, [114]) defended a Kantian-like view according to which the consequent of an implication should not contain concepts not already included in its antecedent. Given such a tenet, addition is clearly no good. Some philosophers ([55, 124, 136]), rather than questioning a specific step in Lewis' argument, put the blame on the possibility of freely chaining such inferential steps together. The resulting notions of entailment are only restrictedly transitive. Finally, a few commentators ([23, 44]) detected an equivocation in Lewis' use of "entails", while others ([46, 137]) accepted the independent proof, but denied the conclusion that it shows that every impossible proposition implies anything.

A completely different reply to Lewis came in the late 1950's from the American logicians Alan R. Anderson and Nuel D. Belnap, who, developing ideas of W. Ackermann ([1]), introduced the systems of *relevant logic* E and R ([3, 4]). Anderson and Belnap ([2]) suggest that the argument by Lewis is a *fallacy of ambiguity*: it equivocates over the meaning of disjunction. There is no single disjunction that allows both addition and the disjunctive syllogism; rather, there is an *intensional* relevant disjunction for which the disjunctive syllogism is valid but addition is not; and an *extensional* truth-functional disjunction for which addition, but not the disjunctive syllogism, holds. Although the connection was only implicit in Anderson and Belnap's paper, the extensional "or" roughly corresponds to additive disjunction, while the intensional "or" corresponds to multiplicative disjunction.

The placing of Anderson and Belnap's R within the framework of substructural logics is most apparent for the implication-negation fragment: it is the corresponding fragment of LK without the weakening rules [3] (or, identifying \ and / with $\rightarrow$, the calculus $InFL_{ec}$). However, the full logic corresponding to LK without weakening (InFL$_{ec}$) is not R but rather a weaker logic, studied intensively by Robert K. Meyer in his PhD thesis [103] and known in relevant circles as LR, that fails to derive the critical half of the distributive law:

$$\alpha \wedge (\beta \vee \gamma) \rightarrow (\alpha \wedge \beta) \vee (\alpha \wedge \gamma).$$

For a partly satisfactory solution to the problem of finding a suitable Gentzen-style formulation for R, instead, logicians had to wait until 1982, when Belnap ([10]) introduced the formalism of display logic.

An algebraic semantics for R in terms of involutive and distributive FL$_{ec}$ algebras, suggested by J.M. Dunn, was already available in 1966; later, Urquhart introduced an operational semantics for the implication fragment of R and, finally, in the early 1970's, a series of papers by R. Routley and R.K. Meyer launched a long-awaited Kripke-style relational semantics for both R and an array of kindred systems (see [4] and [117]). This new research trend triggered the introduction of additional relevant systems, mostly weaker than E or R, motivated by natural semantical conditions. Since the Routley-Meyer evaluation clauses for conjunction and disjunction validate distribution, such non-distributive logics as LR have remained out of the limelight over the last decades as far as relevant logics are concerned.

Fuzzy logics

The first examples of logics failing to admit the structural rule of contraction occurred in the setting of many-valued logics. In the 1920's, Jan Łukasiewicz introduced logics with n truth values for every finite $n > 2$, as well as an infinite-valued logic Ł over the closed real unit interval $[0, 1]$, where 0 represents absolute falsity, 1 represents absolute truth, and values between 0 and 1 can be thought of as intermediate "degrees of truth." The truth functions corresponding to negation and implication are

$$\neg x = 1 - x \quad \text{and} \quad x \rightarrow y = \min(1, 1 - x + y).$$

In classical logic, there are several equivalent ways to define disjunction using negation and implication; for example, $\alpha \vee \beta$ is classically equivalent to both $\neg \alpha \rightarrow \beta$ and $(\alpha \rightarrow \beta) \rightarrow \beta$. However, in Ł:

$$x \oplus y = \neg x \rightarrow y = \min(1, x+y) \quad \text{and} \quad x \vee y = (x \rightarrow y) \rightarrow y = \max(x, y).$$

Similarly, there are two possibilities for conjunction:

$$x \cdot y = \max(0, x + y - 1) \quad \text{and} \quad x \wedge y = \min(x, y).$$

An axiomatization for Ł was proposed by Łukasiewicz, but although a completeness proof was obtained by Wajsberg in the 1930's, published proofs, by Rose and Rosser [116] and Chang [28] (introducing *MV-algebras*, see Example 4.9), appeared only in the late 1950's.

In subsequent years, the study of Łukasiewicz logic has become ever more intertwined with research into *fuzzy logics*. Following the approach promoted by Hájek in his influential monograph [67], the conjunction connective · of fuzzy logics is interpreted by a continuous t-norm (commutative associative increasing binary function on [0, 1] with unit 1), the implication connective → by its residuum, and falsity constant 0 by 0. Other connectives, 1 interpreted by 1, ∧ by min, and ∨ by max, are definable. Fundamental examples of continuous t-norms are the Łukasiewicz conjunction $\max(0, x+y-1)$, Gödel t-norm $\min(x, y)$, and the product t-norm xy (product of reals), which give rise, respectively, to Łukasiewicz logic Ł, Gödel-Dummett logic G ([43, 62]), as well as a relative newcomer, product logic P ([68]).

Common generalizations of these logics have included (in chronological order) Hájek's *basic logic* BL ([67]), Esteva and Godo's *monoidal t-norm logic* MTL ([48]), and Metcalfe and Montagna's *uninorm logic* UL ([101]). The variety of UL-algebras $\mathcal{UL}$ consists of semilinear bounded commutative FL-algebras, the variety $\mathcal{MTL}$ consists of UL-algebras satisfying $1 \approx \top$ and $0 \approx \bot$, and the variety $\mathcal{BL}$ of BL-algebras consists of MTL-algebras satisfying $x \wedge y \approx x \cdot (x \to y)$. The importance of these logics and their algebras is supported by the fact that BL has been shown to be the logic of continuous t-norms ([32]), MTL, the logic of left-continuous t-norms ([81]), and UL, the logic of left-continuous uninorms ([101]). Algebraically, this means that the varieties $\mathcal{BL}$, $\mathcal{MTL}$, and $\mathcal{BL}$, are generated not only by their totally ordered members but also by their so-called "standard" members of the form $\langle [0, 1], \min, \max, \cdot, \to, 1, 0, \bot, \top \rangle$.

Perhaps surprisingly given their origins, it has emerged that many of these logics and their accompanying classes of algebras have an elegant presentation as Gentzen-style proof systems. The catch is that instead of sequents, they are formulated using *hypersequents*, introduced by Avron in [5] and consisting of finite multisets of sequents. The monograph [102] provides a comprehensive account of hypersequent (and other) calculi for fuzzy logics and their applications.

Set theory

A particularly intriguing motivation for dropping contraction rules arises from Curry's 1942 proof that the law of absorption, corresponding to the sequent $\Rightarrow (\alpha \rightarrow (\alpha \rightarrow \beta)) \rightarrow (\alpha \rightarrow \beta)$, plays a crucial role in set-theoretic paradoxes ([35]). At the turn of the last century, Russell showed that naive set theory yields, via unrestricted comprehension, a formula provably equivalent to its own negation. This is a contradiction if the underlying logic derives the law of excluded middle and other classical principles. Some mathematicians, including Brouwer, began therefore to nurture the belief that an intuitionistically correct set theory would escape Russell's paradox. However, as Curry demonstrated, the following variant of Russell's paradox follows from intuitionistically acceptable principles. Let α be an arbitrary sentence in the language of the theory, and let $C = \{x \mid x \in x \rightarrow \alpha\}$. Then:

1. $C \in C \rightarrow (C \in C \rightarrow \alpha)$ definition of C;
2. $(C \in C \rightarrow \alpha) \rightarrow C \in C$ definition of C;
3. $(C \in C \rightarrow (C \in C \rightarrow \alpha)) \rightarrow (C \in C \rightarrow \alpha)$ law of absorption;
4. $C \in C \rightarrow \alpha$ 1, 3, modus ponens;
5. $C \in C$ 2, 4, modus ponens;
6. α 4, 5, modus ponens.

Hence any set theory which contains an unrestricted comprehension axiom and the (intuitionistically correct) law of absorption is bound to be trivial. From a substructural point of view, however, the excluded middle and the law of absorption are equally vicious: the former requires a use of right contraction, while the latter presupposes an application of left contraction. Intuitionistic logic is not substructural enough to accommodate naive set theory.

The first logician who explored the possibility of reconstructing set theory on a nonclassical basis was Skolem, who devoted a series of papers to the subject in the late 1950's and early 1960's (see, *e.g.*, [123]). Skolem thought that infinite-valued Łukasiewicz logic Ł was a plausible candidate: Russell's paradoxical sentence is equivalent to its own negation, but this causes no problem in Ł where any formula whose value is 0.5 has this property. Moreover, absorption and other contraction-related principles do not hold for Ł.

However, weakening also plays a role in producing the set-theoretic paradoxes. A result by Grishin ([66]), in fact, indicates that if we add the extensionality axioms to classical logic without contraction (a weaker system than Ł), contraction can be recovered. Subsequent research on logical bases for naive set theories has therefore focused on systems in the vicinity of linear logic although the systems FL_{ew} and $InFL_{ew}$, interesting for being decidable at the first order level, have also been investigated extensively by Ono and Komori in [111].

Linear lògic

Further motivation for dropping structural rules arose from the constructive approach to logic. Since Heyting, followers of this approach have focused on the notion of proof, stressing, however, that what matters is not *whether* a given formula is provable, but *how* it is proved. A formula may therefore be identified with its set of proofs, so that a proof of α from the assumptions $\alpha_1, ..., \alpha_n$ – seen as a method for converting proofs of $\alpha_1, ..., \alpha_n$ into a proof of α – amounts to a function $f(x_1, ..., x_n)$ which associates to elements $a_i \in A_i$, the element $f(a_1, ..., a_n) \in A$.

This idea is already implicit in the so-called Brouwer-Heyting-Kolmogorov interpretation of intuitionistic logic. In the 1960's, W. A. Howard ([80]) added to this interpretation the identification of intuitionistic natural deduction proofs with terms of typed lambda calculus. A proof of the formula α is associated with a term of type α, and it then becomes possible to spell out the computational content of the inference rules in the $\{\wedge, \rightarrow\}$-fragment of the intuitionistic natural deduction calculus:

- if t and s are terms of respective types α and β, then $\langle t, s \rangle$ (the *pairing* of t and s) is a term of type $\alpha \wedge \beta$;
- if t is a term of type $\alpha \wedge \beta$, then $\pi_1(t)$ and $\pi_2(t)$ (the first and second *projections* of t) are terms of respective types α and β;
- if x is a variable of type α and t is a term of type β, then $\lambda x.t$ (the *abstraction* of t w.r.t. x) is a term of type $\alpha \rightarrow \beta$;
- if t and s are terms of respective types $\alpha \rightarrow \beta$ and α, then ts (the *application* of t to s) is a term of type β.

The ensuing correspondence between intuitionistic natural deduction proofs and terms in the lambda calculus with projection and pairing functors can be seen as a fully-fledged isomorphism (and is indeed referred to as the *Curry-Howard isomorphism*) in that there is a perfect match between the notions of conversion, normality, and reduction introduced in the two frameworks.

In the light of the Curry-Howard isomorphism, it was acknowledged that the problem of finding a "semantics of proofs" for a given constructive logic and the problem of providing lambda calculus (or, for that matter, functional programming) with a semantic interpretation were two sides of the same coin. In Dana Scott's *domain theory*, a first attempt to accomplish this task, a type α was interpreted by means of a particular topological space. On the other hand, Jean-Yves Girard ([59]) introduced for this purpose in the mid 1980's, the notion of a *coherent space* – a set A equipped with a reflexive and symmetric relation R^A, called the *coherence relation* of the space. Now, a type (*alias* formula) α can be associated with a coherent space $\mathbf{A} = \langle A, R^{\mathbf{A}} \rangle$, and a term t of type α (*alias* a proof of α) with a *clique* of $\mathbf{A}$, *i.e.*, with a subset $B \subseteq A$ of pairwise coherent elements of A. Similarly, compound formulas and their proofs can be interpreted using more complex coherent spaces. This semantics of proofs allows for substructural distinctions; for example:

- The space $\mathbf{A} \cdot \mathbf{B}$ is a coherent space whose universe is the cartesian product $A \times B$, and whose coherence relation is given by:

$$(a, b) R^{\mathbf{A} \cdot \mathbf{B}}(a', b') \quad \text{iff} \quad a R^{\mathbf{A}} a' \text{ and } b R^{\mathbf{B}} b'.$$

- The space $\mathbf{A} \wedge \mathbf{B}$ is a coherent space whose universe is the disjoint union $A \uplus B = \{(a, 0) : a \in A\} \cup \{(b, 1) : b \in B\}$, and whose coherence relation is given by:

$$(a, 0) R^{\mathbf{A} \wedge \mathbf{B}}(a', 0) \quad \text{iff} \quad a R^{\mathbf{A}} a';$$
$$(b, 1) R^{\mathbf{A} \wedge \mathbf{B}}(b', 1) \quad \text{iff} \quad b R^{\mathbf{B}} b';$$
$$(a, 0) R^{\mathbf{A} \wedge \mathbf{B}}(b, 1) \quad \text{for any } a \in A, b \in B.$$

Omitting details, let us just mention that Girard introduces a coherent space $\mathbf{A} \to \mathbf{B}$ corresponding to intuitionistic implication and a coherent space $\mathbf{A} \multimap \mathbf{B}$ corresponding to a new kind of implication, which he terms *linear* implication. If we let $!\mathbf{A}$ be the space whose universe is the set of finite cliques of $\mathbf{A}$, and whose coherence relation is given by

$$c R^{!\mathbf{A}} c' \quad \text{iff} \quad \text{there exists a clique } c'' \text{ of } \mathbf{A} \text{ such that } c, c' \subseteq c'',$$

then we get the fundamental property that the space $\mathbf{A} \to \mathbf{B}$ is isomorphic to the space $!\mathbf{A} \multimap \mathbf{B}$; in other words, the semantics of coherent spaces yields a decomposition of intuitionistic implication into a new kind of implication, linear implication, and a new unary operator, $!$. A new kind of logic, *linear logic*, had been born.

How can we make intuitive sense of this logic? One option is to view formulas as concrete resources that once consumed in a deduction to get some conclusion, cannot be recycled or reused. Formulas of the form $!\alpha$, on the other hand, represent "ideal" resources that can be reused at will. Thus, while the availability of an intuitionistic implication $\alpha \to \beta$ means that using as many α's as I might need I can get one β, the availability of a linear implication $\alpha \multimap \beta$ expresses the fact that using just one α I can get one β – something that squares perfectly with the coherent space isomorphism pointed out above. We can also view the other compound formulas of linear logic as concrete resources: for example, $\alpha \cdot \beta$ expresses the availability of *both* resource α and resource β, while $\alpha \wedge \beta$ expresses the availability of *any* one of these resources.

In his seminal 1987 paper, Girard introduces a sequent calculus for this new logic that corresponds (with a quite different syntax) to InFL_e^B (*i.e.*, LK without weakening and contraction, with the split connectives) extended with the following rules for the unary connective ! (*of course!*) (? (*why not?*) can be defined dually as $?\alpha = \neg!\neg\alpha$):

$$\frac{\alpha, \Gamma \Rightarrow \Delta}{!\alpha, \Gamma \Rightarrow \Delta} \ (! \Rightarrow) \qquad \frac{\Gamma \Rightarrow \Delta}{!\alpha, \Gamma \Rightarrow \Delta} \ (!\mathrm{WL}) \qquad \frac{!\alpha, !\alpha, \Gamma \Rightarrow \Delta}{!\alpha, \Gamma \Rightarrow \Delta} \ (!\mathrm{CL}) \qquad \frac{!\Gamma \Rightarrow \alpha}{!\Gamma \Rightarrow !\alpha} \ (\Rightarrow !)$$

where $!\Gamma$ is obtained by prefixing ! to all formulas in Γ.

The importance of the exponentials is fully realized if we take into account the fact that Girard was not interested in setting up a logic weaker than classical or intuitionistic logic: he rather wanted a logic that permits a better analysis of proofs through a stricter control of structural rules. Exponentials are there precisely to recapture the deductive power of weakening and contraction, an aim that is attained – in a sense – by showing that both classical logic and intuitionistic logic can be embedded into linear logic.

6   The interplay of algebra and logic

As we have seen in earlier sections, although ordered algebras and logics have typically emerged from distinct traditions, they may nevertheless be related via algebraization. In this final section, we illustrate the benefits of this correspondence with some examples where techniques in one field may be used to solve problems in the other. Needless to say, we do not aim for completeness, not even for the few topics covered here; rather our intention is to share some of the core ideas involved in the proofs and provide a pointer to key references.

6.1   Completeness

One of the most surprising and revealing features of recent work on the correspondence between ordered algebras and logic is the encroachment of algebraic methods on that typically most syntactic of endeavors: establishing cut elimination for Gentzen systems. To be more precise (since cut elimination in proof-theoretic parlance usually implies giving an algorithm for removing cuts from derivations), these methods, originating independently in the work of Maehara ([95]) and Okada ([106–108]), establish the admissibility of cut for the cut-free system. Or to put matters yet another way, completeness is proved for the cut-free system with respect to some class of algebras. Such results do not supplant constructive proofs (where the elimination algorithm may be fundamental), but suffice for the kind of benefits, such as establishing decidability or interpolation (see below), that a cut-free Gentzen system affords a class of ordered algebras.

The algebraic approach to completeness has undoubtedly made Gentzen systems more attractive to algebraists, and promises to lead – through the combination with other algebraic techniques – to insights and proofs unobtainable solely by syntactic means. Indeed, algebraic methods have been employed by Terui ([133]) to give a semantic characterization of extensions of the sequent calculus RL for residuated lattices with (certain forms of) structural rules that admit cut elimination. Also, Ciabattoni, Galatos, and Terui ([30]) have combined algebraic and syntactic techniques to obtain an algorithm that converts Hilbert-style axioms of a certain form into structural rules for sequent and hypersequent calculi that preserve cut elimination.

Here our aim will be more modest. We will give the main ideas of the proof of the completeness of cut-free RL with respect to residuated lattices, taking elements of our presentation from [9, 82], and [133]. As a starting point, let us consider a construction of residuated lattices that has proved invaluable in several contexts. Let $\mathbf{M} = \langle M, \cdot, 1 \rangle$ be a monoid, and for each $X, Y \subseteq M$, define:

$$X \cdot Y = \{x \cdot y \mid x \in X \text{ and } y \in Y\};$$
$$X \backslash Y = \{y \in M \mid X \cdot \{y\} \subseteq Y\};$$
$$Y / X = \{y \in M \mid \{y\} \cdot X \subseteq Y\}.$$

Then the "powerset algebra" $\wp(\mathbf{M}) = \langle \wp(M), \cap, \cup, \cdot, \backslash, /, \{1\} \rangle$ is easily seen to be a residuated lattice.

This is just the first step of the construction, however. Next, a special kind of map on $\wp(M)$ is used to refine this algebra according to the job at hand. A *nucleus* on the powerset $\wp(M)$ is a map $\gamma : \wp(M) \to \wp(M)$ satisfying $X \subseteq \gamma(X)$, $\gamma(\gamma(X)) \subseteq \gamma(X)$, $X \subseteq Y$ implies $\gamma(X) \subseteq \gamma(Y)$, and $\gamma(X) \cdot \gamma(Y) \subseteq \gamma(X \cdot Y)$.

Lemma 6.1. *If* $\mathbf{M}$ *is a monoid and* γ *is a nucleus on* $\wp(M)$, *then*

$$\wp(\mathbf{M})_\gamma = \langle \gamma(\wp(M)), \cap, \cup_\gamma, \cdot_\gamma, \backslash, /, \gamma(\{1\}) \rangle$$

is a complete residuated lattice with $X \cup_\gamma Y = \gamma(X \cup Y)$ *and* $X \cdot_\gamma Y = \gamma(X \cdot Y)$.

Now we turn our attention to the sequent calculus RL and residuated lattices. The idea is to construct a special example of the latter such that (similar to the Lindenbaum-Tarski algebra construction), validity in this algebra corresponds to cut-free derivability in RL. Let $\mathbf{Fm}^*$ be the free monoid generated by the formulas of RL; that is, the elements Fm^* of $\mathbf{Fm}^*$ are finite sequences of formulas, multiplication is concatenation, and the unit element is the empty sequence. Intuitively, we build our algebra from sets of sequences of formulas that "play the same role" in cut-free derivations in RL. We define:

$$[\Gamma_1\_\Gamma_2 \Rightarrow \alpha] = \{\Gamma \in Fm^* \mid \Gamma_1, \Gamma, \Gamma_2 \Rightarrow \alpha \text{ is cut-free derivable in RL}\};$$

$$\mathcal{D} = \{[\Gamma_1\_\Gamma_2 \Rightarrow \alpha] \mid \Gamma_1, \Gamma_2 \in Fm^* \text{ and } \alpha \in Fm\};$$

$$\gamma(X) = \bigcap\{Y \in \wp(Fm^*) \mid X \subseteq Y \subseteq \mathcal{D}\}.$$

Then γ is a nucleus on $\wp(Fm^*)$ and hence the algebra $\wp(\mathbf{Fm}^*)_\gamma$ is a residuated lattice.

We define an evaluation for this algebra by $e(p) = \gamma(\{p\})$ and prove by induction on formula complexity that for each $\alpha \in Fm$:

$$\alpha \in e(\alpha) \subseteq [\_ \Rightarrow \alpha].$$

Now consider a sequent $\alpha_1, \ldots, \alpha_n \Rightarrow \beta$ such that $\alpha_1 \cdot \ldots \cdot \alpha_n \leq \beta$ holds in all residuated lattices. In particular, it holds in $\wp(\mathbf{Fm}^*)_\gamma$, so $e(\alpha_1) \cdot \ldots \cdot e(\alpha_n) \subseteq e(\beta)$. But then, since $\alpha_i \in e(\alpha_i)$ for $i = 1 \ldots n$ and $e(\beta) \subseteq [\_ \Rightarrow \beta]$,

$$\alpha_1 \cdot \ldots \cdot \alpha_n \in [\_ \Rightarrow \beta].$$

I.e., $\alpha_1, \ldots, \alpha_n \Rightarrow \beta$ is cut-free derivable in RL.

The core idea of proofs of the form outlined above is to show the completeness of a system by establishing the admissibility of a particular rule for that system: in this case, cut. This idea applies also to other completeness results and gives corresponding generation results for classes of algebras. In particular, elimination of the "density rule" of Takeuti and Titani ([130]) has been used by Metcalfe and Montagna ([101], see also [102]) to show the completeness of certain fuzzy logics with respect to algebras based on the rational unit interval $[0, 1] \cap \mathbb{Q}$, or, reformulating this algebraically, the generation of certain varieties of semilinear commutative bounded FL-algebras by their dense totally ordered members. First it is shown that adding the density rule to any axiomatic extension of a Hilbert system UL for semilinear commutative bounded FL-algebras gives a system that is complete with respect to the dense totally ordered members of the corresponding variety. For the second step, axiomatizations are reformulated as hypersequent calculi and it is shown that in certain cases, applications of the hypersequent version of the density rule can be eliminated (constructively, similarly to cut elimination) from derivations.

6.2 Decidability

Decidability problems – determining whether there exists an effective method for checking membership of some class – have long played a prominent role in both logic and algebra, bridging the gap between abstract presentations and computational methods.   Perhaps most significant are the validity problem for a logic L – can we decide whether $\vdash_L \alpha$ holds for any formula α? – and, for a class of algebras $\mathcal{K}$, the decidability of the equational theory – can we decide whether $\vdash_{Eq(\mathcal{K})} \alpha \approx \beta$ holds for any equation $\alpha \approx \beta$? Of course, decidability of the validity problem for an algebraizable logic implies decidability of the equational theory for the corresponding class of algebras and vice versa.

Intriguingly, when tackling these problems for substructural logics and classes of residuated lattices, methods from both fields, logic and algebra, appear to be essential. Let us consider first a strategy for logics that makes use of cut-free Gentzen systems. We have already seen that cut elimination for LJ and LK facilitates a simple proof of decidability of the validity problem for propositional classical logic and propositional intuitionistic logic, and consequently, the equational theory of Boolean algebras and Heyting algebras. The proof for the sequent calculus RL and neighboring systems admitting cut elimination such as FL, FL_e, FL_w, and FL_{ew} is even easier. We simply observe that sequents occurring in any cut-free derivation for these systems must get smaller as we progress upwards in the tree, so proof search is finite. The proof for FL_{ec} (a reworking of a proof for $InFL_{ec}$ by Meyer [103]) is more complicated, but follows a similar pattern: we introduce a restricted version of the calculus and show that the number of sequents occurring in cut-free derivations for the restricted system must be finite (see, e.g., [52] for details).

Cut-free sequent calculi are powerful tools but there remain many substructural logics, not to mention interesting classes of algebras, that are (at least so far) lacking in this department. In some cases, however, more complicated structures can be used to obtain decidability results. For example, display calculi, mentioned in the remarks on relevance logics, with more structural connectives have been used to establish decidability for the equational theory of various varieties of distributive residuated lattices ([21, 22, 88]). Also, hypersequent calculi, mentioned above in connection with fuzzy logics, have been used to establish decidability for various semilinear varieties (see [102] for details).

A complementary algebraic approach to establishing decidability stems from the familiar observation that if a finitely axiomatizable logic is complete with respect to a class of finite algebras – the so-called *finite model property* – then we can check whether a given formula holds in the one-element members, then the two-element members, and so on, and at the same time search for a derivation of height one, height two, etc. Hence the validity problem is decidable. From an algebraic perspective, we say that a class of algebras has the *finite model property* FMP if every equation that fails to hold in the class, fails in a finite member of the class. However, when the forthcoming algebraic method applies, we actually get something stronger, the *strong finite model property* SFMP: any *quasi-equation* that fails to hold in the class, fails in a finite member of the class. If a finitely axiomatizable class of algebras has the FMP, then its equational theory is decidable, and if it has the SFMP, then its quasi-equational theory (and in fact, for quasivarieties of residuated lattices, its universal theory) is decidable.

The SFMP corresponds in turn to an embedding property known to researchers such as McKinsey and Tarski in the 1940's ([100]), explored in particular by Evans ([49]), and developed extensively for residuated lattices and related structures by Blok and Van Alten ([15, 17, 135]). Given an algebra $\mathbf{A} = \langle A, \langle f_i^{\mathbf{A}} \mid i \in I \rangle \rangle$ of any type and $B \subseteq A$, a *partial subalgebra* $\mathbf{B}$ of $\mathbf{A}$ is the partial algebra $\langle B, \langle f_i^{\mathbf{B}} \mid i \in I \rangle \rangle$ where for $i \in I$, k-ary f_i, and $b_1, \ldots, b_k \in B$,

$$f_i^{\mathbf{B}}(b_1, \ldots, b_k) = \begin{cases} f_i^{\mathbf{B}}(b_1, \ldots, b_k) & \text{if } f_i^{\mathbf{B}}(b_1, \ldots, b_k) \in B \\ \text{undefined} & \text{otherwise.} \end{cases}$$

An *embedding* of a partial algebra $\mathbf{B}$ into an algebra $\mathbf{A}$ of the same type is a 1-1 map $\varphi : B \to A$ such that $\varphi(f_i^{\mathbf{B}}(b_1, ..., b_k)) = f_i^{\mathbf{A}}(\varphi(b_1), ..., \varphi(a_k))$ whenever $f_i^{\mathbf{B}}(b_1, ..., b_k)$ is defined.

A class $\mathcal{K}$ of algebras of the same type has the *finite embeddability property* (FEP for short) if every finite partial subalgebra of some member of $\mathcal{K}$ can be embedded into some finite member of $\mathcal{K}$. It is easily seen that if a (quasi)variety $\mathcal{K}$ has the FEP, then $\mathcal{K}$ has the SFMP. Moreover, for quasivarieties of finite type such as (quasi)varieties of residuated lattices we have an equivalence between the two properties.

The FEP is easily established for the variety of Heyting algebras $\mathcal{HA}$ (the subvariety of $\mathcal{CFL}$ satisfying the additional equations $xy \approx x \wedge y$ and $x \wedge 0 \approx 0$; see Example 4.6). Following McKinsey and Tarski's proof ([100]), let **B** be a finite partial subalgebra of some $\mathbf{A} \in \mathcal{HA}$. Then the lattice **D** generated by $B \cup \{0, 1\}$ is a finitely generated distributive lattice and hence finite, even though this might not be true of the Heyting algebra finitely generated by **B**. Since the meet operation in any finite distributive lattice is residuated, **D** can be made into a Heyting algebra. Moreover, the partially defined residuum operation of **B** coincides (where defined) with the residuum of the meet of **D**, so **B** can be embedded into this algebra.

This proof, however, relies on the fact that the lattice reduct of a member of the variety is distributive, and that the meet coincides with the product. A more complicated construction has been introduced by Blok and Van Alten ([15, 17, 135]) that establishes the FEP for numerous subvarieties of $\mathcal{FL}$ (and many other classes of algebras) obeying some kind of integrality or idempotency property. In particular, this construction was used by Ono (private communication, see also [87]) to establish the decidability of various semilinear varieties corresponding to fuzzy logics; a simplified presentation may be found in [29].

For varieties of residuated lattices such as $\mathcal{RL}$ and $\mathcal{CRL}$ that lack integrality and idempotence properties, (versions of) the following algebra based on the integers provides a good candidate for a counterexample:

$$\mathbf{Z} = \langle \mathbb{Z}, \min, \max, +, \to, 0 \rangle,$$

where $x \to y = -x + y$. (Refer to Example 4.8.) Consider the quasi-equation:

$$(1 \leq x \ \& \ x \cdot y \approx 1) \quad \Rightarrow \quad (x \approx 1).$$

It is easy to see that this holds in all finite residuated lattices, but fails in **Z**. So the SFMP and hence the FEP fails for $\mathcal{RL}$ and $\mathcal{CRL}$.

Finally, what of classes of algebras for which these algebraic or syntactic methods do not suffice? Proofs that the universal theories of (commutative) residuated lattices are undecidable may be lifted from the corresponding proofs for full linear logic of [93]. Urquhart proved that certain varieties of distributive residuated lattices are undecidable ([134]). Finally, many interesting problems are open. In particular, it is unknown whether the variety of cancellative residuated lattices (even adding commutativity and/or integrality) is decidable, or the variety of semilinear (commutative) residuated lattices. A selection of results with references to the first (perhaps implicit) proof is given in Table 2, omitting references when the result is folklore.

Variety	Name	Equational Theory	Universal Theory
Residuated lattices	$\mathcal{RL}$	decidable	undecidable [82]
Commutative $\mathcal{RL}$	$\mathcal{CRL}$	decidable	undecidable [93]
Distributive $\mathcal{RL}$	$\mathcal{DRL}$	decidable [88]	undecidable [51]
Distributive $\mathcal{CRL}$	$\mathcal{CDRL}$	decidable [22]	undecidable [51]
Idempotent $\mathcal{CDRL}$	$\mathcal{CIdDRL}$	undecidable [134]	undecidable [134]
Integral $\mathcal{RL}$	$\mathcal{IRL}$	decidable	decidable [15]
Integral $\mathcal{CRL}$	$\mathcal{CIRL}$	decidable	decidable [15]
Semilinear $\mathcal{RL}$	$\mathcal{SemRL}$		
Semilinear $\mathcal{CRL}$	$\mathcal{CSemRL}$		
MTL-algebras	$\mathcal{MTL}$	decidable [87]	decidable [87]
Cancellative $\mathcal{RL}$	$\mathcal{CanRL}$		
Cancellative $\mathcal{CRL}$	$\mathcal{CCanRL}$		
ℓ-groups	$\mathcal{LG}$	decidable [79]	undecidable [60]
MV-algebras	$\mathcal{MV}$	decidable	decidable
Abelian ℓ-groups	$\mathcal{AbLG}$	decidable [78]	decidable [78]
Heyting algebras	$\mathcal{HA}$	decidable [56]	decidable [56]
Boolean algebras	$\mathcal{BA}$	decidable	decidable

Table 2. (Un)decidability of some subvarieties of $\mathcal{RL}$ and $\mathcal{FL}$.

6.3 Amalgamation and interpolation

Amalgamation is an important categorical property of classes of algebras (and more generally in model theory, structures) that guarantees that under certain conditions, two members of the class that contain a common subalgebra can be regarded as subalgebras of a third member so that their intersection contains the common subalgebra. Close relationships between amalgamation and fundamental logical properties, including the *Robinson property*, *Beth definability*, and various forms of *interpolation*, are well known and much studied ([6, 37, 94, 96–98, 115]). Indeed, a broad and quite bewildering number of notions have been introduced to match exactly algebraic and logical concepts occurring in this area. We refer to [52] and [86] for a guide to some of the choices on offer. Here we prefer to illustrate this quite general relationship between logic and algebra by focussing on one particularly useful example: a connection between amalgamation and the *deductive interpolation property*.

A variety $\mathcal{V}$ has the *amalgamation property* AP if for all $\mathbf{A}, \mathbf{B}, \mathbf{C} \in \mathcal{V}$ and embeddings i and j of $\mathbf{A}$ into $\mathbf{B}$ and $\mathbf{C}$, respectively, there exist $\mathbf{D} \in \mathcal{V}$ and embeddings h, k of $\mathbf{B}$ and $\mathbf{C}$, respectively, into $\mathbf{D}$ such that $h \circ i = k \circ j$.

Let us write $var(K)$ for the variables occurring in some expression (formula, equation, set of equations, etc.) K. A variety $\mathcal{V}$ is said to have the *Deductive Interpolation Property* DIP if whenever $\Sigma \vdash_{Eq(\mathcal{V})} \varepsilon$, there exists a set of equations Π with $var(\Pi) \subseteq var(\Sigma) \cap var(\varepsilon)$ such that $\Sigma \vdash_{Eq(\mathcal{V})} \Pi$ and $\Pi \vdash_{Eq(\mathcal{V})} \varepsilon$.

Theorem 6.2. *A variety of commutative residuated lattices has the* AP *iff it has the* DIP.

The preceding result is stated and proved in [53]. However, it is a consequence of the general fact that for varieties whose members have one nullary operation that generates a one-element subalgebra and all operations of finite arity, the AP and the DIP are equivalent in the presence of the congruence extension property. This being said, the result paves the way for an intriguing approach to proving amalgamation for varieties of commutative residuated lattices: namely, we show that a related interpolation property holds for the corresponding logic. Let us say that a logic L of a variety of commutative residuated lattices has the *Craig interpolation property* CIP if whenever $\vdash_L \alpha \to \beta$, there exists a formula γ with $var(\gamma) \subseteq var(\alpha) \cap var(\beta)$ such that $\vdash_L \alpha \to \gamma$ and $\vdash_L \gamma \to \beta$. In fact, in the context of commutative residuated lattices, the CIP is a strictly stronger property, a consequence of the *local deduction theorem* (part (1) of the following theorem).

Theorem 6.3. *Suppose that a variety $\mathcal{V}$ of commutative residuated lattices is an equivalent algebraic semantics for a logic* L. *Then:*

1. $T \cup \{\alpha\} \vdash_L \beta$ *iff* $T \vdash_L (\alpha \wedge 1)^n \to \beta$ *for some* $n \in \mathbb{N}$.
2. *If* L *has the* CIP, *then* $\mathcal{V}$ *has the* DIP *and hence the* AP.

Proof. 1 is provable either by a simple induction on the height of a derivation of $T \cup \{\alpha\} \vdash_L \beta$ or as an immediate consequence of Corollary 4.28 (1). For 2, it is suffices by algebraizability to prove the logical counterpart of the DIP for L. Suppose that $T \vdash_L \alpha$. Then $\{\beta_1 \wedge \ldots \wedge \beta_n\} \vdash_L \alpha$ for some $\{\beta_1, \ldots, \beta_n\} \subseteq T$ and by (1), $\vdash_L (\beta_1 \wedge \ldots \wedge \beta_n \wedge 1)^n \to \alpha$ for some $n \in \mathbb{N}$. If L has the CIP, then $\vdash_L (\beta_1 \wedge \ldots \wedge \beta_n \wedge 1)^n \to \gamma$ and $\vdash_L \gamma \to \alpha$ for some formula γ with $var(\gamma) \subseteq var(\beta_1 \wedge \ldots \wedge \beta_n) \cap var(\alpha)$. But then, again by (1), $\{\beta_1 \wedge \ldots \wedge \beta_n\} \vdash_L \gamma$ and $\{\gamma\} \vdash_L \alpha$ as required. $\square$

Theorem 6.4 ([109]). FL_e *has the* CIP.

Proof. For convenience (basically to reduce the number of cases considerably), we will make use of a slightly different calculus. Let us use (just for this proof) Γ and Δ to denote finite *multisets* rather than sequences of formulas; Γ, Δ and () now denote the multiset union of Γ and Δ, and the empty multiset, respectively. Sequents are then ordered pairs of finite multisets of formulas and we obtain a calculus – easily seen to prove the same sequents as FL_e – simply by reinterpreting the calculus FL with this new definition. Let us make a couple more cosmetic changes. We remove the redundant exchange rules, write $\alpha \rightarrow \beta$ for $\alpha \backslash \beta$ and drop the rules for /, to obtain a calculus that we denote FL_e^m. It then suffices to prove the following:

If $\vdash_{FL_e^m} \Gamma, \Delta \Rightarrow \alpha$, then there exists a formula β with $var(\beta) \subseteq var(\Gamma) \cap var(\Delta, \alpha)$ such that $\vdash_{FL_e^m} \Gamma \Rightarrow \beta$ and $\vdash_{FL_e^m} \Delta, \beta \Rightarrow \alpha$.

We proceed by induction on the height of a cut-free derivation d of $\Gamma, \Delta \Rightarrow \alpha$ in FL_e^m. Suppose first for the base case that $\Gamma, \Delta \Rightarrow \alpha$ is an instance of (ID). If Γ is (α) and Δ is (), let β be α. If Γ is () and Δ is (α), let β be 1. Also, if $\Gamma, \Delta \Rightarrow \alpha$ is an instance of ($\Rightarrow 1$), then Γ and Δ are () and α is 1, so let β be 1. The case for 0 is similar.

For the inductive step, we must consider the last application of a rule in d. Let us just treat the paradigmatic case of implication. Suppose first that α is $\alpha_1 \rightarrow \alpha_2$ and d ends with:

$$\frac{\begin{array}{c} \vdots \\ \Gamma, \Delta, \alpha_1 \Rightarrow \alpha_2 \end{array}}{\Gamma, \Delta \Rightarrow \alpha_1 \rightarrow \alpha_2} \ (\Rightarrow\rightarrow)$$

Then by the induction hypothesis, there exists β with $var(\beta) \subseteq var(\Gamma) \cap var(\Delta, \alpha_1 \rightarrow \alpha_2)$ such that $\Gamma \Rightarrow \beta$ and $\Delta, \beta, \alpha_1 \Rightarrow \alpha_2$ are derivable in FL_e^m. Hence also, by ($\Rightarrow\rightarrow$), the sequent $\Delta, \beta \Rightarrow \alpha_1 \rightarrow \alpha_2$ is derivable in FL_e^m.

Suppose now that d ends with:

$$\frac{\begin{array}{cc} \vdots & \vdots \\ \Gamma_1, \Delta_1 \Rightarrow \gamma_1 & \Gamma_2, \Delta_2, \gamma_2 \Rightarrow \alpha \end{array}}{\Gamma_1, \Gamma_2, \gamma_1 \rightarrow \gamma_2, \Delta_1, \Delta_2 \Rightarrow \alpha} \ (\rightarrow\Rightarrow)$$

There are two subcases. First, suppose that Γ is Γ_1, Γ_2 and Δ is $\gamma_1 \rightarrow \gamma_2, \Delta_1, \Delta_2$. Then by the induction hypothesis twice, there exist β_1, β_2 such that the following sequents are derivable in FL_e^m:

$$\Gamma_1 \Rightarrow \beta_1 \qquad \Gamma_2 \Rightarrow \beta_2 \qquad \Delta_1, \beta_1 \Rightarrow \gamma_1 \qquad \Delta_2, \gamma_2, \beta_2 \Rightarrow \alpha$$

where $var(\beta_1), var(\beta_2) \subseteq var(\Gamma) \cap var(\Delta, \alpha)$. Let β be $\beta_1 \cdot \beta_2$. Then we have the following required derivations:

$$
\cfrac{
\cfrac{\vdots}{\Gamma_1 \Rightarrow \beta_1} \quad \cfrac{\vdots}{\Gamma_2 \Rightarrow \beta_2}
}{\Gamma_1, \Gamma_2 \Rightarrow \beta_1 \cdot \beta_2} \;(\Rightarrow \cdot)
\qquad
\cfrac{
\cfrac{
\cfrac{\vdots}{\Delta_1, \beta_1 \Rightarrow \gamma_1} \quad \cfrac{\vdots}{\Delta_2, \gamma_2, \beta_2 \Rightarrow \alpha}
}{\Delta_1, \Delta_2, \gamma_1 \to \gamma_2, \beta_1, \beta_2 \Rightarrow \alpha} \;(\to \Rightarrow)
}{\Delta_1, \Delta_2, \gamma_1 \to \gamma_2, \beta_1 \cdot \beta_2 \Rightarrow \alpha} \;(\cdot \Rightarrow)
$$

Now suppose that Γ is $\Gamma_1, \Gamma_2, \gamma_1 \to \gamma_2$ and Δ is Δ_1, Δ_2. Here we must be a bit more careful. Considering the derivable sequent $\Gamma_1, \Delta_1 \Rightarrow \gamma_1$, we associate Γ_1 with γ_1, and obtain by the induction hypothesis, a formula β_1 with $var(\beta_1) \subseteq var(\Gamma) \cap var(\Delta, \alpha)$ such that the following sequents are FL_e^m-derivable:

$$\Delta_1 \Rightarrow \beta_1 \qquad \Gamma_1, \beta_1 \Rightarrow \gamma_1.$$

For the derivable sequent $\Gamma_2, \Delta_2, \gamma_2 \Rightarrow \alpha$, we apply the induction hypothesis to obtain β_2 with $var(\beta_2) \subseteq var(\Gamma) \cap var(\Delta, \alpha)$ such that the following sequents are FL_e^m-derivable:

$$\Gamma_2, \gamma_2 \Rightarrow \beta_2 \qquad \Delta_2, \beta_2 \Rightarrow \alpha.$$

Let β be $\beta_1 \to \beta_2$. Then we have the following required derivations:

$$
\cfrac{
\cfrac{
\cfrac{\vdots}{\Gamma_1, \beta_1 \Rightarrow \gamma_1} \quad \cfrac{\vdots}{\Gamma_2, \gamma_2 \Rightarrow \beta_2}
}{\Gamma_1, \Gamma_2, \gamma_1 \to \gamma_2, \beta_1 \Rightarrow \beta_2} \;(\to \Rightarrow)
}{\Gamma_1, \Gamma_2, \gamma_1 \to \gamma_2 \Rightarrow \beta_1 \to \beta_2} \;(\Rightarrow \to)
\qquad
\cfrac{
\cfrac{\vdots}{\Delta_1 \Rightarrow \beta_1} \quad \cfrac{\vdots}{\Delta_2, \beta_2 \Rightarrow \alpha}
}{\Delta_1, \Delta_2, \beta_1 \to \beta_2 \Rightarrow \alpha} \;(\to \Rightarrow)
$$

$\qquad\qquad\qquad\qquad\qquad\qquad\qquad\qquad\qquad\qquad\qquad\qquad$ $\square$

Corollary 6.5. $\mathcal{CRL}$ *admits the* DIP *and therefore the* AP.

This proof method works also for proving amalgamation for $\mathcal{CIRL}$ and related varieties, as well as $\mathcal{BA}$ and $\mathcal{HA}$, but fails in the absence of a suitable Gentzen system. Indeed, we remark that there are many open problems regarding interpolation and amalgamation to be resolved for classes of residuated lattices. In particular, although $\mathcal{RL}$ has the Craig interpolation property, it is unknown whether it has the amalgamation or deductive interpolation properties. Table 3 summarizes the known landscape for a selection of subvarieties of $\mathcal{RL}$ and $\mathcal{FL}$.

Variety	Name	CIP	DIP	AP
Residuated lattices	$\mathcal{RL}$	yes	?	?
Commutative $\mathcal{RL}$	$\mathcal{CRL}$	yes	yes	yes
Integral $\mathcal{CRL}$	$\mathcal{CIRL}$	yes	yes	yes
Semilinear $\mathcal{CRL}$	$\mathcal{CSemRL}$	no	?	?
MTL-algebras	$\mathcal{MTL}$	no	?	?
ℓ-groups	$\mathcal{LG}$	no	no	no
MV-algebras	$\mathcal{MV}$	no	yes	yes
Abelian ℓ-groups	$\mathcal{AbLG}$	no	yes	yes
Heyting algebras	$\mathcal{HA}$	yes	yes	yes
Boolean algebras	$\mathcal{BA}$	yes	yes	yes

Table 3. Amalgamation and interpolation properties for some subvarieties of $\mathcal{RL}$ and $\mathcal{FL}$.

References

[1] W. ACKERMANN, *Begrundung einer strengen Implikation*, J. Symbolic Logic, **21**, 2 1956, 113–128.

[2] A. R. ANDERSON, and N. D. BELNAP, *Tautological entailments*, Philosophical Studies **13** 1962, 9–24.

[3] A. R. ANDERSON, and N. D. BELNAP, "Entailment. The logic of relevance and necessity", Princeton University Press, 1975.

[4] A. R. ANDERSON, N. D. BELNAP and J. M. DUNN, "Entailment: The Logic of Relevance and Necessity II", Princeton University Press, 1992.

[5] A. AVRON, *A Constructive Analysis of RM*, J. Symbolic Logic **52**, 4 (1987), 939–951.

[6] P. D. BACSISH, *Amalgamation properties and interpolation theorems for equational theories*, Algebra Universalis **5** (1975), 45–55.

[7] R. BALBES and P. DWINGER, "Distributive Lattices", University of Missouri Press, 1974.

[8] J. BARNES, *Logical form and logical matter*, In: "Logica, Mente e Persona", A. Alberti (ed.), Olschki, Firenze, 1990, 7–119.

[9] F. BELARDINELLI, P. JIPSEN and H. ONO, *Algebraic aspects of cut elimination*, Studia Logica **77** (2004), 209–240.

[10] N. D. BELNAP, *Display logic*, Journal of Philosophical Logic **11**, (1982), 375–417.

[11] G. BIRKHOFF, "Lattice Theory", American Mathematical Society Colloquium Publications, American Mathematical Society, Providence, R.I., 1940.

[12] W. J. BLOK and B. JÓNSSON, *Equivalence of consequence operators*, Studia Logica **83** (2006), 91–110.

[13] W. BLOK and D. PIGOZZI, *Algebraizable Logics*, Memoirs of the AMS, Vol. 396, American Mathematical Society, Providence, Rhode Island, 1989.

[14] W. J. BLOK and J. REBAGLIATO, *Algebraic semantics for deductive systems*, Studia Logica **73** (2003), 1–29.

[15] W. J. BLOK and C. J. VAN ALTEN, *The finite embeddability property for residuated lattices, pocrims and BCK-algebras*, Algebra Universalis **48** (2002), 253–271.

[16] W. J. BLOK and C. J. VAN ALTEN, "Residuated Ordered Structures in Algebra and Logic", Talk by W. Blok presented at Vanderbilt University, 2002.

[17] W. J. BLOK and C. J. VAN ALTEN, *On the finite embeddability property for residuated ordered groupoids*, Trans. Amer. Math. Soc. **357** (2005), 4141–4157.

[18] K. BLOUNT, "On the Structure of Residuated Lattices", Ph.D. thesis, Vanderbilt University, Nashville, Tennessee, 1999.

[19] K. BLOUNT and C. TSINAKIS, *The structure of residuated lattices*, International Journal of Algebra and Computation **13** (2003), 437–461.

[20] G. BOOLOS, *Frege's theorem and the Peano postulates*, Bulletin of Symbolic Logic **1** (1995), 317–326.

[21] R. T. BRADY, *The Gentzenization and decidability of RW*, Journal of Philosophical Logic **19** (1990), 35–73.

[22] R. T. BRADY, *The Gentzenization and decidability of some contractionless relevant logics*, Journal of Philosophical Logic **20** (1991), 97–117.

[23] D. J. BRONSTEIN, *The meaning of implication*, Mind **45, 178** (1936), 157–180.

[24] S. BURRIS and H. P. SANKAPPANAVAR, "A Course in Universal Algebra", Graduate Texts in Mathematics, Vol. 78, Springer, 1981, available online,
http://www.math.uwaterloo.ca/snburris/htdocs/ualg.html.

[25] A. CANTINI, "I Fondamenti della Matematica", Loescher, Torino, 1979.

[26] E. CASARI, *Comparative logics and Abelian l-groups*, Logic Colloquium '88, R. Ferro *et al.*, North Holland, Amsterdam, 1989, 161–190.

[27] J. CERTAINE, "Lattice-Ordered Groupoids and Some Related Problems", Ph.D. thesis, Harvard University, Cambridge, MA, 1945.

[28] C.C. CHANG, *A new proof of the completeness of the Lukasiewicz axioms*, Transactions of the AMS **93** (1959), 74–90.

[29] A. CIABATTONI, G. METCALFE and F. MONTAGNA, *Algebraic and proof-theoretic characterizations of truth stressers for MTL and its extensions*, Fuzzy Sets and Systems **161**, (2010), 369–389.

[30] A. CIABATTONI, N. GALATOS and K. TERUI, *From axioms to analytic rules in nonclassical logics*, Proceedings of LICS 2008, (2008), 229–240.

[31] R. CIGNOLI, I. D'OTTAVIANO and D. MUNDICI, "Algebraic Foundations of Many-valued Reasoning", Trends in Logic, Vol. 7, Kluwer, Dordrecht, 2000.

[32] R. CIGNOLI, F. ESTEVA, L. GODO, L. and A. TORRENS, *Basic Fuzzy Logic is the logic of continuous t-norms and their residua*, Soft Computing **4** (2000), 106–112.

[33] C. CONSTANTINESCU, *Some properties of spaces of measures*, Atti Sem. Mat. Fis. Univ. Modena **35** (1989), 1–286.

[34] J. CORCORAN, *Aristotle's natural deduction system*, In: "Ancient Logic and Its Modern Interpretations", J. Corcoran (ed.), Reidel, Dordrecht, 1974, 85–131.

[35] H. B. CURRY, *The inconsistency of certain formal logics*, J. Symbolic Logic **7** (1942), 115–117.

[36] H. B. CURRY, *The inferential approach to logical calculus*, Logique et Analyse, n.s., 3 (1960), 119–136.

[37] J. CZELAKOWSKI, and D. PIGOZZI, *Amalgamation and interpolation in abstract algebraic logic*, In: "Models, Algebras, and Proofs (Bogota 1995)", X. Caicedo and C. H. Montenegro (eds.), Lecture Notes in Pure and Applied Mathematics, Vol. 203, Marcel Dekker, Inc., 1999, 187–265.

[38] B. A. DAVEY and H. A. PRIESTLEY, "Introduction to Lattices and Order", Vol. 2, Cambridge University Press, New York, 2002, xii+298, 0-521-78451-4.

[39] R. P. DILWORTH, *Abstract residuation over lattices*, Bulletin of the American Mathematical Society **44** (1937), 262–268.

[40] R. P. DILWORTH, *Non-commutative residuated lattices*, Transactions of the American Mathematical Society **46** (1939), 426–444.

[41] R. P. DILWORTH, "The structure and arithmetical theory of non-commutative residuated lattices", PhD thesis, California Institute of Technology, 1939.

[42] K. DOŠEN, *A historical introduction to substructural logics*, In: "Substructural Logics", K. Došen, and P. Schroeder-Heister (eds.), Oxford Univ. Press, 1993, 1–30.

[43] M. DUMMETT, *A propositional calculus with denumerable matrix*, J. Symbolic Logic **24** (1959), 97–106.

[44] A. E. DUNCAN-JONES, *Is strict implication the same as entailment?*, Analysis **2** (Unknown Month 1934), 1970-1978.

[45] H. M. EDWARDS, *The genesis of ideal theory*, Archive for History of Exact Sciences **23** (1980), 321–378.

[46] A. F. EMCH, *Implication and deducibility*, J. Symbolic Logic **1** (1936), 26–35.

[47] M. ERNÉ, *Adjunctions and Galois connections: Origins, history and development*, In: "Galois Connections and Applications", K. Denecke, M. Erné, and S.L. Wismath (eds.), Kluwer, Dordrecht, 2004, 1–138.

[48] F. ESTEVA and L. GODO, *Monoidal t-norm based logic: towards a logic for left-continuous t-norms*, Fuzzy Sets and Systems **124** (2001), 271–288.

[49] T. EVANS, *Some connections between residual finiteness, finite embeddability and the word problem*, Journal of London Mathematics Society **2** (1969), 399–403.

[50] M. FRIEDMAN, *Kant's theory of geometry*, Philosophical Review **94** (1985), 455–506.

[51] N. GALATOS, *The undecidability of the word problem for distributive residuated lattices*, In: "Ordered algebraic structures", Martinez and Jorge (eds.), Kluwer, Dordrecht, 2002, 231–243.

[52] N. GALATOS, P. JIPSEN, T. KOWALSKI and H. ONO, "Residuated Lattices: An Algebraic Glimpse at Substructural Logics", Studies in Logic and the Foundations of Mathematics, Vol. 151, Elsevier, Amsterdam, 2007.

[53] N. GALATOS and H. ONO, *Algebraization, parametrized local deduction theorem and interpolation for substructural logics over* **FL**, Studia Logica **83** (2006), 279–308.

[54] N. GALATOS and C. TSINAKIS, *Equivalence of closure operators: an order-theoretic and categorical perspective*, J. Symbolic Logic **74** (2009), 780–810.

[55] P. T. GEACH, *Entailment*, Proceedings of the Arist. Society, s.v. **32** (1958), 157–172.

[56] G. GENTZEN, *Untersuchungen über das logische Schließen I, II*, Mathematische Zeitschrift **39** (1935), 176–210, 405–431.

[57] G. GENTZEN, *Die Widerspruchfreiheit der reinen Zahlentheorie*, Mathematische Annalen **112** (1936), 493–565.

[58] G. GIERZ, K. H. HOFMANN, K. KEIMEL, J. LAWSON, M. MIS-LOVE and D. SCOTT, "A Compendium of Continuous Lattices", Springer-Verlag, 1980.

[59] J-Y. GIRARD, *Linear logic*, Theoretical Computer Science **50** (198), 1–102.

[60] A. M. W. GLASS and Y. GUREVICH, *The word problem for lattice-ordered groups*, Trans. Amer. Math. Soc. **280** (1983), 127–138.

[61] V. GLIVENKO, *Sur quelques points de la logique de M. Brouwer*, Bulletin Academie des Sciences de Belgique **15** (1929), 183–188.

[62] K. GÖDEL, *Zum intuitionistischen Aussagenkalkül*, Anzeiger der Akad. der Wiss. Wien **32** (1932), 65–66.

[63] I. GRATTAN-GUINNESS, "The Search for Mathematical Roots 1870-1940", Princeton University Press, Princeton, 2000.

[64] G. GRÄTZER, "General Lattice Theory", 2nd ed., Birkhäuser Verlag, Basel, 1998. New appendices by the author with B.A. Davey, R. Freese, B. Ganter, M. Greferath, P. Jipsen, H.A. Priestley, H. Rose, E. T. Schmidt, S. E. Schmidt, F. Wehrung and R. Wille.

[65] G. GRÄTZER, "Universal Algebra (paperback)", Springer, 2nd ed., 2008.

[66] V. N. GRIŠIN, *Predicate and set-theoretical calculi based on logic without the contraction rule*, Mathematical USSR Izvestiya **18** (1982), 41–59; English transl. Izvéstia Akademii Nauk SSSR **45** (1981), 47–68.

[67] P. HÁJEK, "Metamathematics of Fuzzy Logic", Trends in Logic, Vol. 4, Kluwer, Dordrecht, 1998.

[68] P. HÁJEK, L. GODO and F. ESTEVA, *A complete many-valued logic with product-conjunction*, Archive for Mathematical Logic **35** (1996), 191–208.

[69] T. HALPERIN, "Boole's Logic and Probability", In: "Studies in Logic and the Foundations of Mathematics", Vol. 85, North Holland, Amsterdam, 1986.

[70] T. HALPERIN, "Boole's algebra isn't Boolean algebra", In: A Boole Anthology, J. Gasser (ed.), Kluwer, Dordrecht, 2000.

[71] J. HART, L. RAFTER, L. and C. TSINAKIS, *The Structure of Commutative Residuated Lattices*, International Journal of Algebra and Computation **12** (2002), 509–524.

[72] T. L. HEATH, "A History of Greek Mathematics", Oxford University Press, Oxford, 1921.

[73] R. HECK, *The development of arithmetic in Frege's* Grundgesetze der Arithmetik, J. Symbolic Logic **58**, (1993), 579–601.

[74] A. HEYTING, *Die formalen Regeln der intuitionistischen Logik*, Sitzungsberichte der Preuischen Akademie der Wissenschaften, Phys.-mathem. Klasse (1930), 42–56.

[75] D. HILBERT, *Die Grundlagen der Mathematik*, Abhandlungen aus dem Seminar der Hamburgischen Universität **6** (1928), 65–85.

[76] D. HILBERT and P. BERNAYS, "Grundlagen der Mathematik", Springer, 1934 (first volume), 1939 (second volume).

[77] J. HINTIKKA, *Kant on the mathematical method*, In: Kant Studies Today, L. W. Beck (ed.), Open Court, La Salle, 1969, 117–140.

[78] N. G. HISAMIEV, *Universal theory of lattice-ordered Abelian groups*, Algebra i Logika Sem. **5** (1966), 71–76.

[79] W. C. HOLLAND and S. H. MCCLEARY, *Solvability of the word problem in free lattice-ordered groups*, Houston Journal of Mathematics **5** (1979), 99–105.

[80] W. A. HOWARD, *The formulae-as types notion of construction*, In: "To H.B. Curry: Essays on Combinatory Logic, Lambda Calculus, and Formalism", J. R. Hindley and J. Seldin (eds.), Academic Press, New York, 1980, 479–490.

[81] S. Jenei and F. Montagna, *A Proof of Standard Completeness for Esteva and Godo's Logic MTL*, Studia Logica **70**, (2002), 183–192.

[82] P. JIPSEN and C. TSINAKIS, *A survey of residuated lattices*, In: "Ordered Algebraic Structures", Jorge Martinez (ed.), Kluwer, Dordrecht, 2002, 19–56.

[83] I. KAPLANSKI, "Fields and Rings", Chicago Lectures in Mathematics, Chicago University Press, Chicago, 1965.

[84] K. KEIMEL, *Some trends in lattice-ordered groups and rings*, In: "Lattice Theory and its Applications", K. A. Baker and R. Wille (eds.), Heldermann Verlag, Berlin, 1995, 131–161.

[85] O. KETONEN, *Untersuchungen zum Prädikatenkalkül*, Annales Academiae Scientiarum Fennicae, **I, 23** (1944).

[86] H. KIHARA and H. ONO, *Interpolation Properties, Beth Definability Properties and Amalgamation Properties for Substructural Logics*, Journal of Logic and Computation, to appear.

[87] T. KOWALSKI and H. ONO, *Fuzzy logics from substructural perspective*, Fuzzy Sets and Systems **161** (2010), 301–310.

[88] M. KOZAK, *Distributive full Lambek calculus has the finite model property*, Studia Logica **91** (2009), 201–216.

[89] G. KREISEL, *Perspectives in the philosophy of pure mathematics*, In: "Logic, Methodology and Philosophy of Science IV", P. Suppes *et al.* (eds.), North Holland, Amsterdam, 1973, 255–277.

[90] J. LAMBEK, *The mathematics of sentence structure*, American Mathematical Monthly **65** (1958), 154–170.

[91] J. LEAR, "Aristotle and Logical Theory", Cambridge University Press, Cambridge, 1999.

[92] C. I. LEWIS and C. H. LANGFORD, "Symbolic Logic", The Century Co., New York, 1932.

[93] P. LINCOLN, J. MITCHELL, A. SCEDROV and N. SHANKAR, *Decision problems for propositional linear logic*, Ann. Pure Appl. Logic **56** (1992), 239–311.

[94] J. MADARÁSZ, *Interpolation and amalgamation: Pushing the limits. Part I*, Studia Logica **61** (1998), 311–345.

[95] S. MAEHARA, *Lattice-valued representation of the cut-elimination theorem*, Tsukuba Journal of Mathematics **15** (1991), 509–521.

[96] L. L. MAKSIMOVA, *Craig's theorem in superintuitionistic logics and amalgamable varieties of pseudo-Boolean algebras*, Algebra i Logika **16** (1977), 643–681.

[97] L. L. MAKSIMOVA, *Interpolation properties of superintuitionistic logics*, Studia Logica **38** (1979), 419–428.

[98] L. L. MAKSIMOVA, *Interpolation theorems in modal logics and amalgamable varieties of topological Boolean algebras*, Algebra i Logika **18** (1979), 556–586.

[99] R. N. MCKENZIE, G. F. MCNULTY and W. F. TAYLOR, "Algebras, Lattices, Varieties", Vol. 1, Wadsworth & Brooks/Cole, Monterey, California, 1987.

[100] J. C. C. MCKINSEY and A. TARSKI, *On closed elements in closure algebras*, Ann. of Math. **47** (1946), 122–162.

[101] G. METCALFE and F. MONTAGNA, *Substructural Fuzzy Logics*, J. Symbolic Logic **72** (2007), 834–864.

[102] G. METCALFE, N. OLIVETTI and D. GABBAY, "Proof Theory for Fuzzy Logics", Applied Logic, Vol. 36, Springer, 2008.

[103] R. K. MEYER, "Topics in modal and many-valued logic", Doctoral dissertation, University of Pittsburgh, 1966.

[104] T. NAKANO, *Rings and partly ordered systems*, Mathematische Zeitschrift **99** (1967), 355–376.

[105] E. J. NELSON, *Intensional relations*, Mind **39** (1930), 440–453.

[106] M. OKADA, *Phase semantics for higher order completeness, cut-elimination and normalization proofs (extended abstract)*, Electronic Notes in Theoretical Computer Science **3** (1996), 22 pp. (electronic).

[107] M. OKADA, *Phase semantic cut-elimination and normalization proofs of first- and higher-order linear logic*, Theoretical Computer Science **227** (1999), 333–396.

[108] M. OKADA and K. TERUI, *The finite model property for various fragments of intuitionistic linear logic*, J. Symbolic Logic **64** (1999), 790–802.

[109] H. ONO, *Proof-theoretic methods for nonclassical logic: an introduction*, In: "Theories of Types and Proofs", M. Takahashi, M. Okada, M. Dezani-Ciancaglini (eds.), MSJ Memoirs, Vol. 2, Mathematical Society of Japan, 1998, 207–254.

[110] H. ONO, *Logics without contraction rule and residuated lattices I*, Festschrift on the occasion of R.K. Meyer's 65th birthday, forthcoming.

[111] H. ONO and Y. KOMORI, *Logics without the contraction rule*, J. Symbolic Logic **50** (1985), 169–201.

[112] F. PAOLI, "Substructural Logics: A Primer", Trends in Logic, Vol. 13, Kluwer, Dordrecht, 2002.

[113] F. PAOLI, M. SPINKS and R. VEROFF, *Abelian logic and the logic of pointed lattice-ordered varieties*, Logica Universalis **2** (2008), 209–233.

[114] W. T. PARRY, *Ein Axiomensystem für eine neue Art von Implikation*, Ergebnisse eines Mathematischen Kolloquiums (1933), 5–6.

[115] D. PIGOZZI, *Amalgamations, congruence-extension, and interpolation properties in algebras*, Algebra Universalis **1** (1972), 269–349.

[116] A. ROSE and J. B. ROSSER, *Fragments of many-valued statement calculi*, Transactions of the American Mathematical Society **87** (1958), 1–53.

[117] R. ROUTLEY, V. PLUMWOOD, R. K. MEYER and R. T. BRADY, "Relevant Logics and Their Rivals", Ridgeview, Atascadero, 1982.

[118] V. V. RAMA RAO, *On a common abstraction of Boolean rings and lattice-ordered groups I*, Monatshefte für Mathematik **73** (1969), 411–421.

[119] G. RESTALL, "An Introduction to Substructural Logics", Routledge, London, 2000.

[120] W. SCHARLAU, "Richard Dedekind 1831-1981, Festschrift 1981", Vieweg, Braunschweig, 1982.

[121] K. D. SCHMIDT, *A common abstraction of Boolean rings and lattice ordered groups*, Compositio Mathematica **54** (1985), 51–62.

[122] K. D. SCHMIDT, *Minimal clans: a class of ordered partial semigroups including Boolean rings and partially ordered groups*, In:

"Semigroups: Theory and Applications", Springer, Berlin, 1988, 300–341.

[123] T. SKOLEM, *Studies on the axiom of comprehension*, Notre Dame Journal of Formal Logic **4, 3** (1963), 162–179.

[124] T. J. SMILEY, *Entailment and deducibility*, Proceedings of the Arist. Society **s.v. 33** (1959), 233–249.

[125] T. J. SMILEY, *What is a syllogism?*, Journal of Philosophical Logic **2** (1973), 136–154.

[126] M. H. STONE, *The theory of representations for Boolean algebras*, Trans. Amer. Math. Soc. **40** (1936), 37–111.

[127] K. L. N. SWAMY, *Dually residuated lattice-ordered semigroups*, Mathematische Annalen **159** (1965), 105–114.

[128] K. L. N. SWAMY, *Dually residuated lattice-ordered semigroups II*, Mathematische Annalen **160** (1965), 64–71.

[129] K. L. N. SWAMY, *Dually residuated lattice-ordered semigroups III*, Mathematische Annalen **167** (1966), 71–74.

[130] G. TAKEUTI and T. TITANI, *Intuitionistic fuzzy logic and intuitionistic fuzzy set theory*, J. Symbolic Logic **49** (1984), 851–866.

[131] A. TARSKI, *Foundations of the calculus of systems* (Engl. transl.), In: "Logic, Semantics, Metamathematics", Oxford University Press, New York, 1956, 342–383.

[132] A. TARSKI, *On the concept of logical consequence* (Engl. transl.), In: "Logic, Semantics, Metamathematics", Oxford University Press, New York, 1956, 409–420.

[133] K. TERUI, *Which structural rules admit cut elimination? – an algebraic criterion*, J. Symbolic Logic **72** (2007), 738–754.

[134] A. URQUHART, *The undecidability of entailment and relevant implication*, J. Symbolic Logic **49** (1984), 1059–1073.

[135] C. J. VAN ALTEN, *The finite model property for knotted extensions of propositional linear logic*, J. Symbolic Logic **70** (2005), 84–98.

[136] G. H. VON WRIGHT, "Logical Studies", Routledge, London, 1957.

[137] P. G. J. VREDENDUIJN, *A system of strict implication*, J. Symbolic Logic **4, 2** (1939), 73–76.

[138] M. WARD, *Structure residuation*, Annals of Mathematics **39** (1938), 558–568.

[139] M. WARD and R. P. DILWORTH, *Residuated lattices*, Proceedings of the National Academy of Sciences **24** (1938), 162–164.

[140] M. WARD and R. P. DILWORTH, *Residuated lattices*, Transactions of the AMS **45** (1939), 335–354.

[141] A. M. WILLE, "Residuated Structures with Involution", Shaker Verlag, Aachen, 2006.

[142] R. WÓJCICKI, "Theory of Logical Calculi", Kluwer, 1988.

[143] O. WYLER, *Clans*, Compositio Mathematica **17** (1965–66), 172–189.

2

PROBABILITY

The social entropy process: Axiomatising the aggregation of probabilistic beliefs

George Wilmers

1 Introduction

The present work stems from a desire to combine ideas arising from two historically different schemes of probabilistic reasoning, each having its own axiomatic traditions, into a single broader axiomatic framework, capable of providing general new insights into the nature of probabilistic inference in a multiagent context. In the present sketch of our work we first describe briefly the background context, and we then present a set of natural principles to be satisfied by any general method of aggregating the partially defined probabilistic beliefs of several agents into a single probabilistic belief function. We will call such a general method of aggregation a social inference process. Finally we define a particular social inference process, the Social Entropy Process (abbreviated to **SEP**), which satisfies the principles formulated earlier. **SEP** has a natural justification in terms of information theory, and is closely related to the maximum entropy inference process: indeed it can be regarded as a natural extension of that inference process to the multiagent context.

By way of comparison, for any appropriate set of partial probabilistic beliefs of an *isolated* individual the well-known maximum entropy inference process, **ME**, chooses a probabilistic belief function consistent with those beliefs. We conjecture that **SEP** is the only "natural" social inference process which extends **ME** to the multiagent case, always under the assumption that no additional information is available concerning the expertise or other properties of the individual agents.[1]

Proofs of the results in the present paper, while reasonably straightforward, have mostly not been included, but will be included in a more detailed version of this work which will appear elsewhere.

---

[1] This condition is sometimes known as the Watts Assumption (see [12]).

In order to fix notation let $S = \{\alpha_1, \alpha_2, \ldots \alpha_J\}$ denote some fixed finite set of mutually exclusive and exhaustive outcomes or atomic events. A probability function w on S is a function $w : S \to [0, 1]$ such that $\Sigma_{j=1}^{J} w(\alpha_j) = 1$. Slightly abusing notation we will identify w with the vector of values $< w_1 \ldots w_J >$ where w_j denotes $w(\alpha_j)$ for $j = 1 \ldots J$. If such a w represents the subjective belief of an individual **A** in the outcomes of S we refer to w as **A**'s *belief function*. All other more complex events considered are equivalent to disjunctions of the α_j and are represented by the Greek letters θ, ϕ, ψ etc. A probability function w is assumed to extend so as to take values on complex events in the standard way, *i.e.* for any θ

$$w(\theta) = \sum_{\alpha_j \models \theta} w(\alpha_j)$$

where $\models$ denotes the classical notion of logical implication. Conditional probabilities are defined in the usual manner. We note that in this paper the use the term "belief function" will always denote a probability function in the above sense.

The first scheme referred to above is the notion of an *inference process* first formulated by Paris and Vencovská some twenty years ago (see [11–13], and [14]). The problematic of Paris and Vencovská is that of an isolated individual **A** whose belief function is in general not completely specified, but whose set of beliefs is instead regarded as a set of constraints **K** on the possible values which the vector $< w_1 \ldots w_J >$ may take. The constraint set **K** therefore defines a certain region of the Euclidean space $\mathbb{R}^J$, denoted by $\mathbf{V_K}$, consisting of all vectors $< w_1 \ldots w_J >$ which satisfy the constraints in **K** together with the conditions that $\sum_{j=1}^{J} w_j = 1$ and that $w_j \geq 0$ for all j. It is assumed that the constraint sets **K** which we consider are consistent (*i.e.* $\mathbf{V_K}$ is non-empty), and are such that $\mathbf{V_K}$ has pleasant geometrical properties. More precisely, the exact requirement on a set of constraints **K** is that the set $\mathbf{V_K}$ forms a non-empty closed convex region of Euclidean space. Throughout the rest of this paper **all constraint sets to which we refer will be assumed to satisfy this requirement**, and we shall refer to such constraint sets as *nice* constraint sets.[2] Paris and Vencovská ask the question: given any such **K**, by what rational principles should **A** choose his

---

[2] This formulation ensures that conditions such as

$$w(\theta) = \frac{1}{3}, \; w(\phi \mid \psi) = \frac{4}{5}, \; \text{and} \; w(\psi \mid \theta) \leq \frac{1}{2},$$

where θ, ϕ, and ψ are Boolean combinations of the α_j's, are all permissible in **K** provided probabilistic belief function w consistent with **K** *in the absence of any other information*?

A set of constraints **K** as above is often, slightly misleadingly, called a knowledge base. A rule **I** which for every such **K** chooses such a $w \in \mathbf{V_K}$ is called an *inference process*. Given **K** we denote the belief function w chosen by **I** by **I(K)**. The question above can then be reformulated as: what self-evident general principles should an inference process **I** satisfy? This question has been intensively studied over the last twenty years and much is known. In particular in [11] Paris and Vencovská found an elegant set of principles which uniquely characterise the maximum entropy inference process,[3] **ME**, which is defined as follows: given **K** as above, **ME(K)** chooses that unique probability distribution w which maximises the Shannon entropy of w,

$$-\sum_{j=1}^{J} w_j \log w_j$$

subject to the condition that $w \in \mathbf{V_K}$. Although some of the principles used to characterise **ME** may individually be open to philosophical challenge, they are sufficiently convincing overall to give **ME** the appearance of a gold standard, in the sense that no other known inference process satisfies an equally convincing set of principles.[4]

An apparently rather different problematic of probabilistic inference has been much studied in decision theoretic literature. Given possible outcomes $\alpha_1, \alpha_2, \ldots \alpha_J$ as before, let $\{\mathbf{A}_i \mid i = 1 \ldots m\}$ be a finite set of agents each of whom possesses his own particular probabilistic belief function $w^{(i)}$ on the set of outcomes, and let us suppose that these $w^{(i)}$ have already been determined. How then should these individual belief functions be aggregated so as to yield a single probabilistic belief function v which most accurately represents the *collective* beliefs of the agents? We call such an aggregated belief function a *social belief function*, and a general method of aggregation a *pooling operator*. Again we

---

that they are consistent. Here a conditional constraint such as $w(\psi \mid \theta) \leq \frac{1}{2}$ is interpreted as $w(\psi \wedge \theta) \leq \frac{1}{2} w(\theta)$ which is always a well-defined linear constraint, albeit trivial when $w(\theta) = 0$.

[3] This characterisation considerably strengthens earlier work of Shore and Johnson in [16].

[4] Other favored inference processes which satisfy many, but not all, of these principles are the minimum distance inference process, **MD**, the limit centre of mass process, $\mathbf{CM}^{\infty}$, all **Renyi** inference processes, and the remarkable **Maximin** process of Hawes [8]. (See [12, Paris] for a general introduction to inference processes, and also Hawes [8], especially the comparative table in Chapter 9, for an excellent résumé of the current state of knowledge concerning this topic).

can ask: what principles should a pooling operator satisfy? In this framework various plausible principles have been investigated extensively in the literature, and have in particular been used to characterise two popular, but very different pooling operators **LinOp** and **LogOp**. **LinOp** takes v to be the arithmetic mean of the $w^{(i)}$, *i.e.*

$$v_j = \frac{1}{m} \sum_{i=1}^{m} w_j^{(i)} \quad \text{for each} \ \ j = 1 \ldots J$$

whereas **LogOp** chooses v to be the normalised geometric mean given by:

$$v_j = \frac{(\prod_{i=1}^{m} w_j^{(i)})^{\frac{1}{m}}}{\sum_{k=1}^{J} (\prod_{i=1}^{m} w_k^{(i)})^{\frac{1}{m}}} \quad \text{for each} \ \ j = 1 \ldots J.$$

Various continua of other pooling operators related to **LinOp** and **LogOp** have also been investigated. However the existing axiomatic analysis of pooling operators, while technically simpler than the analysis of inference processes, is also more ambiguous and perhaps less intellectually satisfying in its conclusions than the analysis of inference processes developed within the Paris-Vencovská framework ; in the former case one arrives at rival, apparently plausible, axiomatic characterisations of various pooling operators, including in particular **LinOp** and **LogOp**, without any very convincing foundational criteria for deciding, within the limited context of the framework, which operator is justified, if any. (See *e.g.* [1,2,4–7,15] for further discussion of the axiomatics of pooling operators).

In the present paper we seek to provide an axiomatic framework to extend the Paris-Vencovská notion of inference process to the multiagent case, thereby encompassing the framework of pooling operators as a very special case. Thus we consider, for any $m \geq 1$, a set **M** consisting of m individuals $\mathbf{A}_1 \ldots \mathbf{A}_m$, each of whom possesses his own set of consistent constraints, respectively $\mathbf{K}_1 \ldots \mathbf{K}_m$, on his possible belief function on the set of outcomes $\alpha_1, \alpha_2, \ldots \alpha_J$. (Note that we are only assuming that the beliefs of each individual are consistent, not that the beliefs of different individuals are jointly consistent). We shall refer to such a set **M** of individuals as a *college*.

A *social inference process*, $\mathfrak{F}$, is a function which chooses, for any such $m \geq 1$ and $\mathbf{K}_1 \ldots \mathbf{K}_m$, a probability function on $\alpha_1, \alpha_2, \ldots \alpha_J$, denoted by $\mathfrak{F}(\mathbf{K}_1 \ldots \mathbf{K}_m)$, which we refer to as the *social belief function* defined by $\mathfrak{F}(\mathbf{K}_1 \ldots \mathbf{K}_m)$. Note that, trivially, provided that when $m = 1$ $\mathfrak{F}(\mathbf{K}_1) \in \mathbf{V}_{\mathbf{K}_1}$ for all $\mathbf{K}_1$, $\mathfrak{F}$ marginalises to an inference process. On the other hand, in the special case where $\mathbf{V}_{\mathbf{K}_i}$ is a singleton for all

$i = 1 \ldots m$, $\mathfrak{F}$ marginalises to a pooling operator. The new framework therefore encompasses naturally the two classical frameworks described above.

Again we can ask: what principles would we wish such a social inference process $\mathfrak{F}$ to satisfy *in the absence of any further information?* Is there any social inference process $\mathfrak{F}$ which satisfies them? If so, to which inference process and to which pooling operator does such an $\mathfrak{F}$ marginalise? It turns out that merely by posing these questions in the right framework, and by making certain simple mathematical observations, we can gain considerable insight. It is however essential to note that our standpoint is strictly that of a logician: we insist on the absoluteness of the qualification above that we are given no further information than that stated in the problem. In particular we are given no information about the expertise of the individuals or about the independence of their opinions. This insistence on sticking to a problem where the available information is rigidly defined is absolutely essential to our analysis, just as it is in the analysis of inference processes by Paris and Vencovská and their followers. We make no apology for the fact that such an assumption is almost always unrealistic: in order to tackle difficult foundational problems it is necessary to start with a general but *precisely defined* problematic. As has in essence been pointed out by Paris and Vencovská, unless one is prepared to make certain assumptions which precisely delimit the probabilistic information under consideration, even the classical notion of an inference process becomes incoherent. Indeed failure to define precisely the information framework lies behind several so-called paradoxes of reasoning under uncertainty.[5]

2 An axiomatic framework for a social inference process

The underlying idea of a social inference process is not new. (See *e.g.* [17]). However, to the author's knowledge, the work which has been done hitherto has largely been pragmatically motivated, and has not considered foundational questions. This is possibly due in part to a rather tempting reductionism, which would see the problem of a finding a social inference process as a two stage process in which a classical inference process is first chosen and applied to the constraints $\mathbf{K}_i$ of each agent i to yield a belief function $w^{(i)}$ appropriate to that agent, and a pooling operator is then chosen and applied to the set of $w^{(i)}$ to yield a social belief

---

[5] The interested reader may consult [13] for a detailed analysis of this point in connection with supposed paradoxes arising from "representation dependence".

function.[6] Of course from this reductionist point of view a social inference process would not be particularly interesting foundationally, since
we could hardly expect an analysis of such social inference processes to
tell us anything fundamentally new about collective probabilistic reasoning.

Our approach is however completely different. We reject the two stage
approach above on the grounds that the classical notion of an inference
process applies to an isolated single individual, and is valid only on the
assumption that that individual has absolutely no knowledge or beliefs
other than those specified by his personal constraint set. Indeed the preliminary point should be made that in the case of an isolated individual
A, whereas **A**'s constraint set **K** is subjective and personal to that individual, the *passage* from **K** to **A**'s assumed belief function w via an
inference process should be made using rational or normative principles,
and should therefore be considered to have an *intersubjective* character.
We should not confuse the epistemological status of w with that of **K**. By
hypothesis **K** represents the sum total of **A**'s beliefs; *ipso facto* **K** also
represents, in general, a description of the extent of **A**'s *ignorance*. Thus
while w may be regarded as the *belief function which best represents* **A**'s
subjective beliefs, it must not be confused with those beliefs themselves,
since in the passage from **K** to w it is clear that certain "information"
has been discarded;[7] thus, while w is determined by **K** once an inference
process is given, neither **K** nor $\mathbf{V_K}$ can be recaptured from w. As a trivial
example we may note that specifying that **A**'s constraint set **K** is empty,
i.e. that **A** claims total ignorance, is informationally very different from
specifying that **K** is such that $\mathbf{V_K} = \{< \frac{1}{J}, \frac{1}{J} \dots \frac{1}{J} >\}$, although the
application of **ME**, or of any other reasonable inference process, yields
$w = < \frac{1}{J}, \frac{1}{J} \dots \frac{1}{J} >$ in both cases.

From this point of view the situation of an individual who is a member of a college whose members seek to *collaborate* together to elicit a
social belief function seems quite different from that of an isolated individual. Indeed in the former context it appears more natural to assume as
a normative principle that, if the social belief function is to be optimal,
then each individual member $\mathbf{A}_i$ *should be deemed* to choose his personal
belief function $w^{(i)}$ so as to take account of the information provided by
the other individuals, in such a way that $w^{(i)}$ is *consistent* with his own

---

[6] We note that by no means are all authors reductionist in this sense: in particular although their
concerns are somewhat different from ours, neither [17] nor [9] make such an assumption.

[7] The word "information" is used here in a different sense from that of Shannon information.

belief set $\mathbf{K}_i$, while being as informationally close as possible to the social belief function $\mathfrak{F}(\mathbf{K}_1 \dots \mathbf{K}_m)$ which is to be defined. We will show in Section 3 that this key idea is indeed mathematically coherent and can be used to define a particular social inference process with remarkable properties. Notice however that it is not necessary to assume that a given $\mathbf{A}_i$ subjectively or consciously holds the particular personal belief function $w^{(i)}$ which is attributed to him by the procedure above: such an $w^{(i)}$ is viewed as nothing more than the belief function which $\mathbf{A}_i$ *ought rationally to hold*, given the personal constraint set $\mathbf{K}_i$ which represents his own beliefs, together with the *extra information* available to him by virtue of his knowledge of the constraint sets of the remaining members of the college. Just as in the case of an isolated individual, the passage from $\mathbf{A}_i$'s actual subjective belief set $\mathbf{K}_i$ to his notional subjective belief function $w^{(i)}$ has an intersubjective or normative character: however the calculation of $w^{(i)}$ now depends not only on $\mathbf{K}_i$ but on the belief sets of all the other members of the college.

Considerations similar to the above give rise to the following radical but attractive principle for a social inference process to satisfy:

The Collegial Principle

A social inference process $\mathfrak{F}$ satisfies the Collegial Principle (abbreviated to *Collegiality*) if for any $m \geq 1$ and $\mathbf{A}_1 \dots \mathbf{A}_m$ with respective constraint sets $\mathbf{K}_1 \dots \mathbf{K}_m$, if for some $k < m$ $\mathfrak{F}(\mathbf{K}_1 \dots \mathbf{K}_k)$ is consistent with $\mathbf{K}_{k+1} \cup \mathbf{K}_{k+2} \cup \dots \cup \mathbf{K}_m$, then

$$\mathfrak{F}(\mathbf{K}_1 \dots \mathbf{K}_m) = \mathfrak{F}(\mathbf{K}_1 \dots \mathbf{K}_k) \qquad \Box$$

Collegiality may be interpreted as stating the following: if the social belief function v generated by some subset of the college is consistent with the individual beliefs of the remaining members, then v is also the social belief function of the whole college. In particular this means that adding to the college a new individual whose constraint set is empty will leave the social belief function unchanged.

We now introduce a number of other desirable principles for a social inference process $\mathfrak{F}$ to satisfy. Several of these are obvious transfers of familiar symmetry axioms from the theory of inference processes or from social choice theory.

The Equivalence Principle

If for all $i = 1 \dots m$ $\mathbf{V}_{\mathbf{K}_i} = \mathbf{V}_{\mathbf{K}'_i}$ then

$$\mathfrak{F}(\mathbf{K}_1 \dots \mathbf{K}_m) = \mathfrak{F}(\mathbf{K}'_1 \dots \mathbf{K}'_m) \qquad \Box$$

.

Otherwise expressed the Equivalence Principle states that substituting constraint sets which are equivalent in the sense that the set of probability functions which satisfy them is unchanged will leave the values of $\mathfrak{F}$ invariant. This principle is a familiar one adopted from the theory of inference processes (*cf.* [12]). In this paper we shall always consider only social inference processes (or inference processes) which satisfy the Equivalence Principle. For this reason we may occasionally allow a certain sloppiness of notation in the sequel by identifying a constraint set **K** with its set of solutions $\mathbf{V_K}$ where the meaning is clear and this avoids an awkward notation. In particular if $\boldsymbol{\Delta}$ is a non-empty closed convex set of belief functions then we may write $\mathbf{ME}(\boldsymbol{\Delta})$ to denote the unique $w \in \boldsymbol{\Delta}$ which maximises the Shannon entropy function.

The Anonymity Principle

This principle states that $\mathfrak{F}(\mathbf{K}_1 \ldots \mathbf{K}_m)$ depends only on the multiset of constraint sets $\{\mathbf{K}_1 \ldots \mathbf{K}_m\}$ and not on the characteristics of the individuals with which the $\mathbf{K}_i$'s are associated nor the order in which the $\mathbf{K}_i$'s are listed.   □

In order to ensure that $\mathfrak{F}$ behaves like an inference process for the case $m = 1$ we need the following axiom.

The Consistency Axiom

For the case when $m = 1$

$$\mathfrak{F}(\mathbf{K}_1) \in \mathbf{V_{K_1}}$$

for any constraint set $\mathbf{K}_1$.   □

It is immediate from Consistency, Anonymity and Collegiality that $\mathfrak{F}$ satisfies the "Unanimity" property that for any $\mathbf{K}$,

$$\mathfrak{F}(\mathbf{K} \ldots \mathbf{K}) = \mathfrak{F}(\mathbf{K}).$$ □

Another immediate consequence of the above axioms is:

Lemma 2.1. *If $\mathfrak{F}$ satisfies Consistency and Collegiality, then if* $\mathbf{K}_1 \ldots \mathbf{K}_m$ *are such that*

$$\bigcap_{i=1}^{m} \mathbf{V_{K_i}} \neq \emptyset$$

then

$$\mathfrak{F}(\mathbf{K}_1 \ldots \mathbf{K}_m) \in \bigcap_{i=1}^{m} \mathbf{V_{K_i}}.$$

The following principle is again a familiar one satisfied by classical inference processes (*cf.* [12]):

Let σ denote a permutation of the atoms of S. Such a σ induces a corresponding permutation on the coordinates of probability distributions $< w_1 \ldots w_J >$, and on the corresponding coordinates of variables occurring in the constraints of constraint sets $\mathbf{K}_i$, which we denote below with an obvious notation.

The Atomic Renaming Principle

For any permutation σ of the atoms of S, and for all $\mathbf{K}_1 \ldots \mathbf{K}_m$

$$\mathfrak{F}(\sigma(\mathbf{K}_1) \ldots \sigma(\mathbf{K}_m)) = \sigma(\mathfrak{F}(\mathbf{K}_1 \ldots \mathbf{K}_m)). \qquad \square$$

Our next axiom goes to the heart of certain basic intuitions concerning probability. For expository reasons we will consider first the case when $m = 1$, in which case we are essentially discussing a principle to be satisfied by a classical inference process. First we introduce some fairly obvious terminology.

Let w denote $\mathbf{A}_1$'s belief function. Since we are considering the case when $m = 1$ we will drop the superscript from $w^{(1)}$ for ease of notation. For some non-empty set of atoms $\{\alpha_{j_1} \ldots \alpha_{j_t}\}$ let ϕ denote the event $\bigvee_{r=1}^t \alpha_{j_r}$. Suppose that $\mathbf{K}$ denotes a set of constraints on the variables $w_{j_1} \ldots w_{j_t}$ which defines a non-empty closed convex region of t-dimensional Euclidean space with $\sum_{r=1}^t w_{j_r} \leq 1$ and all $w_{j_r} \geq 0$. We shall refer to such a $\mathbf{K}$ as a *nice set of constraints about* ϕ.

Now let $\hat{w}_r$ denote $w(\alpha_{j_r} \mid \phi)$ for $r = 1 \ldots t$, with the $\hat{w}_r$ undefined if $w(\phi) = 0$. Then $\hat{w} = < \hat{w}_1 \ldots \hat{w}_t >$ is a probability distribution provided that $w(\phi) \neq 0$. Let $\mathbf{K}$ be a nice set of constraints on the probability distribution $\hat{w}$: we shall refer to such a $\mathbf{K}$ as a *nice set of constraints conditioned on* ϕ. In line with our previous conventions we shall consider such $\mathbf{K}$ to be trivially satisfied in the case when $w(\phi) = 0$.

The following principle captures a basic intuition about probabilistic reasoning which is valid for all standard inference processes:

The Locality Principle (for an inference process)

If $\mathbf{K}_1$ is a nice set of constraints *conditioned on* ϕ, and $\mathbf{K}_1^*$ is a nice set of constraints *about* $\neg\phi$, then for every event θ

$$\mathfrak{F}(\mathbf{K}_1 \cup \mathbf{K}_1^*)(\theta \mid \phi) = \mathfrak{F}(\mathbf{K}_1)(\theta \mid \phi)$$

provided that $\mathfrak{F}(\mathbf{K}_1 \cup \mathbf{K}_1^*)(\phi) \neq 0$ and $\mathfrak{F}(\mathbf{K}_1)(\phi) \neq 0$. $\qquad \square$

Let us refer to the set of all events which logically imply the event ϕ as *the world of* ϕ. Then the Locality Principle may be roughly paraphrased as saying that if $\mathbf{K}_1$ contains only information about the *relative* probabilistic beliefs between events in the world of ϕ, while $\mathbf{K}_1^*$ contains only information about beliefs concerning events in the world of $\neg\phi$, then the values which the inference process $\mathfrak{F}$ calculates for probabilities of events conditioned on ϕ should be unaffected by the information in $\mathbf{K}_1^*$, except in the trivial case when belief in ϕ is forced to take the value 0. Put rather more more succinctly: beliefs about the world of $\neg\phi$ should not affect beliefs conditioned on ϕ. Note that we cannot expect to satisfy a strengthened version of this principle which would have belief in the events in the world of ϕ unaffected by $\mathbf{K}_1^*$ since the constraints in $\mathbf{K}_1^*$ may well affect belief in ϕ itself. In essence the Locality Principle asserts that *ceteris paribus* rationally derived *relative probabilities between events inside a "world"* are unaffected by information about what happens strictly outside that world.

As an additional justification for the above principle we may also note the following:

Theorem 2.2. *The inferences processes* **ME**, **CM**$^\infty$, **MD** (*minimum distance*), *together with all* **Renyi** *inference processes* (*see e.g.* [8]), *all satisfy the Locality Principle.*

The Locality Principle is in essence a generalisation of the Relativisation Principle of Paris [12] and the Homogeneity Axiom of Hawes [8], and the above theorem is very similar to results proved previously, especially to results in [8]. It follows from Theorem 2.2 that if we reject the Locality Principle, then we are in effect forced to reject not just **ME**, but also all the currently most favoured inference processes.

An interesting aspect of the Locality Principle is that the justification given above appears no less cogent when we attempt to generalise it to a collective situation. If we accept the arguments in favour of the Locality Principle in the case of a single individual then it is hard to see why we should reject analogous arguments in the case of a social belief function which is derived by considering the beliefs of m individuals each of whom has constraint sets of the type considered above. Accordingly we may formulate more generally.

The Locality Principle (for a social inference process)

For any $m \geq 1$ let **M** be a college of m individuals $\mathbf{A}_1 \ldots \mathbf{A}_m$. If for each $i = 1 \ldots m$ $\mathbf{K}_i$ is a nice set of constraints conditioned on ϕ, and

$\mathbf{K}_i^*$ is a nice set of constraints about $\neg\phi$, then for every event θ

$$\mathfrak{F}(\mathbf{K}_1 \cup \mathbf{K}_1^*, \ldots, \mathbf{K}_m \cup \mathbf{K}_m^*)(\theta \mid \phi) = \mathfrak{F}(\mathbf{K}_1, \ldots, \mathbf{K}_m)(\theta \mid \phi)$$

provided that $\mathfrak{F}(\mathbf{K}_1 \cup \mathbf{K}_1^*, \ldots, \mathbf{K}_m \cup \mathbf{K}_m^*)(\phi) \neq 0$
and $\mathfrak{F}(\mathbf{K}_1, \ldots \mathbf{K}_m)(\phi) \neq 0$.   □

At this point we make a simple observation. In the special case when for each i the constraint sets $\mathbf{K}_i \cup \mathbf{K}_i^*$ are such as to completely determine $\mathbf{A}_i$'s belief function, so that the task of $\mathfrak{F}$ reduces to that of a pooling operator, it is easy to construct an example to show that if the Locality Principle is to be satisfied then that pooling operator cannot be **LinOp**. On the other hand the pooling operator **LogOp** is perfectly consistent with the Locality Principle, as we shall see in the final section of this paper. Related facts concerning **LinOp** and **LogOp** have been widely noted in the literature on pooling operators; what we are noting that is new here is the compelling nature of certain arguments in favour of the Locality Principle in the far broader context of a social inference process. Nonetheless, as remarked above, the Locality Principle is violated by the widely used pooling operator **LinOp**, a fact which appears to us to cast serious doubt on the intrinsic plausibility of **LinOp** as a pooling operator.

Our final axiom relates to a hypothetical situation where several exact copies of a college are amalgamated into a single college.

A *clone* of a member $\mathbf{A}_i$ of $\mathbf{M}$ is a member $\mathbf{A}_{i'}$ whose set of belief constraints on his belief function is identical to that of $\mathbf{A}_i$: *i.e.* $\mathbf{K}_i = \mathbf{K}_{i'}$. Suppose now that each member $\mathbf{A}_i$ of $\mathbf{M}$ is replaced by k clones of $\mathbf{A}_i$, so that we obtain a new college $\mathbf{M}^*$ with km members. $\mathbf{M}^*$ may equally be regarded as k copies of $\mathbf{M}$ amalgamated into a single college; so since the social belief function associated with each of these copies of $\mathbf{M}$ would be the same, we may argue that surely the result of amalgamating the copies into a single college $\mathbf{M}^*$ should again yield the same social belief function. This argument generates the following:

The Proportionality Principle

For any integer $k > 1$

$$\mathfrak{F}(\mathbf{K}_1 \ldots \mathbf{K}_1, \mathbf{K}_2 \ldots \mathbf{K}_2, \ldots, \mathbf{K}_m \ldots \mathbf{K}_m) = \mathfrak{F}(\mathbf{K}_1, \mathbf{K}_2, \ldots, \mathbf{K}_m)$$

where in the expression on the left there are exactly k copies of each $\mathbf{K}_i$.   □

The proportionality principle looks rather innocent. Nevertheless as we shall see at the end of the next section a slight generalisation of the

same idea formulated as a limiting version has some surprising consequences.

3   The Social Entropy Process (SEP)

In this section we introduce a social inference process, **SEP**, which satisfies all the principles introduced in the previous section, and which extends both the inference process **ME** and the pooling operator **LogOp**.

In order to avoid problems with our definition of **SEP** however, we
are forced to add a slight further restriction to the set of m constraint sets
$\mathbf{K}_1 \ldots \mathbf{K}_m$ which respectively represent the beliefs sets of the individuals
$\mathbf{A}_1 \ldots \mathbf{A}_m$. We assume in this section that the constraints are such that
there exists at least one atom α_{j_0} such that no constraint set $\mathbf{K}_i$ forces
α_{j_0} to take belief 0. In the special case when each $\mathbf{K}_i$ specifies a unique
probability distribution, the condition corresponds to that necessary to
ensure that **LogOp** is well-defined.

In order to motivate the definition of **SEP** heuristically, let us imagine
that the college $\mathbf{M}$ decide to appoint an independent chairman $\mathbf{A}_0$, whom
we may suppose to be a mathematically trained philosopher, and whose
only task is to aggregate the beliefs of $\mathbf{A}_1 \ldots \mathbf{A}_m$ into a social belief
function v according to strictly rational criteria, but ignoring any personal
beliefs which $\mathbf{A}_0$ himself may hold. He must then convince the members
of $\mathbf{M}$ that his method is optimal.

$\mathbf{A}_0$ decides that he will choose a social belief function $v = <v_1 \ldots v_J>$
in such a manner as to minimise the average informational distance be-
tween $< v_1 \ldots v_J >$ and the m belief functions $w^{(i)} = < w_1^{(i)} \ldots w_J^{(i)} >$
of the members of $\mathbf{M}$, where the $w^{(i)}$ are each simultaneously chosen in
such a manner as to minimise this quantity subject to the relevant sets of
belief constraints $\mathbf{K}_i$ of each of the members of the college.

Using the standard cross-entropy measure of informational distance
this idea amounts to minimising the function

$$\frac{1}{m} \sum_{i=1}^{m} \sum_{j=1}^{J} v_i \log \frac{v_i}{w_j^{(i)}}$$

subject to all the constraints. In the above the usual convention is ob-
served that $v_i \log \frac{v_i}{w_j^{(i)}}$ takes the value 0 if $v_i = 0$ and the value $+\infty$ if
$w_j^{(i)} = 0$ and $v_i \neq 0$.

A little algebraic manipulation establishes that minimising the above
expression subject to all the constraints is equivalent to *first* choosing the

$w^{(i)}$ subject to the $\mathbf{K}_i$ so as to maximise the function

$$\sum_{j=1}^{J}(\prod_{i=1}^{m} w_j^{(i)})^{\frac{1}{m}}$$

and then, if this maximum value attained is say M, setting

$$v_j = \frac{1}{M}(\prod_{i=1}^{m} w_j^{(i)})^{\frac{1}{m}}$$

for each $j = 1 \ldots J$. Notice that the function being maximised above is just a sum of geometric means. Since this function is bounded and continuous and the space over which it is being maximised is by assumption closed, a maximum value M is certainly attained. Moreover it is easy to see that

Lemma 3.1. *Given $\mathbf{K}_1 \ldots \mathbf{K}_m$ and M defined as above then $0 < M \leq 1$.*

Furthermore the value $M = 1$ occurs if and only if for every $j = 1 \ldots J$ and for all $i, i' \in \{1 \ldots m\}$ $w_j^{(i)} = w_j^{(i')}$. Hence given $\mathbf{K}_1 \ldots \mathbf{K}_m$ the following are equivalent:

1. *$M = 1$*
2. *Every $w^{(1)} \ldots w^{(m)}$ which generates the value M satisfies $w^{(1)} = \ldots = w^{(m)} = v$.*
3. *The constraints $\mathbf{K}_1 \ldots \mathbf{K}_m$ are jointly consistent: i.e there exists some belief function which satisfies all of them.*

Now it is obvious from the above that chairman $\mathbf{A}_0$'s proposed method of choosing v will not in general result in a uniquely defined social belief function. Indeed if $\bigcap_{i=1}^{m} \mathbf{V}_{\mathbf{K}_i} \neq \emptyset$ then *any* point v in this intersection, if adopted as the belief function of each member, will generate the maximum possible value for M of 1 and so will be a possible candidate for a social belief function v. Moreover even if $\bigcap_{i=1}^{m} \mathbf{V}_{\mathbf{K}_i} = \emptyset$ the process above may not result in a unique choice of either the $w^{(i)}$ or of v.

Chairman $\mathbf{A}_0$ now reasons as follows: if the result of the above operation of minimising the average cross-entropy does not result in a unique solution for v, then the best rational resource which he has left is to choose that v which has maximum entropy from the set of possible v previously obtained. Chairman $\mathbf{A}_0$ reasons that by adopting this procedure he is treating the set of v defined by minimising the average cross-entropy of college members *as if* it were the set of belief functions defined by his own beliefs, and then choosing a belief function from that set by applying the **ME** inference process.

However in order to show that this procedure is well-defined chairman $\mathbf{A}_0$ needs to prove a number of lemmas.

Definition 3.2. For constraint sets $\mathbf{K}_1 \ldots \mathbf{K}_m$ we define

$$M_{\mathbf{K}_1 \ldots \mathbf{K}_m} = \text{Max} \left\{ \sum_{j=1}^{J} \left(\prod_{i=1}^{m} w_j^{(i)} \right)^{\frac{1}{m}} \mid w^{(i)} \in \mathbf{V}_{\mathbf{K}_i} \text{ for all } i = 1 \ldots m \right\}$$

and

$$\Gamma(\mathbf{K}_1 \ldots \mathbf{K}_m)$$

$$= \left\{ < w^{(1)} \ldots w^{(m)} > \in \bigotimes_{i=1}^{m} \mathbf{V}_{\mathbf{K}_i} \mid \sum_{j=1}^{J} \left(\prod_{i=1}^{m} w_j^{(i)} \right)^{\frac{1}{m}} = M_{\mathbf{K}_1 \ldots \mathbf{K}_m} \right\}.$$

$\square$

By the earlier discussion, each point $< w^{(1)} \ldots w^{(m)} >$ in $\Gamma(\mathbf{K}_1 \ldots \mathbf{K}_m)$ gives rise to a uniquely determined corresponding social belief function v whose j'th coordinate is given by

$$v_j = \frac{1}{M_{\mathbf{K}_1 \ldots \mathbf{K}_m}} \left(\prod_{i=1}^{m} w_j^{(i)} \right)^{\frac{1}{m}}.$$

We will refer to the v thus obtained from $< w^{(1)} \ldots w^{(m)} >$ as

$$\mathbf{LogOp}(w^{(1)} \ldots w^{(m)})$$

and we let

$$\Delta(\mathbf{K}_1 \ldots \mathbf{K}_m)$$
$$= \{\mathbf{LogOp}(w^{(1)} \ldots w^{(m)}) \mid < w^{(1)} \ldots w^{(m)} > \in \Gamma(\mathbf{K}_1 \ldots \mathbf{K}_m)\}$$

$\Delta(\mathbf{K}_1 \ldots \mathbf{K}_m)$ is thus the candidate set of possible social belief functions from which Chairman $\mathbf{A}_0$ wishes to make his final choice by selecting the point in this set which has maximum entropy.

The following structure theorem for $\Gamma(\mathbf{K}_1 \ldots \mathbf{K}_m)$, which depends strongly on the concavity properties the geometric mean function and of sums of such functions, guarantees that Chairman $\mathbf{A}_0$'s plan is realisable.

Theorem 3.3. *For fixed constraint sets* $\mathbf{K}_1 \ldots \mathbf{K}_m$

(i) *For any two points* $< w^{(1)} \ldots w^{(m)} >$ *and* $< \bar{w}^{(1)} \ldots \bar{w}^{(m)} >$ *in* $\Gamma(\mathbf{K}_1 \ldots \mathbf{K}_m)$ *there exists real numbers* $\mu_1 \ldots \mu_J \in \mathbb{R}$ *such that*

$$\bar{w}_j^{(i)} = w_j^{(i)}(1 + \mu_j)$$

for all $i = 1 \ldots m$ *and* $j = 1 \ldots J$.

(ii) $\Gamma(\mathbf{K}_1 \ldots \mathbf{K}_m)$ *is a compact non-empty convex set.*
(iii) $\Delta(\mathbf{K}_1 \ldots \mathbf{K}_m)$ *is a compact non-empty convex set.*
(iv) *The map*

$$\mathbf{LogOp}: \ \Gamma(\mathbf{K}_1 \ldots \mathbf{K}_m) \ \rightarrow \ \Delta(\mathbf{K}_1 \ldots \mathbf{K}_m)$$

is a continuous bijection.

$\square$

Now since $\Delta(\mathbf{K}_1 \ldots \mathbf{K}_m)$ is a closed convex set by 3.3(iii) and since the entropy function

$$-\sum_{j=1}^{J} v_j \log(v_j)$$

is strictly concave over this set, the set contains a unique point v at which the entropy function achieves its maximum value. It follows at once that the following formal definition of **SEP** defines, for every $\mathbf{K}_1 \ldots \mathbf{K}_m$ satisfying the conditions of this section, a unique social belief function.

Definition 3.4. The Social Entropy Process, **SEP**, is the social inference process defined by

$$\mathbf{SEP}(\mathbf{K}_1 \ldots \mathbf{K}_m) \ = \ \mathbf{ME}(\Delta(\mathbf{K}_1 \ldots \mathbf{K}_m)).$$

Theorem 3.5. **SEP** *satisfies the seven principles of the previous section: Equivalence, Anonymity, Atomic Renaming, Consistency, Collegiality, Locality, and Proportionality.*

It is worth remarking that Theorem 3.3(i) provides a simple sufficient condition for $\Delta(\mathbf{K}_1 \ldots \mathbf{K}_m)$ to be a singleton and thus for the application of **ME** in the definition of **SEP** to be redundant:

Theorem 3.6. *If $\mathbf{K}_1 \ldots \mathbf{K}_m$ are such that for each $j = 1 \ldots J$ except possibly at most one there exists some i with $1 \leq i \leq m$ such that $\mathbf{K}_i$ forces $w_j^{(i)}$ to take a unique value, then $\Delta(\mathbf{K}_1 \ldots \mathbf{K}_m)$ is a singleton. In particular this occurs if for some i $\mathbf{V}_{\mathbf{K}_i}$ is a singleton.*

An interesting characteristic of **SEP** is that the **ME** second stage of the defining process, which is included in order to force the choice of a social belief function to be unique in cases when this would not otherwise hold, can actually be eliminated by insisting that the social inference process satisfies a variant of the axiom of proportionality. Such an argument counters a possible objection that the invocation of maximum entropy at the second stage of the definition is somewhat artificial. To be precise it is possible to substitute the following procedure to define **SEP**.

We define a member i of the college to be an *ignorant fanatic* if $\mathbf{V_{K_i}}$ consists of the single point $< \frac{1}{J}, \frac{1}{J} \ldots \frac{1}{J} >$. Now starting with a college $\mathbf{M}$ of m individuals and constraint sets $\mathbf{K}_1 \ldots \mathbf{K}_m$ as before, let us form for any $k \in \mathbb{N}$, a new college $\mathbf{M}_k^*$ of $km + 1$ members, consisting of a single ignorant fanatic together with k copies of $\mathbf{M}$. Now one would hope that for a well-behaved social inference process, in applying the social inference process to $\mathbf{M}_k^*$ the effect of the ignorant fanatic would become negligible as $k \to \infty$, in which case by proportionality in the limit we should get the same answer as we would get by applying the social inference process to $\mathbf{M}$.

Pleasingly this is exactly what happens for **SEP**. In fact, if we accept the above principle it turns out that we need never invoke the **ME** stage in the definition of **SEP** at all, since by 3.6 the first stage of the definition of **SEP** already guarantees the uniqueness of the social belief function for each $\mathbf{M}_k^*$, owing to the presence of the ignorant fanatic, while the limit of these social belief functions turns out to be $\mathbf{SEP(K}_1 \ldots \mathbf{K}_m)$. We restate this as the following theorem:

Theorem 3.7. *With the notation as above the set Δ_k of solutions for the social belief functions for $\mathbf{M}_k^*$ corresponding to the first stage of applying* **SEP** *consists, for each k, of a set containing a single probability distribution, say $v^{[k]}$, and furthermore*

$$\lim_{k \to \infty} v^{[k]} = \mathbf{SEP(K}_1 \ldots \mathbf{K}_m).$$

Since the left hand side of the above identity does not involve maximum entropy in its definition, we can argue that this shows that the invocation of maximum entropy in the second stage of the original definition of **SEP** is indeed entirely natural.

A suggestive way of interpreting this result is as follows. In order to calculate the social belief function v for $\mathbf{M}$, chairman $\mathbf{A}_0$ first minimizes the sum of the cross entropies as in the first stage of the calculation of **SEP**. If this results in a unique solution then that is taken as the social belief function. If the result is not unique then $\mathbf{A}_0$ adds his own casting constraint set $\mathbf{K}_0$ as that of an ignorant fanatic[8] and recalculates, while diluting his own effect as much as possible by imagining that there are k clones of each of the other members of the college, and that $k \to \infty$. The resulting inference process is just **SEP**.

---

[8] Of course in this context it may be preferable to replace the designation "ignorant fanatic" by "impartial chair with leadership qualities".

In conclusion I am grateful to Alena Vencovská for her helpful comments on some of the ideas presented here, and I also wish to thank Hykel Hosni and Franco Montagna for their careful editorial suggestions. The sole responsibility for any errors lies however with the author.

References

[1] R. M.COOKE, "Experts in Uncertainty: Opinion and Subjective Probability in Science", Environmental Ethics and Science Policy Series, Oxford University Press, New York, 1991.

[2] S. FRENCH, *Group consensus probability distributions: A critical survey*, In: J. M. Bernardo, M. H. De Groot, D. V. Lindley, and A. F. M. Smith (eds.), "Bayesian Statistics", Elsevier, North Holland, 1985, 183–201.

[3] A. GARG, T. S. JAYRAM, S. VAITHYANATHAN and H. ZHU, *Generalized opinion pooling*, In: "Proceedings of the 8th Intl. Symp. on Artificial Intelligence and Mathematics", 2004.

[4] C. GENEST, *A conflict between two axioms for combining subjective distributions*, J. Royal Statistical Society **46** (1984), 403–405.

[5] C. GENEST and C. G. WAGNER, *Further evidence against independence preservation in expert judgement synthesis*, Aequationes Mathematicae **32** (1987), 74–86.

[6] C. GENEST and J. V. ZIDEK, *Combining probability distributions: A critique and an annotated bibliography*, Statistical Science **1** (1986), 114–135.

[7] C. GENEST, K. J. MCCONWAY and M. J. SCHERVISH, *Characterization of externally Bayesian pooling operators*, Ann. Statist. **14** (1986), 487–501.

[8] P. HAWES, "An Investigation of Properties of Some Inference Processes", PhD Thesis, Manchester University, MIMS eprints, 2007, available from http://eprints.ma.man.ac.uk/1304/

[9] W. B. LEVY and H. DELIC, *Maximum entropy aggregation of individual opinions*, IEEE Trans. Systems, Man, Cybernetics **24** (1994), 606–613.

[10] J. MYUNG, S. RAMAMOORTI, A. D. BAILEY,JR., *Maximum entropy aggregation of expert predictions*, Management Science, **42** (1996), 1420–1436.

[11] J. B. PARIS and A. VENCOVSKÁ, *A note on the inevitability of maximum entropy*, International Journal of Approximate Reasoning, **4** (1990), 183–224.

[12] J. B. PARIS, "The Uncertain Reasoner's Companion - A Mathematical Perspective", Cambridge University Press, Cambridge, UK, 1994.

[13] J. B. PARIS and A. VENCOVSKÁ, *In defence of the Maximum Entropy Inference Process*, International Journal of Approximate Reasoning **17** (1997), 77–103.

[14] J. B. PARIS, *Common sense and maximum entropy*, Synthese, **117** (1999), 75–93.

[15] D. M. PENNOCK and M. P. WELLMAN, *Graphical models for groups: belief aggregation and risk sharing*, Decision Analysis **2** (2005), 148–164.

[16] J. E. SHORE and R. W. JOHNSON, *Axiomatic derivation of the principle of maximum entropy and the principle of minimum cross-entropy*, IEEE Transactions on Information Theory, IT-26 (1980), 26–37.

[17] D. OSHERSON and M. VARDI, *Aggregating disparate estimates of chance*, Games and Economic Behavior, July 2006, 148–173.

[18] C. WAGNER, *Aggregating subjective probabilities: Some limitative theorems*, Notre Dame J. Formal Logic **25** (1984), 233–240.

[19] T. S. WALLSTEN, D. V. BUDESCU, I. EREV and A. DIEDERICH, *Evaluating and combining subjective probability estimates*, Journal of Behavioral Decision Making **10** (1997).

Conditional probability in the light of qualitative belief change

David Makinson

Introduction

In this paper we explore ways in which purely qualitative belief change in the AGM tradition can throw light on options in the treatment of conditional probability. First, by helping see why we sometimes need to go beyond the ratio rule defining conditional from one-place probability. Second, by clarifying criteria for choosing between various non-equivalent accounts of the two-place functions. Third, by suggesting novel forms of conditional probability, notably screened and hyper-revisionary. Finally, we show how qualitative uncertain inference suggests another, very broad, class of 'proto-probability' functions.

To keep the main text reader-friendly, most of the verifications and historical remarks are placed in an extended Appendix, whose sections run parallel to the main text.

1 Why go beyond the ratio rule?

Kolmogorov's axioms for one-place probability functions are simple, natural and easy to work with, and the ratio definition of conditional probability is convenient to use (see Appendix). They have become standard. So why go beyond them?

The reasons advanced in the literature are of two main kinds: a metaphysical complaint and a pragmatic appeal for greater expressiveness. We outline them in this section, suggesting that while the metaphysical grounds do not stand up to scrutiny, there is a real need for greater expressive capacity. In the following section, we show how a comparison with the situation in qualitative belief revision makes the need all the more evident.

1.1 Metaphysical *vs* pragmatic considerations

It is commonly felt (see Appendix) that all probability is at bottom conditional anyway, and we should bring this out from the very beginning by integrating it into our formal treatment. From a subjective perspective: a probability judgement is always made given a whole lot of background information, and so is in some sense conditional on that information. From a frequency perspective: probability is some sort of limiting frequency of a type of item in a set, and if we enlarge or diminish the set, the frequency will in general change.

However, the argument has its limitations. From the subjective standpoint, it is not at all clear whether all kinds of background information can, or should, be treated as events of the probability calculus. From both perspectives, the argument may involve an infinite regress. This is most easily seen in the field-of-sets mode. Suppose we do take probability as a two-place function $p : F^2 \to [0, 1]$ where F is a field of subsets of a set S. This still depends on the choice of S. Making it into a three-place function $p : F^3 \to [0, 1]$ will not help, as that still depends on S, taking us one step further in an infinite regress. The only way to eliminate all such dependence is take the domain to be the *universal* class. But practising probabilists never do this, and if done, it might as well be done from the beginning with one-place functions.

Historically, the argument is reminiscent of an early way of looking at classical first-order logic, according to which universal quantifications $\forall x \phi(x)$ are at bottom always conditionals, since their range depends on the choice of domain of discourse. On this view, the dependency should be made explicit from the outset by always quantifying over the entire universe, writing the restricted generalizations as $\forall x [Dx \to \phi(x)]$ where D is the intended domain. Such a view had some currency for a while despite the difficulties of talking about a universal set (so the universe was thought of as a class rather than a set). But we have become accustomed to working with the simpler mode of representing universal quantification without running into difficulty, and the philosophical worries have simply withered away.

The historical precedent carries a methodological lesson. Even if all quantification or probability can be said to be in some sense conditional, this does not imply that the conditionality should always be brought into the formalism of the theory itself. It may sometimes be better treated as part of the business of applying the theory to a specific problem.

Thus, it would seem that the metaphysical reasons for always taking conditional probability as primitive are less than compelling. Nevertheless, a pragmatic consideration remains. When conditional probability is defined by the ratio rule, it has limited expressive capacity. We would sometimes like to allow propositions that have been accorded zero probability to serve as conditions for the probability of other propositions. This is impossible when $p(x|a)$ is put as $p(a \wedge x)/p(a)$, for it is undefined when $p(a) = 0$.

The most famous example of this expressive gap is due to Borel. Suppose a point is selected at random from the surface of the earth. What is the probability that it lies in the Western hemisphere, given that it lies on the equator? The condition of lying on the equator has probability 0 under the random selection, but we would be inclined to regard the question as meaningful and even as having $1/2$ for its answer. Other examples are given by *e.g.* [40].

This complaint is more modest than the metaphysical one, pointing to a gap rather than alleging a defect. But it is much more productive. There is no escaping its basic point: it would sometimes be helpful to have a more general conception that covers what we will be calling *the critical zone* – the case where the condition a is consistent but of zero probability – and we should try to formulate it.

There are, of course, quite trivial ways of making the ratio definition cover the critical zone. One, due to Carnap, is to declare that the zone is empty: whenever $p(x) = 0$ then x is inconsistent. This is known as the regularity condition. It has the immediate effect that the *ratio definition* of $p(x|a)$ as $p(a \wedge x)/p(a)$ covers all instances of the right argument a except when a is inconsistent. For inconsistent a, one can then either leave $p(x|a)$ undefined, or take it to have value 1 for all values of the left argument x.

However, as remarked *e.g.* by [38], this is more like a way of avoiding than solving the problem. It abolishes by fiat the distinction between logical impossibility and total improbability. Moreover, as noted by [16] (page 229), it leads to an internal inelegance. Let p be a proper one-place function satisfying Carnap's regularity condition, and consider the two-place function $p(\cdot|\cdot)$ determined by the ratio definition. Now take a contingent proposition a with $1 \neq p(a) \neq 0$, and form the left projection $p_a(\cdot)$ alias $p(\cdot|a)$ of the two-place function. By the definition of left projections (see Appendix) we have $p_a(x) = p(x|a)$ so substituting $\neg a$ for x, we have $p_a(\neg a) = p(\neg a|a) = p(\neg a \wedge a)/p(a) = 0/p(a) = 0$ since $p(a) > 0$. Thus $p_a(\neg a) = 0$ even though $\neg a$ is consistent, violating the regularity condition as applied to p_a. In other words, even when p satisfies the regularity condition, the left projection of its conditionalization under the ratio definition will not do so – which is a discord, to say the least.

Another trivial way of covering the critical zone is to put $p(x|a) = 1$ for every value of x when $p(a) = 0$. This might be called the *ratio/unit definition* of conditional probability. But while this renders the function always-defined, and is very convenient in many contexts, it does not fill the expressive gap satisfactorily. For under the rule, when a is in the critical zone the left projection $p_a(\cdot)$ from right value a of the two-place function is the constant function with value 1, *i.e.* the improper one-place probability function (see Appendix). Although this cannot be described as wrong, it is not very helpful. When *a* is consistent but $p(a) = 0$, hopefully we should be able to conditionalize to something more informative than the unit function; the two-place function should in some sense be essentially conditional.

In mathematical practice one can sometimes 'work around' the problem. The idea is that when *a* is in the critical zone, we should take $p(x|a)$ to be the limit of the values of $p(x|a')$ for a suitable infinite sequence of non-critical approximations a' to a. But this procedure is possible only for suitable domains (notably, fields based directly or indirectly on the real numbers) satisfying appropriate conditions. So while it serves well for some examples (such as the hemisphere/equator one mentioned earlier), it is not a general solution.

1.2   Some notational niceties

In the following sections, we will compare various options for axiomatizing conditional probability in the light of qualitative belief revision. When doing so, we follow certain notational conventions for clarity. In particular, we distinguish $p(x|a)$ from $p(x, a)$, writing:

- $p(x|a)$ with a bar when it is understood as a two-place operation defined from a one-place one by the ratio rule, *i.e.* by putting $p(x|a) = p(a \wedge x)/p(a)$ when $p(a) > 0$, possibly with the extension that puts $p(x|a) = 1$ when $p(a) = 0$ (in which case we call it the ratio/unit rule).
- $p(x, a)$ with a comma when taking p as an undefined (arbitrary or primitive) two-place operation defined over all or part of L^2.

Care will always be taken to specify the arity (number of places) of a function under consideration, either by mentioning it explicitly, or by using place-markers as in $p(\cdot)$, $p(\cdot|\cdot)$, $p(\cdot, \cdot)$.

Throughout, Cn is the operation of classical consequence; we also write $\approx$ for the relation of classical equivalence.

2   Exploring the critical zone

In this section we weigh the significance of the critical zone. We begin by showing how an analogous zone already arises on the qualitative level for AGM belief change, and how this helps bring out the conceptual options underlying different systems for two-place conditional probability. We then review those systems, presenting them in a modular way that makes manifest the rationales for apparently technical choices.

2.1   A leaf from the AGM book

It is instructive to compare the situation for probability change with that for qualitative belief change in the AGM tradition initiated in [1].

There, expansion is one thing, revision another. Let K be any belief set, *i.e.* a set of propositions closed under the operation Cn of classical consequence, *i.e.* $K = Cn(K)$. The *expansion* of K by a is defined simply by putting $K + a = Cn(K \cup \{a\})$. However *revision* is defined by putting $K * a = Cn((K - \neg a) \cup \{a\})$, where $-$ is a suitable contraction operation forming from K a subset that is consistent with a (when a is itself consistent) and satisfying certain regularity conditions.

We thus have two different kinds of change side by side. Again, they differ in the *critical zone* which, in this qualitative context, is the case where we modify the belief set K by a proposition a that is *itself consistent but inconsistent with K*. In this critical zone, expansion creates blow-out to the set of all propositions of the language, while revision forces contraction of the belief set. Outside the critical zone, the two operations coincide. This basic difference should not be obscured by talk of expansion being a special case of revision, which is a sloppy way of saying that the values of the two operations are the same outside the critical zone. Neither operation is a special case of the other.

This basic conceptual difference reflects itself in the different formal properties of expansion and revision. There are principles that hold for expansion but not for revision, and conversely. In particular:

- Expansion never diminishes the initial belief set, *i.e.* $K \subseteq K + a$. This is sometimes known as the principle of *belief preservation*. In contrast, revision will eliminate material from the belief set whenever the input a is in the critical zone.
- When a is inconsistent with K, expansion gives us blow-out: both $a, \neg a \in K + a = Cn(K \cup a) = L$ (the whole language). In contrast for revision, even when a is inconsistent with K, so long as a is itself consistent, so is $K * a$. This property of revision is known as the principle of *(input) consistency preservation*.

The pattern is replicated in the probabilistic context. There too we are looking at two different kinds of operation coinciding outside but differing inside the critical zone which in this context, we recall, is the case where a is consistent but $p(a) = 0$. One is *expansionary*, the other is *revisionary*.

- The expansionary operation is given by the ratio/unit definition. It satisfies a probabilistic analogue of qualitative belief preservation: $p(x|a) = 1$ whenever $p(x) = p(x|\top) = 1$. Expressed with left projections, $p_a(x) = 1$ whenever $p_\top(x) = 1$. In other words, conditionalizing never reduces the corresponding belief set: writing $B(p)$ for $\{x : p(x) = 1\}$ we always have $B(p) \subseteq B(p_a) = \{x : p_a(x) = 1\} = \{x : p(x|a) = 1\}$ (see the Appendix for detailed verification). No juice is lost. In contrast, a revisionary operation would allow for diminution of the associated belief set.

- More specifically, when $p(a) = 0$, the expansionary operation blowsout to the unit function (irrespective of a's own consistency): in that case $p_a(x) = p(x|a) = 1$ for all x, so that $B(p_a) = L$ (see Appendix or a full verification). In contrast, a revisionary conditional probability function would never give us the unit function when the condition a is itself consistent.

These two kinds of conditionalization should not be thought of as competing for the position of 'the right one'. Like expansion and revision in the qualitative context, they can work side by side, as different kinds of conditionalization. But how can the revisionary conception best be expressed?

There are two main approaches to the problem. One is to define a family of revision operations that take one-place probability functions to others. That is the path taken by Grdenfors in a pioneering paper of 1986 (integrated into his book of 1988). The other approach is to define a family of two-place probability functions. That is the path taken in varying manners by [20, 32–35] and others in their wake.

Although different in appearance, the two approaches are intimately related, as hinted by [11] and observed explicitly by [25]. Here, we consider only the approach using two-place probability functions. Our central question is the following: of the differing axiom systems for two-place probability, is there one that is preferable to the others, and on what grounds?

2.2   Bird's-eye view of available systems

The usual presentation of axiom systems for two-place probability functions can be quite confusing. The systems are not always formulated in an intuitively evident manner. They can also be difficult to compare due to differing choices of right domain – sometimes the whole of L, sometimes the consistent propositions in L, sometimes an arbitrary subset of L lying between $\{x : p(x, \top) > 0\}$ and L itself. To facilitate comparison and focus on essentials, we formulate all systems as functions defined on *the whole* of L^2. We also present the systems in a *modular way*, that is, with a common basis and differing in what is added to it.

The leading idea is to exploit Rényi's insight that for 'most' values of the right argument of the two-place function, the left projections should be proper one-place Kolmogorov functions, adding that in the remaining cases it should be the unit function. We obtain modularity by making a different specification of what counts as 'most' for each system.

We begin with the basic *van Fraassen system*, which was formulated (in the field of sets mode) by [40] and [41]. Its axioms concern *all* propositions, rather than special subsets of them. They are the axioms of *right extensionality*, *left projection*, and *product* respectively, for two-place functions $p : L^2 \to [0, 1]$.

(vF1) $p(x, a) = p(x, a')$ whenever $a \approx a'$
(vF2) p_a is a one-place Kolmogorov probability function with $p_a(a) = 1$
(vF3) $p(x \wedge y, a) = p(x, a) \cdot p(y, a \wedge x)$ for all formulae a, x, y

In (vF1), recall that we are using $\approx$ for logical equivalence. Note that (vF2), as formulated here, says that p_a is a one-place Kolmogorov function, but it *does not say whether it is proper or improper* (the unit function). Indeed, the axioms are consistent with p_a being the unit function for every $a \in L$.

Despite their modesty, the van Fraassen axioms have surprisingly many useful consequences. The following were already noticed between [2,3,40,41]. For the convenience of the reader, we recall in the Appendix the brief verifications.

- Left extensionality: $p(x, a) = p(x', a)$ whenever $x \approx x'$.
- When $y \in Cn(x)$ then $p(x, a) \le p(y, a)$.
- When $p(\cdot)$ is defined as $p(\cdot, \top)$, then we have the ratio rule (though not its unit extension to the critical zone, *i.e.* the ratio/unit rule).
- When a is a contradiction, then p_a is the unit function.
- The set Δ of all $a \in L$ such that p_a is the unit function is an ideal. That is, it is closed downwards (whenever $a \in Cn(b)$ and $a \in \Delta$ then $b \in \Delta$) and also closed under disjunction (whenever $a, b \in \Delta$ then $a \vee b \in \Delta$).
- p_a is the unit function iff $p(a, b) = 0$ for all b such that p_b is a proper Kolmogorov function.

In [40, 41] van Fraassen called the $a \in L$ such that p_a is a proper Kolmogorov function *normal*, and the remaining $a \in L$ *abnormal* – of course, modulo the function $p(\cdot, \cdot)$. In that terminology, the set of all abnormal formulae form a non-empty ideal containing the contradictions, and a formula a is abnormal iff $p(a, b) = 0$ for all normal b. Apart from that, they do not tell us anything about which formulae are normal, which abnormal.

Popper's system goes some way to filling the gap. It may be obtained by adding a single axiom, stating that p_a is normal whenever $p(a, \top) > 0$.

(Positive): when $p(a, T) > 0$ then p_a is a proper Kolmogorov function.

More economically but less transparently, Popper's system may also be obtained by instead adding the statement that $p(x, b) \neq 1$ for some x, b (see Appendix), *i.e.* that $p_b(\cdot)$ is not the unit function for some formula b, *i.e.* that $p(\cdot, \cdot)$ is not the unit two-place function.

This still leaves unspecified the status of p_a when a is in the critical zone, *i.e.* consistent but with $p(a, \top) = 0$. The other systems fill this gap in three different ways. *Carnap's system* does so trivially, by declaring that the zone is empty:

(Carnap) When a is consistent then $p(a, \top) > 0$.

This is equivalent to what we would get by staying with one-place functions as primitive, using the familiar ratio/unit definition to generate two-place functions, but declaring that only contradictions can get the value 0.

The *Unit system* fills the gap almost as trivially, by adding instead an axiom saying that any left projection from a point in the critical zone has constant value 1:

(Unit) When a is consistent but $p(a, \top) = 0$, then p_a is the unit function.

This is equivalent to what we would get by keeping one-place functions as primitive and using the ratio/unit definition to generate two-place ones, without requiring that only contradictions can get the value 0.

Hosiasson-Lindenbaum's system (briefly HL) regulates the critical zone by treating its elements just like consistent propositions outside the zone. It adds to the Popper axioms:

(HL) When a is consistent but $p(a, \top) = 0$, then p_a is a proper Kolmogorov probability function.

Thus, in terms of the leading idea mentioned above, 'most propositions' means, respectively:

- The van Fraassen system: an unspecified subset (possibly empty) of the consistent propositions,

- The Popper system: all propositions that are above the critical zone, plus those in an unspecified subset (possibly empty) of it,
- The Unit system: all propositions above the critical zone but no others,
- The Hosiasson-Lindenbaum system: all propositions above or in the critical zone,
- Carnap's system: we can say any of the last three, since the critical zone is declared empty.

It is easy to check that these axiom systems are equivalent to their usual presentations (see Appendix), giving us the sets **Carnap, Unit, HL, Popper, van Fraassen** of functions. The modular arrangement makes it clear at a glance, from their very formulation, what the relations between the systems are. Specifically, we have **Carnap** $=$ **Unit** $\cap$ **HL** $\subset$ **Unit, HL** $\subset$ **Unit**$\cup$**HL** $\subset$ **Popper** $\subset$ **Popper**$\cup\{1(\cdot,\cdot)\} =$ **van Fraassen**, where $1(\cdot,\cdot)$ is the unit two-place function putting $p(x,a)=1$ for all a, x.

The first four relations were established by Leblanc and Roeper ([24, Theorems 4 and 15, Table 5, Figure 15]; also [36, Chapter 3, Section 2]), with however rather laborious verifications from the usual formulations of the systems, and without mentioning the historical role of Hosiasson-Lindenbaum as a key contributor. With the present modular formulation, the inter-relations become trivial, except for the inclusion **van Fraassen** $\subseteq$ **Popper** $\cup \{1(\cdot,\cdot)\}$ and the proper part of the inclusion **Unit** $\cup$ **HL** $\subset$ **Popper**. The former is checked in the Appendix. For the latter, we need a 'mixed' function, failing axioms (Unit) and (HL) but satisfying the Popper axioms. Such a function was already supplied by [24] in the form of a rather enigmatic 64-element table; in the Appendix we provide the same example with an intuitive rule-based formulation. The relations between the classes are pictured in the Hasse diagram of Figure 2.1.

The reader may be surprised that we have not mentioned the axiomatic system of Rényi's [33], also in his later books [34,35]. This is not neglect: his work is indeed capital, providing the leading idea on which most subsequent presentations (including the present one) are based. Rather, Rényi's system takes a form rather different from those above. In effect, he presents a *scheme for a range of axiomatizations*, with the right domain of the function serving as a parameter. For a suitable choice of this parameter (and a little massage) we may obtain the axiomatization of Popper, and likewise of Hosiasson-Lindenbaum. Thus strictly speaking (and taking into account the chronology), Popper's axioms could well be called the Rnyi/Popper postulates. These historical matters are reviewed more fully in the Appendix.

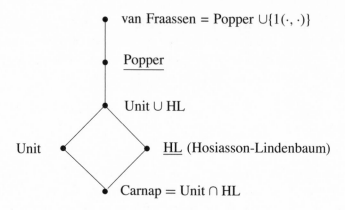

Figure 2.1. Hasse diagram for classes of two-place probability functions.

3 Choosing between systems

Are there any good reasons for preferring one of these systems to another? From our discussion so far, there are two serious contenders
going beyond the ratio/unit account, namely the systems of Hosiasson-
Lindenbaum and of Popper, underlined in the diagram. In this section
we discuss possible criteria for preferring one to the other, coming to the
conclusion that the choice is not a matter of correctness, but of how revisionary we want our revisionary conditional probability to be. We then
give an example of how the difference between the two can sometimes
make a difference to an enterprise using conditional probability.

3.1 Hosiasson-Lindenbaum *vs* Popper

The Hosiasson-Lindenbaum system is not just revisionary it is radically
so, satisfying without reserve the probabilistic counterpart of consistency
preservation. That is, for *every* proposition a, if it is consistent then p_a
is a proper Kolmogorov function. The only values of the right argument
that project to the unit function are the inconsistent ones.

On the other hand Popper's system is 'variably revisionary': it leaves
unspecified the extent to which a function satisfying the axioms is expansionary, and how far it is revisionary. As one extremal case it covers
functions $p(\cdot, \cdot)$ that are purely expansionary, *i.e.* p_a blows out to the
unit function for every a in the critical zone as well as for inconsistent
a. These are the functions satisfying the Unit axiom above. At the other
extreme it covers the Hosiasson-Lindenbaum functions, where p_a never
blows out in the critical zone. In between, it covers many 'mixed' functions, where for certain a, b in the critical zone p_b is the unit function

while p_a is a proper Kolmogorov function. Intuitively, it is in the spirit of [24] who asked rhetorically: "Can't there be *some* statement of L that is 'utterly unbelievable', so unbelievable indeed that should you believe it you'd believe anything, and yet is not truth-functionally false?"

It would be unjustifiably doctrinal to regard one of these policies as right and the other wrong. They are two more options to be added to the traditional one of using the ratio or ratio/unit rule. One option may be appropriate in certain applications, another elsewhere. It is not a matter of choosing once and for all between candidates, but of knowing which candidate to call on for what employment.

Moreover, any Popper function may be transformed into a Hosiasson-Lindenbaum one by suitably expanding the underlying consequence relation. This can be explained formally as follows.

We have already noticed that for any two-place function satisfying the Van Fraassen axioms, and thus *a fortiori* any Popper function $p(\cdot, \cdot)$, the set Δ of all $a \in L$ such that p_a is the unit function is a non-empty ideal. That is, it contains all contradictions, whenever $a \in Cn(b)$ and $a \in \Delta$ then $b \in \Delta$, and whenever $a, b \in \Delta$ then $a \vee b \in \Delta$. Hence the set $\nabla = \{a : \neg a \in \Delta\}$ is a filter, *i.e.* whenever $b \in Cn(a)$ and $a \in \nabla$ then $b \in \nabla$, and whenever $a, b \in \nabla$ then $a \wedge b \in \nabla$). From this in turn it follows that if we define a supraclassical consequence operation Cn' by putting $Cn'(A) = Cn(A \cup \nabla)$ we have: $\bot \in Cn'(a)$ iff $\bot \in Cn(\{a\} \cup \nabla)$ iff $\neg a \in Cn(\nabla) = \nabla$ iff $a \in \Delta$ iff p_a is the unit function. That is, p_a is the unit function iff a is inconsistent modulo Cn'. Moreover, it is easy to check that modulo Cn', the function $p(\cdot, \cdot)$ continues to satisfy all the van Fraassen axioms, given that it satisfied them modulo classical Cn, and so $p(\cdot, \cdot)$ is a Hosiasson-Lindenbaum function modulo Cn'.

In brief: any Popper function (modulo classical Cn) is a Hosiasson-Lindenbaum function modulo a suitably defined supraclassical consequence operation Cn', with the abnormal elements of the critical zone becoming Cn'-inconsistent. Thus the gap between the broader and narrower classes of function is rather less wide than might have been imagined.

Which should we work with? Given that the tighter constraints of HL functions make them easier to handle, it would appear good practice to do so in applications that admit those constraints. Indeed, the following policy suggests itself: (1) When the application does not require any attention to the critical zone, stay with one-place functions as primitive, using the ratio or ratio/unit definition of conditional probability. Otherwise (2) take two-place probability as primitive with the Hosiasson-Lindenbaum axioms if the application admits doing so, otherwise (3) the Popper axioms.

3.2 Does it ever make a difference?

So, does it ever make a substantive difference which kind of essentially conditional probability we use? In some cases it it appears to do so. An example is the theory of 'cores' as set out by [3] using Popper functions, building on ideas of [41] and [2].

Cores were introduced to give a probabilistic account of the intuitive distinction between a broader class of 'plain' beliefs and a narrower one of 'full' beliefs, in such a way that the former, as well as the latter, is closed under classical consequence (and hence under conjunction).

Translating from the field-of-sets mode used by the authors mentioned, a *core* for a Popper function $p : L^2 \to [0, 1]$ is a formula c such that (1) c is normal, that is, the left projection p_c of p from the right value c is a proper Kolmogorov function, and (2) for any consistent formula a logically implying c and any formula b inconsistent with c, $p(b, a \vee b) = 0$.

Plain beliefs modulo p are then identified with those formulae logically implied by at least one core, while *full* beliefs are those implied by every core. The authors show that in the finite case, for any Popper function $p : L^2 \to [0, 1]$ there is a unique strongest core c_0 and a unique weakest one c_1; so that in that case plain beliefs are those formulae logically implied by c_0, while full beliefs are those implied by c_1. Indeed, in the field-of-sets mode we have the same whenever the underlying set is countable and we assume countable additivity.

However, for plain beliefs so defined, there is a difficulty. In the finite case they turn out to be just the formulae x with $p(x, \top) = 1$. In the field-of-sets mode, and assuming countable additivity, this also holds whenever the underlying set is countable. This is given as the 'coincidence lemma' of [2] page 578, and is also an immediate consequence of Lemma 3.1 of [3]. Thus in these contexts, the construal of *plain* belief in terms of cores gives us nothing new, no matter how we choose our Popper function. Nevertheless, as Parikh has urged (personal communication), when we are working in the uncountable case, or in the countable one but without countable additivity, we may not have the same collapse.

It does not seem to have been noticed in the literature that for full beliefs as defined via cores, there is another difficulty. If we work with Hosiasson-Lindenbaum functions rather than the broader class of all Popper functions, it turns out that in every case (finite, countable, uncountable) the full beliefs so construed are just the tautologies – which is hardly what we want. To show this, we need only verify that $\top$ is itself a core. Using the definition above, it suffices to check that $p(\neg\top, \top) = 0$ (which is easy) and that whenever a is consistent while b is inconsistent then $p(b, a \vee b) = 0$. But by the inconsistency of b we have $p(b, a \vee b) = p(b, a)$; and since p is a Hosiasson-Lindenbaum function, its left projection p_a from consistent a is a proper Kolmogorov function, so by the inconsistency of b again, $0 = p_a(b) = p(b, a)$.

Thus the usefulness of cores for defining a formal notion of full belief is not robust between these two axiomatizations of two-place probability. Some might take this as a point against the Hosiasson-Lindenbaum system; the author takes it as one against the edibility of cores.

4   Reverse direction: belief revision in the light of conditional probability

We have been using AGM belief revision to explain why we should take seriously a revisionary reading of two-place probability functions, and to throw light on the options available for them. Insight can also be gained in the other direction. There is a natural map from two-place probability functions satisfying the Hosiasson-Lindenbaum (HL) postulates into (in fact onto) the family of AGM belief revision operations modulo classical consequence. This section is rather more technical than the others; some readers may prefer to skip to the more exciting perspectives of Section 5.

4.1   A map from conditional probability to AGM belief revision

[25] constructed a map from the class of all Gärdenfors probability-revision operations into the class of AGM belief revision operations. The construction below essentially translates it (with some simplifications and an explicit verification of surjectivity) into a map from the class of Hosiasson-Lindenbaum probability functions onto the AGM operations.

We treat AGM belief revision functions as one-place operations $* : L \to 2^L$, with associated current belief sets K. Given any HL function $p : L^2 \to [0, 1]$ as defined in Section 2.2 or equivalently in its Appendix, we construct the associated function $* : L \to 2^L$ and the set $K = B(p)$, as follows.

- The operation $*_p : L \to 2^L$ is defined by putting $*_p(a) = \{x : p(x, a) = 1\}$.
- The set $B(p)$, also called the *top* of p, is defined by putting $B(p) = *_p(\top) = \{x : p(x, \top) = 1\}$.

Then we can show (see Appendix) that for every HL function $p : L^2 \to [0, 1]$:

- $B(p)$ is a consistent belief set.

- The operation $*_p : L \to 2^L$ satisfies the full set of AGM postulates (K∗1) through (K∗8) with respect to $B(p)$.

4.2   Properties of the map: surjective but not injective

The passage from p to $*_p$ is not injective: a counterexample is given in the Appendix. On the other hand, it is surjective for consistent belief sets and under the condition of finiteness (*i.e.* that the propositional language has only finitely many mutually non-equivalent formulae). That is: in such a language, for every consistent belief set K and every revision operation $* : L \to 2^L$ satisfying the AGM postulates with respect to K, there is a HL function $p : L^2 \to [0, 1]$ with $* = *_p$ and $K = B(p)$.

The construction is quite straightforward. Given $*$ and consistent K for such a language, we define $p : L^2 \to [0, 1]$ as follows:

- In the limiting case that a is inconsistent, put $p(x, a) = 1$ for all $x \in L$
- In the principal case that a is consistent, put $p(x, a)$ to be the proportion of $(K * a)$-worlds that are x-worlds.

Here, a *world* is a maximal consistent set of formulae, and an *X-world*, for $X \subseteq L$, is a world Y with $X \subseteq Y$. It is straightforward to verify (see Appendix) that p satisfies the HL postulates, $* = *_p$, and $K = B(p)$ as desired.

Thus, we have a natural surjective (though non-injective) map from the family of all HL conditional probability functions to the family of the AGM revision operations on consistent belief sets. This map helps us see the AGM postulates as reflections of the HL ones. To this extent, the AGM postulates may be said to go back to 1940!

The map and proof may be generalized to cover the Popper functions via the link between the two established in Section 3.1, as follows. In the above construction, we have been taking AGM revision functions as formulated using classical consequence in the background. In fact, as was already made clear in [1], the same theory of belief revision carries through when formulated using arbitrary supraclassical consequence operations satisfying the Tarski closure conditions and disjunction in the premises. So, for a given function $p : L^2 \to [0, 1]$ satisfying the Popper postulates modulo classical Cn, we first see it as a Hosiasson-Lindenbaum function modulo a suitable supraclassical Cn', in the manner described in Section 3.1, checking also that this Cn' satisfies the above conditions. We then construct the same map as above, but with Cn' understood everywhere in place of Cn, and verify in the same manner as before.

5   Alternative forms of Conditionalization and Revision

Existing work on qualitative belief revision can suggest novel forms of conditional probability. In this section we discuss two examples: screened and hyper-revisionary conditionalization. We also explain how recent work on the qualitative part of probabilistically sound inference leads to an interesting notion of proto-probability functions.

5.1   Screened Conditional Probability

Screened revision is a variant form of AGM belief revision. Its basic idea is to see the operation as made up of two steps: a pre-processing step possibly followed by application of an AGM revision. The pre-processor decides the question of whether to revise, and this is done by checking whether the proposed input is consistent with a central part of the belief set under consideration, regarded as a protected subset. If they are mutually inconsistent, the belief set remains unchanged; otherwise, we apply an AGM revision in a manner that protects the privileged material. Clearly, such a composite process will not satisfy all the postulates of AGM revision: to begin with, the postulate of success, $a \in K * a$, may fail. For more details, see [26].

What would a probabilistic analogue of this look like? Roughly speaking, using the language of Leblanc cited earlier, when a is too unbelievable to take seriously as a condition, we put the probability of x on condition a to be just the *unconditioned* probability of x. In other words, for all values of a that are abnormal in the sense on van Fraasen (*i.e.* ones whose left projection p_a is not a proper Kolmogorov function, *cf.* Section 2.2) we require that $p_a = p_\top$ rather than put p_a to be the unit function $1(\cdot)$.

This forces modification of the Van Fraassen axioms. In particular, the axiom (vF2) of left projection must be weakened: we no longer always have $p_a(a) = 1$ since when a is abnormal $p_a(a) = p_\top(a) = p(a, \top) = 0$. In another respect, however, (vF2) can be strengthened: we can require that the left projection from any point is always a proper Kolmogorov function, as we no longer have any use for the unit function. The product axiom (vF3) must also be weakened. To show this, consider any inconsistent a. Unrestricted use of the product axiom would give us that for all x: $p_\top(x) = p_a(x) = p(x, a) = p(x \wedge x, a) = p(x, a) \cdot p(x, a \wedge x) = p(x, a) \cdot p(x, a) = p_a(x) \cdot p_a(x) = p_\top(x) \cdot p_\top(x)$; so that for any x, $p_\top(x)$ is either 0 or 1 which is quite undesirable behaviour.

The question of formulating adequate axiom systems for screened versions of the Popper and Hosiasson-Lindenbaum systems appears to be open. This problem may well be worth investigating, for although screened conditional probability behaves in an unfamiliar way, it is a coherent, intuitively motivated, and possibly useful concept.

5.2   Hyper-revisionary Probability Functions

As is well known, for any van Fraassen function $p(\cdot, \cdot)$ and $a \in L$, if $p(a, \top) > 0$ then $p(x, a)$ is determined by a natural relativization of the ratio rule: $p(x, a) = p(a \wedge x, \top)/p(a, \top)$. Indeed, this equality is almost immediate: the product axiom gives us $p(a \wedge x, \top) = p(a, \top) \cdot p(x, a \wedge \top) = p(a, \top) \cdot p(x, a)$ by right extensionality, permitting division when $p(a, \top) > 0$.

As remarked by Jonny Blamey (personal communication), it may be suggested that this is too conservative, even when we give elements in the critical zone a radically revisionary treatment in the manner of Hosiasson-Lindenbaum. For if a has a very low positive probability say, to fix ideas, $0 < p(a, \top) < 0.01$ then a surprise occurrence of a might sometimes lead us to question whether the function $p(\cdot, \cdot)$ was really right to give $p(a, \top)$ such a small value. We should perhaps move to a function $q(\cdot, \cdot)$ which makes the truth of a less unexpected, *i.e.* puts $q(a, \top)$ well above $p(a, \top)$; and for such a q the value of $q(x, a)$ will be $q(a \wedge x, \top)/q(a, \top)$, which may be quite different from that of $p(a \wedge x, \top)/p(a, \top)$.

This 'hyper-revisionary' proposal has a number of repercussions, some of which may be seen as merits, others as drawbacks.

Philosophically, the proposal drives a wedge between two different ways of 'adopting' a condition a. On the one hand, we may accept it because its truth has been revealed to us; on the other hand, we may entertain it to explore its consequences. The argument above suggests grounds for abandoning $p(\cdot, \cdot)$ when confronted with the truth of a proposition a for which p gives a very low value, but it does not suggest doing so when we merely entertain the truth of a to determine what effect it has on our probabilities. In this way, the hyper-revisionary proposal has the merit of giving formal expression to the contrast between accepting and supposing a condition, which tends to be neglected by the usual treatments of conditional probability.

Pragmatically, there can be no universally fixed cut-off point, like 0.01, at which we should revise rather than apply the relativized ratio rule. Where to draw the line would be a matter of context, purposes and subject matter, balanced in an informal judgement. This may be a source of frustration.

Formally, given the above short derivation of the relativized ratio rule from the product and right extensionality axioms, at least one of the two would have to be given up, or at least restricted in a suitable way. This is quite a loss.

Finally, it is not immediately clear how the new function $q(\cdot, \cdot)$ might be constrained by appropriate conditions. This could be seen as a disappointment – or as a challenge.

What would the qualitative analogue of such hyper-revisionary conditionalization look like? It would be to allow that even when input a is logically consistent with belief set K, we should not always take $K * a$ to be $Cn(K \cup \{a\})$. As well as adding in a, we should perhaps be contracting K, for despite the logical consistency of the two, a may be so implausible in the eyes of K that the exposure of its truth may lead us to an 'agonizing reappraisal' of the latter.

This, of course, is counter to one of the basic postulates of AGM belief revision, which puts $K * a = Cn(K * \{a\})$ in *every case* that a is consistent with K. In brief, AGM does not admit any conflict less than consistency as forcing contraction, just as the standard forms of conditional probability do not allow any improbability other than zero to force us out of the ratio rule.

The formal development of hyper-revisionary qualitative belief change remains open, as does that of hyper-revisonary conditionalization. It would appear that the latter could make contact with the theory of 'error statistics' as developed by Fisher, Neyman and Pearson, which analyses grounds for choosing between statistical hypotheses when faced with evidence that is logically consistent with each but highly improbable given one but not the other. However, we do not attempt to explore this problem in the present paper.

5.3 Proto-probability functions for qualitative inference

In 1996, Hawthorne investigated rules of uncertain inference which, while qualitative, may be given a probabilistic justification, formulating an axiom system that he called Q. All of its axioms are in a natural sense probabilistically sound, although the converse has not yet been settled. The question arises: do we need the full force of the axioms of probability in order to justify the rules of Q, or can it be done with weaker constraints on the 'probability' functions? In this section we observe that considerable weakening is possible. We need only certain modest order-theoretic conditions from among those derivable in the system of van Fraassen, the weakest of those presented in Section 2.2.

First, we recall Hawthorne's axioms. They concern consequence relations $\vdash\!\!\!\sim$ (in words: snake) between formulae of classical propositional logic. There are six Horn rules O1-O6 defining a system O, and one 'almost Horn' rule of 'negation rationality' (NR) whose addition gives Q. As always, Cn is classical consequence and $\approx$ is classical equivalence:

O1. $a \mathrel{|\!\sim} a$;

O2.   When $a \mathrel{|\!\sim} x$ and $y \in Cn(x)$, then $a \mathrel{|\!\sim} y$;

O3.   When $a \mathrel{|\!\sim} x$ and $a \approx b$, then $b \mathrel{|\!\sim} x$;

O4.   When $a \mathrel{|\!\sim} x \wedge y$, then $a \wedge x \mathrel{|\!\sim} y$;

O5.   When $a \mathrel{|\!\sim} x$, $b \mathrel{|\!\sim} x$ and $\neg b \in Cn(a)$, then $a \vee b \mathrel{|\!\sim} x$;

O6.   When $a \mathrel{|\!\sim} x$ and $a \wedge \neg y \mathrel{|\!\sim} y$, then $a \mathrel{|\!\sim} x \wedge y$;

NR.   When $a \vee b \mathrel{|\!\sim} x$ and $\neg b \in Cn(a)$, then either $a \mathrel{|\!\sim} x$ or $b \mathrel{|\!\sim} x$.

O1-O6 are also known as Reflexivity, Right Weakening (RW), Left Classical Equivalence (LCE), Very Cautious Monotony (VCM), Exclusive Or (XOR) and Weak And (WAND), respectively.

As Hawthorne showed, these conditions are probabilistically sound in the sense that for any probability function $p(\cdot, \cdot)$ satisfying Popper's postulates and 'threshold' $t \in [0, 1]$, if we define a relation by putting $a \mathrel{|\!\sim}_{pt} x$ iff $p(x, a) \geq t$, then $\mathrel{|\!\sim}_{pt}$ satisfies all the rules of Q.

Our question is: how much probability is really needed for the job? We show that it can be done with any function into an arbitrary linearly ordered set with greatest and least elements, satisfying certain very weak conditions in which no arithmetic operations appear.

Let D be any non-empty set equipped with a relation $\leq$ that is transitive and complete with a greatest element 1_D and a least element 0_D. A *proto-probability* function into D is any function $p : L^2 \to D$ satisfying the following six conditions:

P1. $p(a, a) = 1_D$;

P2. $p(x, a) \leq p(y, a)$ whenever $y \in Cn(x)$;

P3. $p(x, a) = p(x, b)$ whenever $a \approx b$;

P4. $p(x \wedge y, a) \leq p(y, a \wedge x)$;

P5. $p(x, a) \leq p(x, a \vee b) \leq p(x, b)$
      whenever $p(x, a) \leq p(x, b)$ and $\neg b \in Cn(a)$;

P6. $p(x, a) = p(x \wedge y, a)$ whenever $p(y, a \wedge \neg y) \neq 0_D$.

We call condition (P5) the principle of *disjunctive interpolation*. It is closely related to a principle of 'alternative presumption' of [22, 23] (details in the Appendix).

If we take any proto-probability function $p(\cdot, \cdot)$ and $t \in D$, and define a relation by putting $a \mathrel{|\!\sim}_{pt} x$ iff $p(x, a) \geq t$, then $\mathrel{|\!\sim}_{pt}$ satisfies all the rules of Q. Indeed, each condition (Oi) follows directly from its counterpart (Pi), with (NG) also following from (P5). The verifications are trivial, but given the novelty of the notion of proto-probability, we give them in full in the Appendix.

It is also easy to check that the axioms for proto-probability functions follow from those of van Fraassen, *a fortiori* from the stronger ones of Popper, Hosiasson-Lindenbaum, Carnap, and the Unit system. In fact, they are considerably weaker. Informally, it is clear that the left projection and product axioms of van Fraassen do not hold for all proto-probability functions, even when their top and bottom elements are chosen as the numbers 1,0, since our conditions for the latter make no use of either addition (which is implicit in the left projection axiom) or multiplication (explicit in the product axiom).

For a specific example of a proto-probability function that is not a van Fraassen one, take $p : L^2 \rightarrow [0, 1]$ to be the characteristic function of the classical consequence relation, *i.e.* put $p(x, a) = 1$ when $x \in Cn(a)$, otherwise $p(x, a) = 0$. Clearly, this satisfies conditions P1 through P6, but it fails (vF2) since $p(x \vee \neg x, \top) = 1$ while $p(x, \top) = 0 = p(\neg x, \top)$ for contingent formulae x, so that $p_\top$ is not a Kolmogorov function. The example can be generalized (see Appendix).

Thus, the proto-probability functions are defined by purely order-theoretic conditions that are strictly weaker than the axioms of any of the usual systems for conditional probability, but are strong enough to support the rules defining Hawthorne's system Q of probabilistic inference. In this way, the theory of qualitative uncertain inference, like that of qualitative belief change, provides new perspectives on conditional probability.

Appendix

This Appendix runs parallel to the main text. It contains most of the formal work, verifications, references and historical remarks supporting the main text.

For Section 1. Why go beyond the ratio rule?

The Kolmogorov axioms

There are several modes for presenting the Kolmogorov axioms for one-place probability functions, according to what we take as their domain. It may be a field of sets (most common in mathematics and applications), or equivalently a Boolean algebra (the preferred way of algebraists), or the set of all formulae of a propositional language (whose quotient structure under classical equivalence will be a free Boolean algebra). In this paper we work in the propositional mode, with the following formulation ([27]) of the postulates.

A (one-place) *proper Kolmogorov function* $p : L \to [0, 1]$ is any function defined on the set L of formulae of a language closed under the Boolean connectives, into the real numbers from 0 to 1, such that:

(K1) $p(x) = 1$ for some formula x;

(K2) $p(x) \leq p(y)$ whenever $y \in Cn(x)$;

(K3) $p(x \lor y) = p(x) + p(y)$ whenever $\neg y \in Cn(x)$.

Cn is classical consequence; we also write $\approx$ for classical equivalence. Thus postulate (K1) tells us that 1 is in that range of p; (K2) says that $p(x) \leq p(y)$ whenever x classically implies y; (K3), called the rule of finite additivity, tells us that $p(x \lor y) = p(x) + p(y)$ whenever x is inconsistent with y. It is sometimes extended so as to constrain the probability of countable unions (most easily expressed in the field of sets mode).

As remarked in the text (and observed by several authors, notably [16] and subsequently [11,24]), in comparative contexts it is convenient to regard the *unit function* (*i.e.* the function p that puts $p(x) = 1$ for every $x \in L$) as also being a Kolmogorov function, and we will follow this convention. It can be formalized by the simple expedient of defining a *Kolmogorov function* as one that is *either* a proper Kolmogorov function (*i.e.* satisfies the above postulates) *or* is the unit function. Equivalently, one could weaken axiom (K3) by putting it under the proviso that p is not the unit function. We refer to the unit function as the *improper* Kolmogorov probability function.

The ratio rule

The *ratio rule* for conditional probability uses an arbitrary Kolmogorov function $p : L \to [0, 1]$ to define a two-place function, conventionally written as $p(x|a)$ and read as 'the probability of x given a', defined on $L \times \{a \in L : p(a) > 0\}$ by the rule: $p(x|a) = p(a \land x)/p(a)$ when $p(a) > 0$ and otherwise undefined.

Left projections

We recall the standard concept of the left projection $f_a : X \to Y$ of a two-place function $f : X \times A \to Y$ from point $a \in A$, defined by putting $f_a(x) = f(x, a)$ for all $x \in X$.

For Section 1.1. Metaphysical *vs* pragmatic considerations

Metaphysical considerations

Such metaphysical views have been expressed by a number of probabilists, notably [33] and [9,34] and by some philosophers, *e.g.* [13].

[33, page 286] puts it briefly: "In fact, the probability of an event depends essentially on the circumstances under which the event possibly occurs, and it is a commonplace to say that in reality every probability is conditional". The same idea recurs at greater length in his 1970 (page 35).

[9, page 134] similarly remarks: "Every evaluation of probability is conditional; not only on the mentality or psychology of the individual involved, at the time in question, but also, and especially, on the state of information in which he finds himself at that moment."

More recently, [13] writes: "... given an unconditional probability, there is always a corresponding conditional probability lurking in the background. Your assignment of 1/2 to the coin landing heads super-ficially seems unconditional; but really it is conditional on tacit assumptions about the coin, the toss, the immediate environment, and so on. In fact, it is conditional on your total evidence."

Carnap's regularity condition

Carnap's formulation of the additional 'regularity' condition may be found in his [6, Section 53, Axiom C53-3] and also [8, Chapter 2.7, page 101].

We note in passing that the concept of a 'counterfactual probability function' discussed by [5] (building on [37]) also assumes that the crit-ical zone is empty. That concept, defined in the finite case, is a curious mixture of quantitative and qualitative ingredients. It puts $p(x, a)$, called the counterfactual probability of x given a, to be the proportion of the best a-states of the model that are x-states. The emptiness of the critical zone is assumed to ensure that the denominator is non-zero for consistent formulae a.

For Section 1.2. Some notational niceties

Two-place functions could alternatively be distinguished from one-place ones by different type-faces, *e.g.* lower case for one and upper case for the other. However that convention meshes poorly with the standard notation for left projection, which we also need to use extensively.

For Section 2.1. A leaf from the AGM book

How important is the critical zone?

Our view of the importance of the critical zone contrasts with that of many writers who minimize it. For example [30]: "The problem we have been examining, how to revise one's system of beliefs upon obtaining new evidence that had prior probability 0, is not a problem that has any great practical significance."

Conditional probability in the light of counterfactual conditionals

An argument for going beyond the ratio definition of two-place probability may also be made in terms of counterfactual conditionals rather than belief revision. Indeed, this is way in which it is usually done in philosophical literature going back to [37]. However, in the author's view, the comparison with belief revision affords a clearer view, and also lends itself to the construction of very simple formal maps, as shown in Section 5 and the corresponding part of the Appendix.

Verifications of properties of $B(p)$

We verify the claims made in bullet points about belief sets for probability functions. Let the belief set $B(p)$ corresponding to one-place function p be defined by putting $B(p) = \{x : p(x) = 1\}$. This is also sometimes called the *top* of the function. Write $B + a$ for the qualitative expansion of B by a, i.e. $B + a = Cn(B \cup \{a\})$. With $p_a(\cdot)$ understood as the left projection from a of the conditionalization $p(\cdot|\cdot)$ defined from $p(\cdot)$ by the ratio/unit rule, we want show: (1) in all cases, $B(p) \subseteq B(p)+a \subseteq B(p_a)$ and (2) in the limiting case that $p(a) = 0$ we have belief explosion: $B(p) + a = L = B(p_a)$, where L is the set of all propositions of the language.

For (1), the first inclusion is immediate from the definition of expansion above. To check the second inclusion, note that since $B(p_a)$ is closed under consequence it suffices to show that $a \in B(p_a)$ and $B(p) \subseteq B(p_a)$. The former is immediate since when $p(a) > 0$ then $p_a(a) = 1$ by the ratio definition and the Kolmogorov postulates for one-place probability, and $p_a(a)$ is also 1 when $p(a) = 0$, by the unit part of the ratio/unit definition. For the latter, it suffices to show that whenever $p(x) = 1$ then $p_a(x) = 1$. This is immediate when $p(a) = 0$. When $p(a) > 0$ we have $p_a(x) = p(a \wedge x)/p(a) = p(a)/p(a) = 1$ since the hypothesis $p(x) = 1$ implies that $p(a \wedge x) = p(a)$. For (2), it suffices to show further that when $p(a) = 0$ we have $B(p) + a = L$. But when the hypothesis holds then $p(\neg a) = 1$, so $\neg a \in B(p)$ and thus $B(p) + a \supseteq Cn(\neg a, a) = L$.

For Section 2.2. Bird's-eye view of available systems

Verification of consequences of the van Fraassen axioms

Left extensionality: $p(x, a) = p(x', a)$ whenever $x \approx x'$. *Verification*: By left projection, p_a is either a proper Kolmogorov function or the unit function. In the former case, $p(x, a) = p_a(x) = p_a(x') = p(x', a)$ using the hypothesis. In the latter case, $p(x, a) = p_a(x) = 1 = p_a(x') = p(x', a)$ irrespective of the hypothesis.

When $y = Cn(x)$ then $p(x, a) \leq p(y, a)$. *Verification*: If $y \in Cn(x)$ then $x \approx y \wedge x$ so by left extensionality and product, $p(x, a) = p(y \wedge x, a) = p(y, a) \cdot p(x, a \wedge y) \leq p(y, a)$.

When $p(\cdot)$ is defined as $p(\cdot, \top)$, then we have the ratio rule. *Verification*: Suitably instantiating the product axiom, $p(a \wedge x, \top) = p(a, \top) \cdot p(x, a \wedge \top) = p(a, \top) \cdot p(x, a)$ using right extensionality, so if $p(a, \top) > 0$ we have $p(x, a) = p(a \wedge x, \top)/p(a, \top) = p(a \wedge x)/p(a)$.

When a is a contradiction, then p_a is the unit function. *Verification*: $1 = p_a(a) = p(a, a) \leq p(x, a) = p_a(x)$, using left projection and an inequality already established.

The set Δ of all $a \in L$ such that p_a is the unit function is an ideal. *Verification*: To show that Δ is closed downwards, suppose $a \in \Delta$ and $a \in Cn(b)$. Then $1 = p(b \wedge x, a) = p(b, a) \cdot p(x, a \wedge b) = 1 \cdot p(x, a \wedge b) = p(x, b) = p_b(x)$, using the first supposition, product, first supposition again, second supposition respectively. To show that Δ is closed under disjunction, suppose p_a, p_b are both the unit function. To show that $p_{a \vee b}$ is also the unit function it suffices, by the left projection axiom to show that it is not a proper Kolmogorov function. Suppose it is; we get a contradiction. From the van Fraassen axioms we have $p(\bot, a \vee b) = p(\bot \wedge a, a \vee b) = p(a, a \vee b) \cdot p(\bot, a \wedge (a \vee b)) = p(a, a \vee b) \cdot p(\bot, a) = p(a, a \vee b) \cdot 1 = p(a, a \vee b)$ using the supposition that p_a is the unit function. Likewise $p(\bot, a \vee b) = p(b, a \vee b)$. By the supposition that $p_{a \vee b}$ is a proper Kolmogorov function we have $p(\bot, a \vee b) = 0$ so $p(a, a \vee b) = 0 = p(b, a \vee b)$. By the same supposition, $p(a \vee b, a \vee b) \leq p(a, a \vee b) + p(b, a \vee b) = 0 + 0 = 0$, contradicting the left projection axiom.

Finally, we check that a is abnormal iff $p(a, b) = 0$ for all normal b. *Verification*: From right to left, suppose $p(a, b) = 0$ for all normal b, but a is not abnormal. Then a is normal, so $p(a, a) = 0$, contradicting the left projection axiom. From left to right, suppose a is abnormal and b is normal. Then $a \wedge b$ is abnormal as already established, so $0 = p(\bot, b) = p(\bot \wedge a, b) = p(a, b) \cdot p(\bot, a \wedge b) = p(a, b) \cdot 1 = p(a, b)$ as desired.

Verification of the alternative axiomatization of the Popper system

Assume first the van Fraassen axioms plus (Positive); we need to show that $p(x, b) \neq 1$ for some x, b. By left projection, $p(\top, \top) = 1 > 0$ so by (Positive) $p_\top$ is proper and thus $p(\bot, \top) = 0 \neq 1$ as desired. Now assume the van Fraassen axioms plus $p(x, b) \neq 1$ for some x, b. Suppose $p(a, \top) > 0$; we need to show that p_a is proper, for which it suffices to show that it is not the unit function. First note that $p(\bot, \top) = p(\bot \wedge$

$b, \top) = p(b, \top) \cdot p(\bot, \top \wedge b) = p(b, \top) \cdot p(\bot, b)$; but since $p(x, b) \neq 1$ it follows that p_b is proper so $p(\bot, b) = 0$ and thus $p(\bot, \top) = 0$. But also $p(\bot, \top) = p(\bot \wedge a, \top) = p(a, \top) \cdot p(\bot, a)$, so since $p(a, \top) > 0$ we have $p(\bot, a) = 0$ so that p_a is not the unit function, as desired.

Example of a 'mixed' function

Leblanc and Roeper [24] gave an example of a two-place function satisfying the Popper postulates, whose treatment of formulae with probability zero is a mix of the expansionary and revisionary policies. They presented it rather enigmatically as an $8 \cdot 8$ table (their Table 5). We provide it with a more transparent rule-based presentation, which for convenience we express with a field of sets.

Take the field F of all subsets of the three-element set $S = \{\alpha, \beta, \gamma\}$. For motivation, think of α, β, γ as being of increasing levels of importance beginning from α, which has no importance at all. For $a, x \subset S$, put $p(x, a) = 1$ unless there is some item of positive importance in a and the item of greatest importance in a is not in x. More precisely, we define $p : S^2 \to [0, 1]$, in fact into $\{0, 1\}$ as follows:

1. If $\gamma \in a$ then $p(x, a) = 1$ if $\gamma \in x$, otherwise $p(x, a) = 0$;
2. If $\gamma \notin a$ but $\beta \in a$ then $p(x, a) = 1$ if $\beta \in x$, otherwise $p(x, a) = 0$;
3. If $\gamma \notin a$ but $\beta \notin a$ then $p(x, a) = 1$.

This function is a mix of the two kinds of conditional probability: $p(\{\beta\}, S) = 0 = p(\{\alpha\}, S)$ applying the first clause, but $p(\emptyset, \{\beta\}) = 0$ applying the second while $p(\emptyset, \{\alpha\}) = 1$ by the third. On the other hand, it is straightforward to check that it satisfies the Popper axioms.

Historical development of conditional probability

We review the historical steps in the construction of axioms for two-place probability functions, working backwards from [32]. For ease of comparison, we consider them all in the propositional mode, and treat each as defined on the whole of L^2, but comment on particularities of the original formulations each as we go.

Popper's original postulates for two-place probability functions, contained in an appendix of [32] (recalled *e.g.* in [24] and more accessibly [21]) were in the propositional mode. They reflected a desire for the autonomy of probability theory from logic, abstract algebra and set theory and so avoided any use of concepts from those areas. But if we are happy to use concepts of classical logic in our presentation then, as shown by subsequent writers, Popper's axioms may be given more perspicuously, as in the following formulation of [18], which require, for $p : L^2 \to [0, 1]$, that:

(P0) $p(x, a) \neq 1$ for some formulae a, x;

(P1) $p(x, a) = p(x, b)$ whenever $a \approx b$;

(P2) $p(x, a) = 1$ whenever $x \in Cn(a)$;

(P3) either $p(x \vee y, a) = p(x, a) + p(y, a)$ whenever $\neg(x \wedge y) \in Cn(a)$, or p_a is the unit function;

(P4) $p(x \wedge y, a) = p(y, a) \cdot p(x, y \wedge a)$.

Of course, if we are working in the context of fields of sets, (P1) becomes vacuous. Warning: The term 'Popper function' is sometimes used rather loosely, to refer to almost any primitive two-place probability function defined over the critical zone. For example, [25] use the term to refer to the narrower class of Hosiasson-Lindenbaum functions, defined below.

Our modular presentation takes from Rényi [33–35] his leading idea that for 'most' values of a, the left projection from a will be a proper Kolmogorov function giving a the value 1, and so is very similar in gestalt. But in its details, Rényi's system is rather different from any of those we have considered. Formulated in the field-of-sets mode, it treats the right domain as a *parameter*, allowing it to be chosen as any subset of the left domain that is consistent with the axioms. These axioms are just the product rule and the principle that p_a is a *proper* one-place Kolmogorov function with $p_a(a) = 1$, both formulated under the restriction that all values of the right argument take a value in the restricted right domain. For values of the right argument outside that subset, the probability functions are left undefined. We are thus given a scheme for a *family* of axiom sets, one for each choice of right domain.

This yields the Popper axioms if we constrain the right domain to include $\{a : p(a, S) > 0\}$, where S is the set on which the field is based, and carry out the following editing: (a) put $p(x, a) = 1$ for all a with $p(a, S) = 0$, (b) ensure consistency by allowing in the left projection axiom that p_a may be improper (as in the axiom (vF2) of Section 2.2), (c) for the one-place Kolmogorov functions mentioned in the left projection axiom, weaken Rényi's assumption of countable to finite additivity, and finally (d) translate from the field-of-sets mode to the propositional one.

Hosiasson-Lindenbaum's system [20] concerned what she called 'confirmation' functions, writing them as $c(x, a)$ rather than $p(x, a)$ and working in the propositional mode. This ground-breaking work has been comparatively neglected, despite its accessible and respected place of publication. In particular, the paper is not mentioned in any of [33–35], nor in the wide-ranging discussion of [16] or the comprehensive study of Roeper and Leblanc [36]. Popper does mention Hosiasson-Lindenbaum in passing in [32], but with respect to other questions and without citing her 1940 paper. This contrasts his the explicit acknowledgement (note 12 in new Appendix iv) of the influence of Rényi [33] on with thinking.

Hosiasson-Lindenbaum excluded inconsistent propositions from the right domain. Restoring them, we get the following axioms:

(HL1) $p(x, a) = 1$ whenever $x \in Cn(a)$;

(HL2) $p(x \vee y, a) = p(x, a) + p(y, a)$ whenever $\neg(x \wedge y) \in Cn(a)$, provided a is consistent;

(HL3) $p(x \wedge y, a) = p(x, a) \cdot p(y, a \wedge x)$ for all formulae a, x, y;

(HL4) $p(x, a) = p(x, b)$ whenever $a \approx b$.

Axiom (HL2) thus broadens the conditions under which the left projection of a two-place function satisfies additivity and is thus a proper Kolmogorov function, from the narrower requirement $p(a, \top) > 0$ to the wider one that a is consistent. The system may be obtained fron Rényi's scheme by setting the right domain at the set of all non-empty sets of S and editing by first putting $p(x, \emptyset) = 1$ and then as for Popper's system.

In what respect can it be said that Rényi's formulation was an advance on that of Hosiasson-Lindenbaum? For working mathematicians and statisticians, its use of the field-of-sets mode made application to practical problems more transparent. But at a deeper level, the step forward was *conceptual* – the realization that a rather arbitrary-looking axiom system becomes natural if we build it around the idea that for 'most' values of the right argument, the left projection will be a proper one-place probability function. As Rényi put it: "a conditional probability space is nothing else than a set of ordinary probability spaces which are connected with each other by [the product axiom]" [33, pages 289-290] .

Mini-note: We reverse a correction made by Hailperin [15, page 75] to the effect that since Hosiasson-Lindenbaum's formulation is in the propositional mode, it needs a left companion to (HL4) stating that $p(x, a) = p(y, a)$ whenever $x \approx y$. In fact, this follows from the postulates as given. In the limiting case that a is inconsistent we have $p(x, a) = 1 = p(y, a)$ by (HL1), so suppose a is consistent and $x \approx y$. Then $\neg(x \wedge \neg y) \in Cn(a)$, so by the additivity axiom (HL2) we have $p(x \vee \neg y, a) = p(x, a) + p(\neg y, a)$. But the supposition also gives us $LHS = 1$ by (HL1), so $p(x, a) + p(\neg y, a) = 1$. Moreover, (HL1) and (HL2) imply that $p(\neg y, a) = 1 - p(y, a)$, and so by arithmetic $p(x, a) = p(y, a)$. Essentially this point was already made by Tarski with regard to the earlier axiomatization of [29] (discussed below), and was acknowledged in footnote 1 of that paper.

Hosiasson-Lindenbaum [20] states that her axioms for two-place probability are 'analogous' to still earlier ones of Mazurkiewicz [29]. In fact, they constitute a major simplification and clarification of his quite complex system, which requires the left domain to contain *individual* propositions, while the right one contains consistent *sets* of propositions closed under classical consequence – the two kinds of proposition drawn, moreover, from intersecting and not very clearly defined languages. In his only example, Mazurkiewicz considers a game: the left argument of $p(x, A)$ can be filled by a proposition describing a state of play, while the right one can be occupied by a closed set of propositions containing the rules of the game, the current state of play, and any mathematical apparatus needed for deductions.

In turn, Mazurkiewicz states that he is taking as his starting point the axioms of Bohlmann [4]. However, Bohlmann's postulates are for *one*-place probability in a mode of unanalysed items called events and occurrences, which he supplements with an 'axiom' defining conditional probability by the ratio rule.

For some late nineteenth-century uses of conditional probability (without any attempt at axiomatization) see [14].

Thus our trail into the history of axiomatizations of two-place probability that cover the critical zone appears to end with Mazurkiewicz [29] as first serious attempt, Hosiasson-Lindenbaum [20] as the first really successful one, and Rényi [33] for providing a clear gestalt.

For Section 3.1. Hosiasson-Lindenbaum *vs* Popper

A misguided argument for Popper

We briefly review an interesting, but in the end inconclusive reason that could be given for regarding the Popper account as intrinsically preferable to that of Hosiasson-Lindenbaum.

As well as passing from an unconditional to a conditional function, we often need to strengthen the condition of an already conditional one. It could be useful to be able to express this as an operation taking a two-place function $p(\cdot, \cdot)$ to another two-place function $p_{\wedge b}(\cdot, \cdot)$ by the rule $p_{\wedge b}(x, a) = p(x, a \wedge b)$. But this operation breaks the boundaries of Hosiasson-Lindenbaum functions. It may happen that while a is consistent, $a \wedge b$ is not, in which case for any function $p(\cdot, \cdot)$ satisfying the Hosiasson-Lindenbaum axioms, $(p_{\wedge b})_a$ is the unit function despite the consistency of a, so that $p_{\wedge b}$ does not satisfy axiom (HL).

However, Popper functions face a similar problem. Consider any Popper function $p(\cdot, \cdot)$, and let a be an inconsistent proposition. Then $p_{\wedge a}(a, \top) = p(a, a) = 1 > 0$, while for all values of x we have

$(p_{\wedge a})_a(x) = p_{\wedge a}(x, a) = p(x, a) = 1$ since a is inconsistent, so that $(p_{\wedge a})_a$ is the unit function. These two facts contradict the distinctive Popper axiom (Positive).

The only way to keep our class of two-place functions closed under the 'conjoined condition' operation is to drop (Positive) and retreat to the van Fraassen system. Thus the convenience of being able to strengthen conditions could be seen as a point in favour of the usefulness of that very basic system.

Changing the underlying consequence operation

If one is working in the mode of fields-of-sets or Boolean algebras as carriers for the probability functions, then one can similarly express Popper functions as Hosiasson-Lindenbaum ones by passing to the quotient algebra determined by the same filter as in the propositional case. Essentially this construction was described, under a different light, by [16, page 234] and more explicitly [17, Section 6].

For Section 4.1. A map from conditional probability to AGM revision

We verify the claims made in the text about the map from HL conditional probability functions to AGM revision operations.

For HL functions, the left projection p_a of $p(\cdot, \cdot)$ from a is a proper Kolmogorov one-place probability function whenever a is consistent (Section 2.5), so under that condition we can apply well-known properties of the one-place functions without detailed justification, as well as the HL axioms themselves.

To show that $K = B(p)$ is a belief set, suppose $y \in Cn(K)$; we need to check that $y \in K$. By compactness, $y \in Cn\{\wedge x_i : i \leq n\}$ for some $x_1, ..., x_n \in B(p)$, so each $p(x_i, \top) = 1$, $p(\wedge x_i, \top) = 1$ and thus $p(y, \top) = 1$ so that $y \in B(p)$. To show that $B(p)$ is consistent we need then only note that $p(\bot, \top) = 0$.

Let $p : L^2 \to [0, 1]$ be any HL function. We need to check that the associated function $* : L \to 2^L$ satisfies each of the AGM postulates (K∗1) through (K∗8) with respect to $K = B(p)$. Two general remarks before the details:

- The AGM postulates for revision were first formulated in [10] and a convenient overview may be found in [31], whose presentation we follow. We note in passing that the classic account in [1] focused on contraction, and its axiomatization of revision contains a confusion: it omits postulate (K∗3) below, and treats the definition of revision from contraction via the Harper identity as if it were a postulate.

- We are not verifying satisfaction with respect to an *arbitrary* belief set K, but with respect to the *specific* belief set depending on the choice of p, namely $K = B(p) = \{x : p(x, \top) = 1\}$. This specification is needed for (K∗3) and (K∗4) though not for the other postulates, in which K does not appear in unrevised form.

(K∗1): $K * a = Cn(K * a)$. *Verification*: Same as the above for $B(p) = Cn(B(p))$, but replacing $\top$ by a.

(K∗2): $a \in K * a$. *Verification*: We need $p(a, a) = 1$, immediate from axiom (vF2).

(K∗3): $K * a \subseteq Cn(K \cup \{a\})$. *Verification*: Suppose $y \in LHS$, so that $p(y, a) = 1$. We need to show that $y \in Cn(K \cup \{a\}) = Cn(B(p) \cup \{a\}) = Cn(\{x : p(x, \top) = 1\} \cup \{a\})$, so it suffices to show that $\neg a \vee y \in \{x : p(x, \top) = 1\}$, *i.e.* that $p(\neg a \vee y, \top) = 1$. Now $p(\neg a \vee y, \top) = p(\neg a \vee (a \wedge y), \top) = p(\neg a, \top) + p(a \wedge y, \top)$. But $p(a \wedge y, \top) = p(a, \top) \cdot p(y, a) = p(a, \top)$ since $p(y, a) = 1$. Thus $p(\neg a \vee y, \top) = p(\neg a, \top) + p(a, \top) = p(\top, \top) = 1$ as desired.

(K∗4): $Cn(K \cup \{a\}) \subseteq K * a$ whenever a is consistent with K. *Verification*: Suppose $y \in Cn(K \cup \{a\})$ and a is consistent with K; we need to show $p(y, a) = 1$. By the first supposition, $\neg a \vee y \in Cn(\wedge x_i : i \leq n\}$ for some $x_1, .., x_n \in K = B(p)$ with each $p(x_i, \top) = 1$, so that $p(\wedge x_i, \top) = 1$ and thus $p(\neg a \vee y, \top) = 1$. Hence $p(a, \top) = p(a \wedge (\neg a \vee y), \top) = p(a \wedge y, \top)$. But also we have $p(a \wedge y, \top) = p(a, \top) \cdot p(y, a)$. Putting these together, $p(a, \top) = p(a, \top) \cdot p(y, a)$. But by supposition, $\neg a \notin K = B(p)$ so $p(\neg a, \top) \neq 1$ so $p(a, \top) \neq 0$, so by arithmetic $p(y, a) = 1$ as desired.

(K∗5): $K * a$ is consistent whenever a is consistent. *Verification*: Suppose a is consistent; we need $p(\bot, a) = 0$, which is immediate.

(K∗6): If $a \approx b$ then $K * a \approx K * b$. *Verification*: Suppose $a \approx b$; we need $p(x, a) = 1$ iff $p(x, b) = 1$, again immediate.

(K∗7): $K * (a \wedge b) \subseteq Cn((K * a) \cup \{b\})$. *Verification*: Suppose $x \in LHS$, so that $p(x, a \wedge b) = 1$. It suffices to show that $\neg b \vee x \in K * a$, *i.e.* that $p(\neg b \vee x, a) = 1$. When a is inconsistent, this is immediate, so suppose that a is consistent. From the supposition, $p(\neg b \vee x, a \wedge b) = 1$. Now $p(b \wedge x, a) = p(b \wedge (\neg b \vee x), a) = p(b, a) \cdot p(\neg b \vee x, a \wedge b) = p(b, a) \cdot 1 = p(b, a)$, so since a is consistent we have $p(b \wedge \neg x, a) = 0$ so $p(\neg b \vee x, a) = 1$ as desired.

(K∗8): $Cn((K * a) \cup \{b\}) \subseteq K * (a \wedge b)$ whenever b is consistent with $K * a$. *Verification*: Suppose that $y \in LHS$ and b is consistent with $K * a$; we need to show that $p(y, a \wedge b) = 1$. We have already verified $a \in K * a$, so the second supposition gives us the consistency of a. We now proceed along lines similar to the verification of (K*4). By the first supposition, $\neg b \vee y \in K * a$, i.e. $p(\neg b \vee y, a) = 1$. Hence since a is consistent, $p(b, a) = p(b \wedge (\neg b \vee y), a) = p(b \wedge y, a)$. But also $p(b \wedge y, a) = p(b, a) \cdot p(y, a \wedge b)$. Putting these together, $p(b, a) = p(b, a) \cdot p(y, a \wedge b)$. But by the second supposition again, $\neg b \notin K * a$ so $p(\neg b, a) \neq 1$ and thus $p(b, a) \neq 0$, so by arithmetic $p(y, a \wedge b) = 1$ as desired.

For Section 4.2 properties of the map: surjective but not injective

Failure of injectivity

For the failure of injectivity, it suffices to find two distinct HL functions $p \neq p'$ with $*_p = *_{p'}$, i.e. with $*_p(a) = \{x : p(x, a) = 1\} = \{x : p'(x, a) = 1\} = *_{p'}(a)$ for all $a \in L$, i.e. with $p(x, a) = 1$ iff $p'(x, a) = 1$ for all $a, x \in L$. For simplicity we do this with Boolean algebras rather than propositional languages. Take any finite Boolean algebra with $n \geq 2$ atoms, and two distinct probability distributions f, f' to these atoms with each atom getting a non-zero probability; extend them to one-place probability functions (for simplicity using the same names) on the entire algebra. Noting that every non-zero element of the algebra receives a non-zero probability under each of these functions, we can define two-place functions $p, p' : L^2 \to [0, 1]$ by the ratio rule for non-zero right arguments and putting $p(x, 0) = p'(x, 0) = 1$. These are HL probability functions, in fact they are Carnap functions. Then for all a, x we have $p(x, a) = 1$ iff $p(a \wedge x) = p(a)$ i.e. iff $a \leq x$ and likewise for p', and so $p(x, a) = 1$ iff $p'(x, a) = 1$ as desired.

Surjectivity

Suppose that the language is finite, and let $* : L \to 2^L$ satisfy the AGM postulates with respect to a consistent set K. Define $p(\cdot, \cdot)$ by the rule: $p(x, a) = 1$ for all $x \in L$ in the limiting case that a is inconsistent, while in the principal case that a is consistent $p(x, a)$ is the proportion of $(K * a)$-worlds that are x-worlds. We need to show that (1) p satisfies the HL axioms, (2) $* = *_p$, and (3) $K = B(p)$.

For (1) it is convenient to check the HL axioms in the form given to them by [20] (see Appendix to Section 2.2), as follows.

(HL1) $p(x, a) = 1$ whenever $x \in Cn(a)$. *Verification*: If a is inconsistent then we have $p(x, a) = 1$ by the definition for that case, so we may suppose that a is consistent. By AGM, $a \in K * a$ so if

$x \in Cn(a)$ we have $x \in Cn(K * a) = K * a$. Thus all $(K * a)$-worlds are x-worlds, *i.e.* the proportion of $(K * a)$-worlds that are x-worlds is 1, so $p(x, a) = 1$ as required.

(HL2) $p(x \vee y, a) = p(x, a) + p(y, a)$ whenever a is consistent and $\neg(x \wedge y) \in Cn(a)$. *Verification*: Suppose a is consistent and $\neg(x \wedge y) \in Cn(a)$. By the first supposition, we need to consider proportions, and by the second the proportion of $(K * a)$-worlds that are $(x \vee y)$-worlds is the sum of the proportions of $(K * a)$-worlds that are, separately, x-worlds or y-worlds, and we are done.

(HL3) $p(x \wedge y, a) = p(x, a) \cdot p(y, a \wedge x)$. *Verification*: If a is inconsistent then so is $a \wedge x$ and $LHS = 1 = RHS$. Suppose a is consistent. If $a \wedge x$ is inconsistent then $LHS = 0$ while $RHS = 0 \cdot 1 = 0$ and again we are done. If $a \wedge x$ is consistent then LHS is the proportion of $(K * a)$-worlds that are $(x \wedge y)$-worlds, while RHS is the proportion of $(K * a)$-worlds that are x-worlds multiplied by the proportion of $(K * a \wedge x)$-worlds that are y-worlds. If x is inconsistent with $K * a$ then both LHS and RHS equal 0, so we may suppose that x is consistent with $K * a$. Then by AGM axioms (K$*$7) and (K$*$8) the $(K * a \wedge x)$-worlds are just the $(K * a)$-worlds that are x-worlds. Hence RHS is the proportion of $(K * a)$-worlds that are x-worlds multiplied by the proportion of *those* that are y-worlds, which equals the proportion of $(K * a)$-worlds that are $(x \wedge y)$-worlds, equalling the LHS and we are done.

(HL4) $p(x, a) = p(x, b)$ whenever $a \approx b$. *Verification*: If a is inconsistent then so is b, so $LHS = 1 = RHS$. If a is consistent, then if $a \approx b$ the a-worlds are just the b-worlds, and the proportion of a-worlds that are x-worlds is the same as the proportion of b-worlds that are x-worlds.

To show that (2) $* = *_p$, consider first the principal case that a is consistent, where we need only note that by the definition of $*_p$ we have $x \in *_p(a)$ iff $p(x, a) = 1$ while, by the definition of p, also $p(x, a) = 1$ iff every $(K * a)$-world is an x-world, *i.e.* iff $x \in Cn(K * a) = K * a$. In the limiting case that a is inconsistent, $p(x, a) = 1$ for every x and by the AGM postulates, $x \in K * a$ for every x, so again we are done. Finally, to check (3) that $K = B(p)$ we need only show that $x \in K$ iff $p(x, \top) = 1$. But since K is consistent, the AGM postulates tell us that $K = K * \top$, and the equivalence $p(x, a) = 1$ iff $x \in K * a$ just established may be applied substituting $\top$ for a.

We conjecture that surjectivity fails in the infinite case. Evidently its present proof breaks down there, since one cannot meaningfully speak of proportions of infinite sets, thus blocking the definition of $p(\cdot, \cdot)$ above. Nor is it possible to repair the proof by replacing proportionality by some probability distribution that gives each world a non-zero value. For if the set of formulae is countable, there are continuum many worlds and as is well known, there is no probability distribution on a non-countable set that gives a non-zero value to each element.

For Section 5.3. Proto-probability functions for qualitative inference

The system Q

For further information on systems O and Q see part I of [19].

Disjunctive interpolation

As remarked in the text, the principle of disjunctive interpolation is closely related to a rule discussed by Koopman [22, 23]. Called 'alternative presumption', it states that whenever both $p(x, a \wedge b)$, $p(x, a \wedge \neg b) \leq p(y, c)$ then $p(x, a) \leq p(y, c)$. If we assume that the order $\leq$ is complete (as we do, although Koopman does not), alternative presumption is in fact equivalent to the right half of disjunctive interpolation. *Verification.* To obtain Koopman: by completeness of $\leq$, either $p(x, a \wedge b) \leq p(x, a \wedge \neg b)$ or conversely; in *e.g.* the former case we have by the right part of disjunctive interpolation that $p(x, (a \wedge b) \vee (a \wedge \neg b)) \leq p(x, a \wedge \neg b)$ and by right extensionality and transitivity of $\leq$ we are done. In the converse direction, suppose $p(x, a) \leq p(x, b)$ and $\neg b \in Cn(a)$, we want to show the right hand part of disjunctive interpolation, *i.e.* that $p(x, a \vee b) \leq p(x, b)$. We need only note that $p(x, b) = p(x, (a \vee b) \wedge b)$ and, given the last supposition, that $p(x, a) = p(x, (a \vee b) \wedge \neg b)$, then apply Koopman with a little help from left extensionality, taking (y, c) as (x, b).

Verification that Q is proto-probablistically sound

We check that when we take any proto-probability function $p(\cdot, \cdot)$ and $t \in D$, and define a relation by putting $a \mathrel{|\!\sim_{pt}} x$ iff $p(x, a) \geq t$, then $\mathrel{|\!\sim_{pt}}$ satisfies all the rules of Q. For (O1) we need $p(a, a) \geq t$, which is immediate from (P1). For (O2), we need that when $p(x, a) \geq t$ and $y \in Cn(x)$ then $p(y, a) \geq t$, which is immediate from (P2). For (O3), we need that when $p(x, a) \geq t$ and $a \approx b$ then $p(x, b) \geq t$, which is immediate from (P3). For (O4), we need that when $p(x \wedge y, a) \geq t$ then $p(y, a \wedge x) \geq t$, which is immediate from (P4). For (O5) alias XOR, we need that when $p(x, a) \geq t$, $p(x, b) \geq t$ and $\neg b \in Cn(a)$, then

$p(x, a \vee b) \geq t$. Since the order on D is complete, either $p(x, a) \leq p(x, b)$ or conversely, consider *e.g.* the former. Then using the left half of (P5), $p(x, a) \leq p(x, a \vee b)$ and we are done by transitivity of $\leq$. For (O6) alias WAND, we need that when $p(x, a) \geq t$ and $p(y, a \wedge \neg y) \geq t$ then $p(x \wedge y, a) \geq t$. If $t = 0_D$ then this is immediate, and if $t \neq 0_D$ it is immediate from (P6). It remains to obtain the non-Horn rule (NR) of negation rationality. We need to show that when $\neg b \in Cn(a)$ and $p(x, a \vee b) \geq t$ then either $p(x, a) \geq t$ or $p(x, b) \geq t$. Since the order on D is complete, either $p(x, a) \leq p(x, b)$ or conversely, consider *e.g.* the former. Then using the right half of (P5), $p(x, a \vee b) \leq p(x, b)$ giving $p(x, b) \geq t$ as desired.

Generalizing an example

We can generalize the example given in the text of a proto-probability function that is not a van Fraassen function. Fix any consistent formula c and put $p(x, a) = 1$ when $x \in Cn(a \wedge c)$, else $p(x, a) = 0$. Then $p(\cdot, \cdot)$ satisfies (P1) through (P6), but choosing x as any formula independent of c it fails (vF2) as before.

ACKNOWLEDGEMENTS. Thanks to Horacio Arló Costa, Jonny Blamey, Richard Bradley, Franz Dietrich, Peter Gärdenfors, Jim Hawthorne, Hykel Hosni, Franco Montagna, Rohit Parikh, Wlodek Rabinowicz, Miklos Redei and two referees for remarks on drafts and oral presentations. Also to Hykel, Jacek Malinowski, and Miklos for assistance in obtaining copies of rather inaccessible texts of Bohlmann, Mazurkiewicz and Rényi.

References

[1] C. ALCHOURRÓN, P. GÄRDENFORS and D. MAKINSON, *On the logic of theory change: partial meet contraction and revision functions*, The Journal of Symbolic Logic **50** (1985), 510–530.

[2] H. ARLÓ COSTA, *Bayesian epistemology and epistemic conditionals: on the status of export-import laws*, The Journal of Philosophy **98** (2001), 555–593.

[3] H. ARLÓ COSTA and ROHIT PARIKH, *Conditional probability and defeasible inference*, Journal of Philosophical Logic **34** (2005), 97–119.

[4] G. BOHLMANN, *Die Grundbegriffe der Wahrscheinlichkeitsrechnung in ihrer Anwendung auf die Lebensversicherung*, In: "Atti

del IV Congresso Internazionale dei Matematici, Roma 6-11 Aprile 1908", vol. III, Accademia dei Lincei, Roma, 1909, 244–278.

[5] C. BOUTILIER, *On the revision of probabilistic belief states*, Notre Dame Journal of Formal Logic **36** (1995), 158–183.

[6] R. CARNAP, "Logical Foundations of Probability", Chicago University Press, 1950.

[7] R. CARNAP, "The Continuum of Inductive Methods", Chicago University Press, 1952.

[8] R. CARNAP, *A basic system of inductive logic*, In: "Studies in Inductive Logic and Probability", Carnap and Jeffries (eds.), University of California Press, 1971, 33–165.

[9] B. DE FINETTI, "Theories of Probability", Wiley, New York, 1974.

[10] P. GÄRDENFORS, *The dynamics of belief: contractions and revisions of probability functions* Topoi **5** (1986), 29–37.

[11] P. GÄRDENFORS, "Knowledge in Flux: Modeling the Dynamics of Epistemic States", MIT Press, Cambridge (Mass.), 1988.

[12] W. GOOSENS, *Alternative axioms of elementary probability theory*, Notre Dame Journal of Formal Logic **20** (1979), 227–239.

[13] A. HAJEK, *What conditional probability could not be*, Synthese **137** (2003), 273–323.

[14] T. HAILPERIN, *The development of probability logic from Leibniz to MacColl*, History and Philosophy of Logic **9** (1988), 131–191.

[15] T. HAILPERIN, *Probability logic in the twentieth century*, History and Philosophy of Logic **12** (1991), 71–110.

[16] W. L. HARPER, *Rational belief change, Popper functions and counterfactuals*, Synthese **30** (1975), 221–262. Reprinted with minor editorial changes in: "Foundations of Probability, Statistical Inference, and Statistical theories of Science, Vol. I", W. L. Harper and C. A. Hooker (eds.), Reidel, Dordrecht, 73–115.

[17] W. L. HARPER, *Rational conceptual change*, In: "PSA: Proceedings of the Biennial Meeting of the Philosophy of Science Association 1976, Vol. II" University of Chicago Press, 1976, 462–494.

[18] J. HAWTHORNE, On *the logic of nonmonotonic conditionals and conditional probabilities*, Journal of Philosophical Logic **25** (1996), 185–218.

[19] J. HAWTHORNE and D. MAKINSON, *The quantitative/qualitative watershed for rules of uncertain inference*, Studia Logica **86** (2007), 249–299.

[20] HOSIASSON-LINDENBAUM, *On confirmation*, The Journal of Symbolic Logic **4** (1940), 133–148.

[21] R. KOONS, *Supplement to Defeasible Reasoning*, Stanford Encyclopedia of Philosophy, http://plato.stanford.edu (2009).

[22] B. O. KOOPMAN, *The axioms and algebra of intuitive probability*, The Annals of Mathematics **41** (1940), 269–292.

[23] B. O. KOOPMAN, *The bases of probability*, Bulletin of the American Mathematical Society **46** (1940), 763–774.

[24] H. LEBLANC and P. ROEPER, *On relativizing Kolmogorov's absolute probability functions*, Notre Dame Journal of Formal Logic **30** (1989), 485–512.

[25] S. LINDSTRÖM and W. RABINOWICZ, *On probabilistic representation of non-probabilistic belief revision*, Journal of Philosophical Logic **18** (1989), 69–101.

[26] D. MAKINSON, *Screened revision*, Theoria **63** (1997), 14–23.

[27] D. MAKINSON, "Bridges from Classical to Nonmonotonic Logic", College Publications, London, 2005.

[28] D. MAKINSON, *Logical questions behind the lottery and preface paradoxes: lossy rules for uncertain inference* Synthese, issue commemorating Henry Kyburg, to appear.

[29] S. MAZURKIEWICZ, *Zur Axiomatic der Wahrscheinlichkeitsrechnung*, Comptes rendus des séances de la Societé des Sciences et des Lettres de Varsovie (Sprawozdania z posiedzen Towarzystwa Naukowego Warszawskiego) **25** (1932), 1–4.

[30] V. MCGEE, *Learning the impossible*, In: "Probability and Conditionals: Belief Revision and Rational Decision", E. Eells and B. Skyrms (eds.), Cambridge University Press, 1994, 179–199.

[31] P. PEPPAS, *Belief revision*, Chapter 8 of "Handbook of Knowledge Representation", F. van Harmelen, V. Lifschitz, B. Porter (eds.), Elsevier, Amsterdam, 2007.

[32] K. POPPER, "The Logic of Scientific Discovery", second edition, Basic Books, New York, 1959.

[33] A. RÉNYI, *On a new axiomatic theory of probability*, Acta Mathematica Academiae Scientiae Hungaricae **6** (1955), 268–335.

[34] A. RÉNYI, "Foundations of Probability Theory" Holden-Day, San Francisco, 1970.

[35] A. RÉNYI, "Probability Theory" North-Holland, Amsterdam, 1970.

[36] P. ROEPER and H. LEBLANC, "Probability Theory and Probability Logic" University of Toronto Press, Toronto, 1999.

[37] R. C. STALNAKER, *Probability and conditionals*, Philosophy of Science **37** (1970, 64–80. Reprinted in: "Ifs", W. L. Harper *et al.* (eds.), Reidel, Dordrecht, 1981, 107–128.

[38] W. SPOHN, *The representation of Popper measures*, Topoi **5** (1986), 69–74.

[39] W. SPOHN, *A survey of ranking theory*, In: "Degrees of Belief" , F. Huber *et al.* (eds.), Springer, New York, 2009, 185–228.

[40] B. VAN FRAASSEN, *Representation of conditional probabilities*, Journal of Philosophical Logic **5**, (1976), 417–430.

[41] B. VAN FRAASSEN, *Fine-grained opinion, probability, and the logic of full belief* Journal of Philosophical Logic **24** (1995), 349–377.

Is there a probability theory of many-valued events?

Vincenzo Marra

> Neque omittenda est prophetia Danielis de ultimis mundi temporibus: *Multi pertransibunt et multiplex erit scientia.*
> Francis Bacon, aphorism XCIII in the *Novum Organum*, 1620.

> Ce n'est pas que je n'estime autant que je dois votre invention; mais ce que le chancelier Bacon a dit est bien vrai: *Multi pertransibunt et augebitur scientia.*
> First letter from Fermat to Roberval, August 1636.

1 Prologue

Since Boolean, yes/no events may be denoted by formulæ in classical propositional logic, by analogy one is tempted to regard formulæ in a non-classical propositional logic as linguistic descriptions of non-classical events – whatever the latter are. And since Boolean events may be assigned (subjective) probabilities, one is lured into attempts at defining probabilities of non-classical events.

Ulam had it that, according to Banach, "good mathematicians see analogies between theorems and theories, while the very best ones see analogies between analogies" [31, Report 20]. On this account, mathematicians are analogy-makers. But that is far from being the whole story. Sadly, not all analogies are created equal: a few are fruitful, while the rest are barren.

I offer here the result of my own preliminary efforts to assess the extent to which the analogies in the opening paragraph are likely to be fruitful, given our current state of knowledge. I shall survey some mathematical, logical, and philosophical issues that any serious attempt to weave those analogies into a substantive theory must tackle; many of them seem to be largely unexplored as yet. By way of case study, I will concentrate on the one-variable fragment of a single many-valued propositional logic known as *Gödel logic*. I hope to convince the reader that even in this apparently trivial case much remains to be understood.

Let me begin at the (conventional) beginning.

2 From probability to logic: the classical case

2.1 Events as Propositions

It is a historiographic cliché that the modern theory of probabilities begins in 1654, when Fermat replied to a letter he had received from Pascal. (Pascal's first letter to Fermat is apparently no longer extant.) Here is an excerpt[1] of what Fermat wrote [10, page 288].

> Monsieur,
> If I undertake to make a point with a single die in eight throws, and if we agree after the money is put at stake, that I shall not cast the first throw, it is necessary by my theory that I take $\frac{1}{6}$ of the total sum to be impartial because of the aforesaid first throw. And if we agree after that that I shall not play the second throw, I should, for my share, take the sixth of the remainder that is $\frac{5}{36}$ of the total. If, after that, we agree that I shall not play the third throw, I should to recoup myself, take $\frac{1}{6}$ of the remainder which is $\frac{25}{216}$ of the total. And if subsequently, we agree again that I shall not cast the fourth throw, I should take $\frac{1}{6}$ of the remainder or $\frac{125}{1296}$ of the total, and I agree with you that that is the value of the fourth throw supposing that one has already made the preceding plays.

Fermat and Pascal are discussing the so-called *problem of the points*, that is, the problem of the fair division of a stake between two players whose game is called off before its close. It transpires from their correspondence that the problem was brought to their attention by the Chevalier de Méré. Born Antoine Gombaud and not in fact a Knight at all, the Chevalier de Méré was a writer and an amateur mathematician; he became interested in the problem of points, which already had quite a long history at the time.

True to a scholarly tradition that continues to this day, Fermat and Pascal both took a belligerent stance on research. Their agreement in the passage above does not make it to the next paragraph.

> But you proposed in the last example in your letter (I quote your very terms) that if I undertake to find the six in eight throws and if I have thrown three times without getting it, and if my opponent proposes that I should not play the fourth time, and if he

---

[1] The English translation of all excerpts of the Fermat-Pascal correspondence quoted below is by Vera Sanford, in [28, page 546–565].

wishes me to be justly treated, it is proper that I have $\frac{125}{1296}$ of the entire sum of our wagers. This, however, is not true by my theory [according to which] he who holds the die and agrees to not play his fourth throw should take $\frac{1}{6}$ as his reward. I urge you therefore to write me that I may know whether we agree in the theory, as I believe we do, or whether we differ only in its application.

I am, most heartily, etc.

At the time of this correspondence, Pascal was Fermat's junior by 22 years, and 31 years of age. His reply, dated July 29, 1654, conveys all of the excitement injected into a brilliant young man by a presumably sleepless night spent over a research problem [10, page 289; the emphasis is mine]:

Monsieur,
Impatience has seized me as well as it has you, and *although I am still abed*, I cannot refrain from telling you that I received your letter in regard to the problem of the points *yesterday evening*.

Pascal admits that his proposed solution of $\frac{125}{1296}$ to the second problem discussed in Fermat's letter is wrong, while Fermat's solution of $\frac{1}{6}$ is correct. After stating that he "admires" Fermat's solution, however, he adds at once that Fermat's method for such problems actually was his own first thought, too, but that he went straight on to devise "an abridgement and indeed another method that is much shorter and more neat, which I should like to tell you here in a few words". And, before proceeding, he declares himself pleased that he and his correspondent ultimately do agree, if not on the best method of solution, at least on the final answers [10, page 290]:

I plainly see that the truth is the same at Toulouse and at Paris.
[*Je vois bien que la vérité est la meme à Toulouse et à Paris.*]

As far as I can tell, neither Pascal nor Fermat ever use in their correspondence the word "event", or one of its synonyms. But already by 1718, when the first edition of de Moivre's *Doctrine of Chances* was printed in London [7], the word had become a technical term of the subject; it featured in the subtitle (*A Method of Calculating the Probability of Events in Play*), in the dedication (to Isaac Newton, who befriended the French emigrant de Moivre), in the opening paragraph of the Introduction:

The Probability of an Event is greater, or less, according to the number of Chances by which it may Happen, compar'd with the number of all the Chances, by which it may either Happen or Fail.

The tradition that regards events – in the sense of possible outcomes of idealised experiments – as a key primitive notion in probability theory permeates the subject, from Kolmogorov's *Grundbegriffe* [20, Section 3] through Rényi's *Foundations* [25, Section 1.1], all the way down to many contemporary classroom textbooks. There is, however, a second approach to the notion of event that also has a substantial tradition. It consists in taking propositions as the primitive concept, and in defining events as a derived notion. Thus, in his *Laws of Thought* Boole wrote [3, pages 247–248]:

> [...] it will be advantageous to notice, that there is another form under which all questions in the theory of probabilities may be viewed; and this form consists in substituting for *events* the propositions which assert that those events have occurred, or will occur; and viewing the element of numerical probability as having reference to the *truth* of those *propositions*, not to the *occurrence* of the *events* concerning which they make assertion.

Later, Keynes emphasised the point at the beginning of his treatise [18, page 5].

> With the term *event*, which has taken hitherto so important a place in the phraseology of the subject, I shall dispense altogether. Writers on Probability have generally dealt with what they term the "happening" of "events". In the problems which they first studied this did not involve much departure from common usage. But these expressions are now used in a way which is vague and ambiguous; and it will be more than a verbal improvement to discuss the truth and the probability of *propositions* instead of the occurrence and the probability of *events*.

Keynes asserts that replacing events by propositions is "more than a verbal improvement". How is it so, precisely? Consider, for the sake of argument, the following questions.

(1) What does it mean to say that an event occurs, or that it does not?
(2) Does it make sense to consider arbitrarily many events, or should we restrict ourselves, say, to at most countably many events at any one time?
(3) Is there always, to any given event, another event that occurs precisely when the first one does not? If so, is such an event unique?

Etc., *ad libitum*. Now whereas it is evident that such questions have little import as to whether the proverbial coin will land heads or tails, it is also fair to admit that they might become at least potentially relevant for more advanced topics in probability theory. By contrast, on the Boole-Keynes account the three preceding questions would read as follows.

(1′) What does it means to say that a proposition is true, or that it is not?
(2′) Does it make sense to consider arbitrarily many propositions, or should we restrict ourselves, say, to at most countably many propositions at any one time?
(3′) Is there always, to any given proposition, another proposition that is true precisely when the first one is not? If so, is such a proposition unique?

This shift of perspective is more than a verbal improvement precisely in that our new questions have nothing to do with the theory of probability *per se*. We are now in the realm of *logic*, and that might well give us a head start – after all, logicians have been thinking about propositions for several centuries before the Fermat-Pascal correspondence took place. Questions such as (1′–3′), for instance, have widely accepted answers in logic.

To summarise so far: *The rôle of logic in the theory of probability is to provide a formal model for the notion of event. Specifically, an event is defined as whatever may be denoted by a proposition.*

2.2 Classical logic

Taking stock of available knowledge about the logical analysis of propositions, let us without further ado adopt classical propositional logic as a formal means of modelling events. We start with an arbitrary set of *propositional variables*, or *atomic formulæ*, that are to stand for those propositions that cannot – or that we do not wish to – analyse into simpler constituents. To fix ideas, say we content ourselves with countably many variables:

$$\text{VAR} = \{X_1, X_2, \ldots, X_n, \ldots\} \,.$$

To these we adjoin two symbols $\top$ and $\bot$, say, that are to stand for a proposition that is always true (the *verum*), and one that is always false (the *falsum*), respectively. To construct compound formulæ we use the *logical connectives* $\wedge$ (for conjunction), $\vee$ (for disjunction), $\rightarrow$ (for implication), and $\neg$ (for negation). The usual recursive definition of a formula now reads as follows.

- $\top$ and $\bot$ are formulæ.
- All propositional variables are formulæ.

- If φ and ψ are formulæ, then so are $(\varphi \wedge \psi)$, $(\varphi \vee \psi)$, $(\varphi \rightarrow \psi)$, and $(\neg\varphi)$.
- Nothing else is a formula.[2]

Let us write FORM for the set of formulæ constructed over the countable language VAR. Similarly, if $n \geq 0$ is a non-negative integer, let us write FORM$_n$ for the set of formulæ constructed over the first n propositional variables VAR$_n = \{X_1, \ldots, X_n\}$ only.

With our syntactic conventions in place, we can proceed to devise a formal semantics in order to give formulæ a meaning. To this end, let me quote Frege's celebrated conclusion about the meaning of a proposition [11, page 34].

> We are therefore driven into accepting the truth value [*Wahrheits-wert*] of a sentence[3] as constituting its reference [*Bedeutung*]. By the truth value of a sentence I understand the circumstance that it is true or false. There are no further truth values. For brevity I call the one the True [*das Wahre*], the other the False [*das Falsche*].

Here, the neutral nouns *das Wahre* and *das Falsche* are literal versions of our *verum* and *falsum*, respectively. Accordingly, we consider *assignments of truth values* to propositions, also called *valuations* or *interpretations*. These are functions $w \colon$ FORM $\rightarrow \{\bot, \top\}$ subject to the familiar conditions:

(1) $w(\top) = \top$, $w(\bot) = \bot$;
(2) $w(\varphi \wedge \psi) = \top$ if and only if both $w(\varphi) = \top$ and $w(\psi) = \top$;
(3) $w(\varphi \vee \psi) = \top$ if and only if either $w(\varphi) = \top$ or $w(\psi) = \top$, or both;
(4) $w(\varphi \rightarrow \psi) = \bot$ if and only if $w(\varphi) = \top$ and $w(\psi) = \bot$;
(5) $w(\neg\varphi) = \top$ if and only if $w(\varphi) = \bot$,

for all $\varphi, \psi \in$ FORM. Thus, (2.2) prescribes that $\vee$ is to be interpreted as *inclusive disjunction*; (2.2) that $\rightarrow$ is *material implication*; and so on.

Remark 2.1 (The structure of truth values, I). Above, we are only assuming that the set $\{\bot, \top\}$ is just that – *a two-element set carrying no further structure*. Thus, any isomorphic set could do just as well –

---

[2] Following well-known conventions on the precedence rules among connectives, I will at times omit redundant parentheses from formulæ.

[3] I am glossing over the distinction between propositions and sentences; this is harmless for my purposes here. Readers interested in that distinction may find *e.g.* [22] a useful starting point.

{Gottlob, Frege}, for instance, or {0, 1}. The latter is best avoided *at this stage*, however, lest we let our own notation fool us into thinking that *e.g.* '0 < 1', or '0 + 0 = 0' or like expressions be at all meaningful. Later, though, we will see that there is more than meets the eye here – *cf.* Section 2.3 below.

A second easy observation about these definitions is that a valuation is subject to no restrictions concerning the truth values it assigns to propositional variables. It is also apparent that any valuation w: FORM → {⊤, ⊥} is uniquely determined by such values $w(X_1), \ldots, w(X_n), \ldots$; and this is one way to state the *principle of truth-functionality* (again after Frege) for classical propositional logic.

Partaking in a well-established tradition going back at least to Leibniz, it is often useful to think of a valuation w: FORM → {⊥, ⊤} as a *possible world* in which each proposition $\varphi \in$ FORM either holds (*i.e.* $w(\varphi) = \top$) or fails to hold (*i.e.* $w(\varphi) = \bot$). Because of truth-functionality, then, a classical logician's world may be identified with the collection of atomic sentences that can be truthfully uttered in that world.[4]

Analytic truths, or *tautologies* after Wittgenstein, are now defined as those formulæ $\varphi \in$ FORM that are true in every possible world, *i.e.* such that $w(\varphi) = \top$ for any assignment w: FORM → {⊥, ⊤}. Tautologies are central to logic proper, but they are uninteresting in probability theory. By definition, they are just descriptions of the sure event, an event that always obtains in each world – and thus holds no charms for the forecaster. Nonetheless, it will be useful to briefly recall the completeness theorem for tautologies at this point. Selecting an appropriate collection of formulæ as axioms, and using *modus ponens* as deduction rule, one defines the notion of *provable* or *deducible* formula (I omit the well-known details). Writing ⊢ φ to mean that a formula φ is provable, and writing ⊨ φ to mean that φ is a tautology, one then proves the all-important (soundness and) completeness theorem: *for any $\varphi \in$ FORM, ⊢ φ if and only if ⊨ φ*. A more general version of this result that deals with *synthetic truths* is available. In fact, it is this version that is relevant to probability theory, and to all conceivable realms of applications of classical logic, for that matter: for only synthetic truths may embody specific knowledge, whereas tautologies are unable to reveal anything about any given world. For $\varphi \in$ FORM and an arbitrary subset of formulæ $A \subseteq$ FORM, to write $A ⊢ \varphi$ is to assert that φ is a *syntactic consequence* of A, *i.e.* is deducible via *modus ponens* from the axioms of classical logic augmented by A,

---

[4] I hasten to add that I do not feel bound to take the ontological status of such ethereal worlds very seriously; others do – notably, D. K. Lewis in [21].

the latter being conceived as a set of additional assumptions. On the semantic side, to write $A \models \varphi$ is to assert that φ is a *semantic consequence* of A, *i.e.* is true in each possible world wherein each formula of A is true. Thus, syntactic and semantic consequence generalise provability and tautologousness, respectively. By a *theory* $\Theta \subseteq$ FORM one means a set of formulæ that is deductively closed. Finally, the completeness theorem for theories now reads: *for any $\varphi \in$ FORM and any theory $\Theta \subseteq$ FORM, $\Theta \vdash \varphi$ if and only if $\Theta \models \varphi$.*

After this recapitulation on classical logic, we are ready to pass from *assignments of truth values* to assignments of *degrees of probabilities* to propositions describing events. Before doing so, however, I first discuss a crucial point that underlies the transition from truth to probability.

2.3 The mathematical structure of *verum* and *falsum*

As was discussed in the previous section, if one is only interested in identifying tautologies, then there is no need to impose on the set $\{\bot, \top\}$ of truth values any additional structure. For other purposes, however, this can and must be done.

With hindsight, it is a simple observation that the logical connectives induce on $\{\bot, \top\}$ the structure of a Boolean algebra.[5] For instance, $\wedge$ and $\vee$ induce two binary operations on $\{\bot, \top\}$, that I denote $\wedge$ and $\vee$ again, defined by the tables below.

$\wedge$	$\bot$	$\top$
$\bot$	$\bot$	$\bot$
$\top$	$\bot$	$\top$

$\vee$	$\bot$	$\top$
$\bot$	$\bot$	$\top$
$\top$	$\top$	$\top$

These tables are determined as follows. The result, say, of combining $\top$ and $\top$ with $\wedge$ is $\top$ because $\vdash (\top \wedge \top) \to \top$ and $\vdash \top \to (\top \wedge \top)$ both hold; that is, because $\top \wedge \top$ and $\top$ are *logically equivalent* formulæ, a notion we shall return to in the next section. The operation induced by implication – which, unlike $\wedge$ and $\vee$, is not commutative – deserves special attention. From its defining table, reproduced below,

$\to$	$\bot$	$\top$
$\bot$	$\top$	$\top$
$\top$	$\bot$	$\top$

---

[5] For a good introduction to the theory of Boolean algebras, see [13].

one recognises that $\to$ actually defines an *order* on $\{\bot, \top\}$. Namely, if one sets $\varphi \leq \psi$ whenever $\vdash \varphi \to \psi$, then $\{\bot, \top\}$ becomes totally ordered as $\bot \leq \top$, $\vdash \bot \to \top$ being an instance of the Scholastic principle that *ex falso sequitur quodlibet*. If we therefore insist that the implicit structure induced by connectives on $\{\bot, \top\}$ be explicitly taken into account, then we *can* make sense of expressions such as $\bot \leq \top$, and thus a choice such as $\{0, 1\}$ for our previous set of truth values $\{\bot, \top\}$ begins to ring more defensible.

The well-known fact that $\{\bot, \top\}$ may be naturally endowed with the structure of a Boolean algebra is the beginning of *algebraic logic*, an extensive subject; for an opinionated introduction, see [16]. There is, however, a further fact about the structure of truth values that is possibly less well known than it would deserve, but has great import in the context of this paper. This is the circumstance that $\{\bot, \top\}$ carries an implicit *additive structure*, too. To appreciate this, consider the set of integers $\mathbb{Z} = \{\ldots, -2, -1, 0, 1, 2, \ldots\}$ endowed with its natural order and addition. The resulting structure $(\mathbb{Z}, \leq, +, 0)$ – with 0 the neutral element for addition – is a specimen of an *ordered Abelian* (alias *commutative*) *group*, another subject with a well-developed theory; *cf. e.g.* [14]. Ordered Abelian groups may be regarded as a modern version of the classical Greek theory of magnitudes expounded in Euclid's Elements, most notably in Books V and X (see Heath's English translation [9]). The characteristic feature of ordered Abelian groups is the relation between the order relation and the group-theoretic addition, known as *invariance under translations*: whenever $x \leq y$ holds, then the translated inequality $z + x \leq z + y$ holds too, for any three elements x, y, z of the group. In certain ordered Abelian groups, one may be able to choose a so-called *order unit*. That is an element u with the properties that (i) u is non-negative – meaning that $u \geq 0$ holds – and (ii) any other element x of the group eventually lies below a multiple of u – meaning that

$$x \leq \underbrace{u + u + \cdots + u}_{n \text{ times}} \tag{A}$$

holds for a sufficiently large natural number n depending on x. Choosing an order unit amounts to making a conventional choice for the *unity magnitude*, which is required to enjoy the Archimedean property (A). Returning to the special case of the integers, let us choose the number 1 as an order unit. We thus have the ordered Abelian group with order unit $(\mathbb{Z}, \leq, +, 0, 1)$; let us consider its *unit interval* $\{x \in \mathbb{Z} \mid 0 \leq x \leq 1\} = \{0, 1\}$. Now, while the addition of the whole group does not restrict to an operation on $\{0, 1\}$, because $1 + 1 = 2 \notin \{0, 1\}$, one can consider the *truncated addition*

$$x \oplus y = \begin{cases} x + y & \text{if } x + y \le 1, \\ 1 & \text{otherwise.} \end{cases} \tag{B}$$

Then $\oplus$ indeed is an operation on $\{0, 1\}$. In fact, writing $\{0, 1\}$ in lieu of $\{\bot, \top\}$, $\oplus$ is just the operation $\vee$ induced on $\{0, 1\}$ by disjunction. Moreover, the operation induced on $\{0, 1\}$ by negation also can be defined in terms of $(\mathbb{Z}, \le, +, 0, 1)$; namely,

$$\neg x = 1 - x, \tag{C}$$

where the subtraction is the inverse to the group operation $+$. In turn, the remaining Boolean operations induced on $\{0, 1\}$ by the logical connectives are determined by $\vee$, $\neg$, and 0 (or 1). Observe from (B–C) that the whole structure of ordered Abelian group with order unit of the set of integers is needed to recover the Boolean structure of $\{0, 1\}$. We thus see that $(\mathbb{Z}, \le, +, 0, 1)$ *uniquely determines the structure of Boolean algebra on its unit interval* $\{0, 1\}$. Technically, this is a triviality; conceptually, it hints at a connection between Boolean algebras and ordered Abelian groups with an order unit. That connection was first exposed by C.C. Chang in [4], where he gave an algebraic proof of the completeness of Łukasiewicz infinite-valued logic based on results from the theory of ordered groups. The crux of the connection is the following rather less trivial converse to the preceding statement: *the structure of Boolean algebra of* $\{0, 1\}$ *uniquely determines the whole of* $(\mathbb{Z}, \le, +, 0, 1)$, *among all ordered Abelian groups with an order unit*. Equivalently, since there is just one two-element Boolean algebra up to isomorphism, this means that the *only* such group whose unit interval consists of precisely two elements is $(\mathbb{Z}, \le, +, 0, 1)$ (to within the appropriate notion of isomorphism of ordered groups with an order unit). This fact is a very special case of the definitive result relating ordered groups to the algebraic analysis of Łukasiewicz logic, established by D. Mundici in [23]; for a textbook treatment, see [5].

The above explains in which sense we may, if we so wish, think of addition of truth values – just complete the Boolean algebra $\{\bot, \top\}$ to its unique enveloping ordered Abelian group with order unit $(\mathbb{Z}, \le, +, 0, 1)$. If we follow this route, then it definitely makes sense to rename $\{\bot, \top\}$ as $\{0, 1\}$.

And still, for the theory of probabilities we need even more. For we want to think of $\{0, 1\}$ as a subset of the real numbers $\mathbb{R}$, not just of the integers. That is essential, because degrees of probabilities range in the real unit interval $[0, 1] \subseteq \mathbb{R}$. But in the theory of ordered groups it is well known how to go from $(\mathbb{Z}, \leq, +, 0, 1)$ to the reals in a canonical fashion; I sketch the construction, omitting details. First, every ordered Abelian group can be completed to its *divisible hull*; in our case, the result is that we canonically embed $(\mathbb{Z}, \leq, +, 0, 1)$ into $(\mathbb{Q}, \leq, +, 0, 1)$, the additive group of rationals with its natural order and addition, with order unit again 1. Next, $(\mathbb{Q}, \leq, +, 0, 1)$ can be completed to $(\mathbb{R}, \leq, +, 0, 1)$, the ordered Abelian group of reals with order unit again 1, by a procedure that is similar to the construction of Cauchy sequences in elementary analysis. For the absolute value on the rationals is a *norm* on $(\mathbb{Q}, \leq, +, 0, 1)$, and the latter group is not complete in this norm – there are Cauchy sequences that admit no limit (because the limit, if it existed, would be an irrational number such as $\sqrt{2}$). Its unique *norm-completion* is then precisely $(\mathbb{R}, \leq, +, 0, 1)$.

Remark 2.2 (The structure of truth values, II). The considerations in this section show that, just as we can regard the set $\{\bot, \top\}$ as a Boolean algebra, we can further regard it as the unit interval of the ordered Abelian group of reals $(\mathbb{R}, \leq, +, 0, 1)$ with order unit 1, in which case it is notationally convenient to switch from $\{\bot, \top\}$ to $\{0, 1\}$. And it is in the latter enlarged setting only that it makes sense to consider the probability degrees of formulæ discussed next.

2.4   The axioms of probability assignments

Let us now move on to probability degrees. Fix a theory Θ. A *probability assignment*[6] (relative to Θ) is a function $P \colon \text{FORM} \to \mathbb{R}$ that satisfies the following axioms, for all $\varphi, \psi \in \text{FORM}$.

(K0) $P(\varphi) = P(\psi)$ whenever $\Theta \vdash \varphi \to \psi$ and $\Theta \vdash \psi \to \varphi$.
(K1) $P(\top) = 1$ and $P(\bot) = 0$.
(K2) $P(\varphi) \leq P(\psi)$ whenever $\Theta \vdash \varphi \to \psi$ holds.
(K3) $P(\varphi) + P(\psi) = P(\varphi \wedge \psi) + P(\varphi \vee \psi)$.

Here, (K0) asserts that the probability of a proposition does not depend on the specific wording used to express that proposition. Observe that on the Fregean account of the meaning of a proposition, two propositions that always have the same truth value independently of the specific circumstances (*i.e.* possible world) under which they are considered, must express the same meaning. Say then that $\varphi, \psi \in \text{FORM}$ are *logically equivalent* (*relative to* Θ) – or logically equivalent *tout court* when Θ

---

[6] Often called a (*finitely additive*) *probability measure*, a terminology I prefer to avoid here.

is empty – if $\Theta \vdash \varphi \to \psi$ and $\Theta \vdash \psi \to \varphi$ hold. By completeness for theories, this amounts to saying that φ and ψ embody the very same meaning in different wordings (assuming the assumptions in Θ hold). Accordingly, (K0) asks that they be assigned the same probability degree. Next, (K1) prescribes that 1 and 0 are the probability degrees of *verum* and *falsum*, respectively; thus, (K0) and (K1) together entail that whenever φ is a synthetic truth (relative to Θ), then $P(\varphi) = 1$, and similarly for falsehood. Axiom (K2) dictates that the *order structure* of probability degrees faithfully mirror the *implicative relations* between propositions. Finally, (K3) dictates that the *additive structure* of probability degrees come from the *implicit additive structure of truth values* discussed in Section 2.3. In particular, (K3) becomes the usual condition of finite additivity if $\varphi \wedge \psi$ and $\bot$ are logically equivalent relative to Θ – which is the proposition-theoretic way of stating that the events denoted by φ and ψ are incompatible. In that case, by (K0) and (K1), (K3) reads $P(\varphi) + P(\psi) = P(\bot) + P(\varphi \vee \psi) = P(\varphi \vee \psi)$, which is Kolmogorov's axiom V in [20] rephrased for propositions. Indeed, it is an exercise to show that (K0–K3) are equivalent to the usual finitely additive version of Kolmogorov's axioms.

Remark 2.3 (Valuation-theoretic axiomatisations of probability). In more detail, (K0–K3) compare to Kolmogorov's axioms I–V in [20] for finitely additive assignments of probabilities as follows. I am defining P as a *lattice-theoretic valuation* – (K0) and (K3) – that is *order-preserving* – (K2) – and *normalised* – (K1). Thus I regard probability assignments as a special case of lattice-theoretic valuations. This is expedient here in that it clearly brings out the rôle of implication. By contrast, Kolmogorov's first five axioms I–V do *not* use the set-theoretic counterpart of implication at all, namely, inclusion.[7] It is worth pointing out in passing that (K0–K3) make sense for any Heyting algebra, the algebraic counterpart of intuitionistic logic, and in particular for any finite distributive lattice – because such lattices admit of a unique Heyting implication adjoint to the lattice meet. In this general context, (K0–K3) form an independent system: a valuation need not be order-preserving, nor normalised; for example, the *Euler characteristic* of polyhedra, which can be reformulated in valuation-theoretic terms [26], is not either. For more on the general theory of valuations on distributive lattices, the interested reader can consult [19].

---

[7] His last axiom VI does. This is Kolmogorov's *continuity axiom*: whenever $A_1 \supseteq A_2 \supseteq \cdots$ is a decreasing sequence of sets (modelling events) such that $\bigcap_{i=1}^{\infty} A_i = \emptyset$, then $\lim_{n \to \infty} P(A_n) = 0$. Together with his I–V, this is equivalent to countable additivity.

Remark 2.4 (Order versus Magnitude). From the foregoing, it may be argued that (K2) and (K3) are assumptions of a different nature. The former arises directly from implications that hold between propositions, possibly the most fundamental logical notion of all. By contrast, (K3) arises from the additive structure of truth values discussed in Section 2.3 – and it is at least debatable whether such a structure has the same fundamental logical status as implicative relations alone. Whatever one's attitude to this issue, though, it should be borne in mind that (K0–K3) are *equivalent* to the standard axiomatisation of finitely additive probability assignments. Thus, it is a fact that *the theory of finitely additive probabilities grows out of classical logic by assuming that truth values carry* both *an order-theoretic* and *an additive structure*, whether one states those assumptions explicitly or not. One way to appreciate this point is as follows. Use $\{0, 1\}$ as the set of truth values, and consider an assignment $w \colon$ FORM $\to \{0, 1\}$ along with a theory $\Theta \subseteq$ FORM such that $w(\varphi) = 1$ for all $\varphi \in \Theta$. Then one checks that setting $P(\varphi) = w(\varphi)$ yields a an assignment of probabilities. Thus:

(I) *Each assignment of truth values is an assignment of probability degrees.*

A converse to this statement holds, too. Let me assume for the sake of simplicity that the language is finite, and that Θ is the empty theory. Consider a probability assignment $P \colon$ FORM$_n \to \mathbb{R}$, for some integer $n \geq 0$.

(II) *If the range of* P *is* $\{0, 1\}$, *then* P *is an assignment of truth values.*

To prove this, set $C_i = L_1^i \wedge \cdots \wedge L_n^i$ ($i = 1, \ldots, 2^n$), where L_j^i is either X_j or $\neg X_j$ ($j = 1, \ldots, n$); the formula C_i is called a *conjunctive clause*, and each L_j^i is a *literal*. It is easily checked that $\vdash C_1 \vee \cdots \vee C_{2^n}$, so that $C_1 \vee \cdots \vee C_{2^n}$ is logically equivalent to $\top$; and that $\vdash \neg(C_i \wedge C_{i'})$ whenever $i \neq i'$, so that $C_i \wedge C_{i'}$ is logically equivalent to $\bot$. From this, iterated use of (K3) shows that $1 = P(\top) = \sum_{i=1}^{2^n} P(C_i)$, so that there must be exactly one clause, say C_{i_0}, satisfying $P(C_{i_0}) = 1$, whereas $P(C_i) = 0$ whenever $i \neq i_0$. Further, it is well known that any formula $\varphi \in$ FORM$_n$ is logically equivalent to one that is in *disjunctive normal form*, *i.e.* of the form $C_{i_1} \vee \cdots \vee C_{i_u}$, for distinct clauses $C_{i_k}, k = 1, \ldots, u$. If $i_k = i_0$ for some k, then using (K3) repeatedly as before we conclude that $P(\varphi) = 1$; otherwise, the same argument shows that $P(\varphi) = 0$. Therefore P coincides with the unique assignment of truth values $w \colon$ FORM$_n \to \{0, 1\}$ that evaluates to 1 the clause C_{i_0}, and to 0 all other clauses.

It is because of (I–II) that the "old philosophical thesis that logic is, in some sense, a limiting case of probability" [8, page 1] is actually warranted – provided, though, we are willing to accept 0 and 1 *both* as truth values *and* as elements of the ordered Abelian group $\mathbb{R}$ with order unit 1, at one and the same time.

Remark 2.5 ("Justifications" of probability). As is the case for all well-established mathematical notions, the axioms for probability assignments did not come out of the blue. Between 1654, the year of the Fermat-Pascal correspondence, and 1933, the year that Kolmogorov's *Grundbegriffe* appeared, a large amount of interesting mathematics in measure and probability theory had become available, with significant contributions by Kolmogorov himself.[8] Thus, from the point of view of the conceptual development of mathematics, there is no question that the axiomatisation was already fully justified in 1933. Nonetheless, from other perspectives it is reasonable and even necessary to ask for a different sort of justification. In particular, one possible way to formalise the epistemic state of a rational agent is to use the (subjective) theory of probabilities. But there are competing or more general theories, such as *e.g.* the Dempster-Shafer theory of degrees of belief [27]. In this context, the case for probabilities – or for any other proposal – is to be made against the backdrop of a given theory of knowledge, or at the very least against a set of assumptions concerning the rational agent; and such assumptions have little to do with the inner development of mathematics. For an account of justifications of probability in this sense, see [24, Chapter 3].

3 Tentative steps towards many-valued probability

3.1 Gödel logic

Let us now embark on an attempt at generalising classical probability assignments to a specific non-classical logic.

Gödel (propositional infinite-valued) logic is the many-valued logic of the minimum triangular norm and its residuum, in the sense of Hájek [15, Section 2.1 and Section 4.2]. As such, it is part of a whole array of many-valued logics, all of which are schematic extensions[9] of one fundamental system called (Hájek's) Basic Logic. The two best-known specimen of such extensions are Gödel logic itself, and *Łukasiewicz logic*, for which

---

[8] For much more on this point see the historical account of the modern theory of probabilities in [32].

[9] In accordance with Section 2.2, 'axiom' (of a logic) means 'axiom schema' throughout. Therefore to formalise inference one only needs deduction rules – in this paper, *modus ponens* only – without substitution.

the standard reference is [5]. Gödel logic has precisely the same syntax as classical logic, recalled in Section 2.2. It has, of course, a different semantics: as all many-valued logics in the sense of [15], Gödel logic is meant to be a formal model for propositions whose truth values come in (more than two) degrees – *contra* Frege [*loc. cit.*]. Stronger still: *In a given possible world, each proposition has as truth value some uniquely determined real number in the interval* $[0, 1] \subseteq \mathbb{R}$. There is a remnant of the Fregean assumption, though, in that there exists both a *largest* truth value – 1 for our former $\top$, the *verum* – and a *smallest* truth value – 0 for our former $\bot$, the *falsum*. Further, the *verum* has the same privileged status as in classical logic: tautologies are formulæ that are *absolutely true*, *i.e.* have truth value 1, in every possible world. Truth-functionality is retained as a valid principle. The specific way in which the truth value of a proposition is computed from those of its constituents determines the nature of the specific many-valued logic under consideration.[10] For Gödel logic, a truth-value assignment to formulæ is defined as a function $w \colon \text{FORM} \to [0, 1]$ subject to the conditions:

1. $w(\top) = 1, w(\bot) = 0$;

2. $w(\varphi \wedge \psi) = \min \{w(\varphi), w(\psi)\}$;

3. $w(\varphi \vee \psi) = \max \{w(\varphi), w(\psi)\}$;

4. $w(\varphi \to \psi) = \begin{cases} 1 & \text{if } w(\varphi) \leq w(\psi), \\ w(\psi) & \text{otherwise.} \end{cases}$

5. $w(\neg\varphi) = \begin{cases} 1 \text{ if } w(\varphi) = 0, \\ 0 \text{ if } w(\varphi) > 0, \end{cases}$

for all $\varphi, \psi \in \text{FORM}$. As mentioned, a tautology is a formula that evaluates to 1 under each assignment. For an appropriate Hilbert-style axiomatisation with *modus ponens* as deduction rule, one obtains completeness for tautologies with respect to this many-valued semantics [15, 4.2.17]. In fact, upon defining theories and syntactic and semantic consequences precisely as in the classical case, the stronger completeness result for theories holds, too.[11]

---

[10] I refrain from discussing here in full the general assumptions underlying Hájek's framework. In particular, I am ignoring *residuation* – the crucial adjointness relation between conjunction and implication that I already mentioned in Section 2.4 in connection with Heyting algebras – which not only underpins Basic Logic, but also the more general framework of *residuated lattices*; please see [12] for details.

[11] The reader is warned that this is certainly not the case for all extensions of Basic Logic: for instance, completeness for theories fails in Łukasiewicz logic by [5, 4.6.6].

Remark 3.1 (Intended semantics do not scale down). It is known [15, page 98] that Gödel logic can also be axiomatised as the schematic extension of the intuitionistic propositional calculus by the axiom $(\varphi \rightarrow \psi) \vee (\psi \rightarrow \varphi)$, called "pre-linearity axiom" in [15]. In fact, speaking informally, Gödel logic is the intersection of (schematic extensions of) intuitionistic logic and (schematic extensions of) Basic Logic, and this lends Gödel logic additional interest. Intuitionistic logic has a standard intended semantics known as the *Brouwer-Heyting-Kolmogorov interpretation*. Briefly, this consists in regarding atomic formulæ as denoting problems, compound formulæ as denoting problems that can be constructively reduced to simpler constituent problems (subformulæ), and formal proofs of a formula as constructive solutions to the problem denoted by the formula. (Here, "constructive" is to be read as "witnessed by an explicitly exhibited algorithmic procedure".) Thus, for instance, $\vdash \varphi \wedge \psi$ holds intuitionistically just in case one can exhibit both a constructive solution to φ and one to ψ; $\vdash \varphi \rightarrow \psi$ holds intuitionistically just in case one can exhibit a constructive procedure to reduce any constructive solution to φ to a constructive solution to ψ; etc. (See the standard reference on constructivism [30, page 9–10 and *passim*] for details.) Now suppose we try and apply this semantics to Gödel logic. Then the characteristic pre-linearity axiom $(\varphi \rightarrow \psi) \vee (\psi \rightarrow \varphi)$ reads: *whichever two problems one cares to choose, it is always possible to exhibit either a constructive procedure that reduces the first to the second, or one that reduces the second to the first.* I find it hard to imagine contexts that would warrant such an assumption. Hence: *the Brouwer-Heyting-Kolmogorov interpretation does not scale down well to schematic extensions.* The point has nothing to do with intuitionistic logic in particular; it is perfectly general. Thus, suppose one has a wonderfully convincing informal semantics for (infinite-valued) Łukasiewicz logic. Does one then have the same for three-valued Łukasiewicz logic, or even for classical, two-valued logic? No; at least, not *a priori*. Of course, *formal* semantics do scale down – *e.g.* Boolean algebras are Heyting algebras with special properties, etc.

Let us write $\vdash_\mathscr{G}$ for the syntactic consequence relation in Gödel logic, and let us define logical equivalence as in the classical case. It turns out that over the finite language $\text{VAR} = \{X_1, \ldots, X_n\}$ there is only a finite number of distinct formulæ, up to logical equivalence. In algebraic language, this amounts to saying that the Lindenbaum algebra of Gödel logic over a finitely many propositional variables is finite, or equivalently, that the variety of Gödel algebras is locally finite – every finitely generated algebra is finite. This was first proved in [17, Theorem 1]. Let us write $\mathscr{G}_n$ for the set of equivalence classes of the logical equivalence relation.

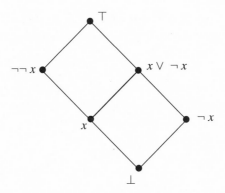

Figure 3.1. All one-variable formulæ in Gödel logic, up to logical equivalence. The order is induced by implication: φ is below ψ if and only if $\varphi \to \psi$ is provable.

Then $\mathcal{G}_n$ can be partially ordered by declaring that

$$[\varphi] \leq [\psi] \text{ if and only if } \vdash_{\mathcal{G}} \varphi \to \psi .$$

This makes $\mathcal{G}_n$ into a (finite) distributive lattice, as is easy to check. Direct computation shows that $|\mathcal{G}_1| = 6$; the Hasse diagram of $\mathcal{G}_1$ is depicted in Figure 3.1, where for convenience X is used in place of X_1. In general, the cardinality $|\mathcal{G}_n|$ can be computed using a recurrence relation first discovered in [17, page 479], and reobtained by different techniques in [6, Section 4.2] and [2, Corollary 1]. The number $|\mathcal{G}_n|$ grows very fast indeed with n: for instance, $|\mathcal{G}_2| = 342$, and $|\mathcal{G}_3| > 10^{11}$. Beyond cardinality, the whole algebraic structure of $\mathcal{G}_n$ is completely understood – see again [2,6] for details.

3.2 The logic of one Gödelian event

Consider the sentence
$$\text{"It rains" .}$$

As is, this sentence is not sufficiently precise to determine a proposition. For instance, when and where is it asserted that there be rain? We may however assume that the sentence is *precisified*[12] in some acceptable manner so as to make it a (classical) proposition. For this, we specify it further in respect of time, place, and so on; say "It rains on February 18, 2010 in the Milanese urban area, etc." Such specifications, along with

---

[12] For a discussion of precisification in the context of vagueness see [29, page 79 and *passim*].

the reader's favourite ones not mentioned here, I shall tacitly assume in the following, whatever they be, provided only they are sharp enough so as to make it clear whether "It rains" is true or false at any given possible world (in the respects under consideration). I will, however, explicitly discuss one further specification. Namely, "It rains" must also be precisified in respect of *quantity of rainfall* – for is a single drop (fallen during February 18, 2010 etc.) to count as rain or not? For this, let us stipulate the following.

(B1) There is an observable feature of rainfall called (*daily*) *precipitation rate* that is measurable in, say, $\frac{\text{mm}}{\text{day}}$ by means of a rain gauge.

(B2) The proposition "It rains" is false (in respect of quantity) if and only if the measured precipitation rate is zero (*i.e.* it lies below the level of precision of the rain gauge we agree to measure precipitation by), and is true otherwise.

(B3) There is another uniquely determined proposition, namely

"It does not rain" ,

that is true precisely when the measured precipitation rate is zero, and is false otherwise.

In the light of (B1–B3) along with our tacit background precisifications, the logic of the single event denoted by "It rains" is very simple. For upon setting[13]

$$X := \text{"It rains"} \quad \text{and} \quad \neg X := \text{"It does not rain"},$$

we see that X and $\neg X$ together *partition* the collection of all events conceivable in our language $\text{VAR}_1 = \{X\}$. Indeed, in any possible world, by (B1–B3) it is either the case that it rains, or that it does not rain (*tertium non datur*), and these two cases cannot occur together (*principium contradictionis*):

$$\vdash (X \vee \neg X) \ , \quad \vdash \neg(X \wedge \neg X) \ . \tag{*}$$

Thus, the distinct formulæ expressible in the language $\{X\}$ are just four, up to logical equivalence in classical logic:

$$\mathscr{B}_1 = \{X, \neg X, \top, \bot\} \ .$$

---

[13] In the remaining part of this paper I write X instead of X_1.

The algebraic structure induced on $\mathscr{B}_1$ by the classical connectives is that of the four-element Boolean algebra – the Boolean algebra freely generated by one generator X, in algebraic language. What probability degrees are we allowed to assign to these (equivalence classes of) formulæ on the basis of (K0–K3)? There is no choice for $\top$ and $\bot$; and one checks that any $P(X), P(\neg X) \in [0, 1]$ will do, provided only

$$P(X) + P(\neg X) = 1 \,. \qquad (**)$$

In short, (**) sums up all there is to say about assigning probability degrees to the single Boolean event "It rains" and its complementary event "It does not rain".

Remark 3.2 (The intrinsic logic of precisified events, I). To repeat the point I am making in the above: *a family of events, as modelled by appropriately precisified sentences, comes endowed with its own intrinsic logic* – provided of course our precisifications are sharp enough to determine it uniquely. In the preceeding example, the logic of the events under consideration happens to be classical logic. In this sense, then, those events are Boolean.

Let me now change the preceding example a little. Consider the sentence

"It rains a lot" .

Question: *How does one precisify the sentence above*? Such a question does not seem to have a single widely accepted answer at present. For the purposes of this paper, let us stipulate the following.

(G4) The proposition "It rains a lot" is true (in respect of quantity) if and only if the measured precipitation rate equals or exceeds $300\frac{\text{mm}}{\text{day}}$, and it is false if and only if the measured precipitation rate is zero.

Along with (G4), we retain assumptions (B1–B3) *verbatim*, renaming them (G1–G3) for convenience.

Claim. The logic of the three propositions "It rains a lot", "It rains", and "It does not rain" as precisified by (G1–G4) is exactly Gödel logic.

To see why the claim is correct, let us reconstruct the Hasse diagram in Figure 3.1 using (G1–G4) only. Set

$X := $ "It rains a lot" , $\neg X := $ "It does not rain" , $\neg\neg X := $ "It rains". $(\star)$

Recall that the second and third proposition in $(\star)$ both behave classically by (G1–G3). Thus the analogues of (*) for $\neg X$ and $\neg\neg X$ hold, namely,

$$\vdash_{\mathscr{G}} (\neg\neg X \vee \neg X) \,, \quad \vdash_{\mathscr{G}} \neg(\neg\neg X \wedge \neg X) \qquad (*')$$

and this is in accordance with the Hasse diagram in Figure 3.1. Now let us analyse the implicative relations between $\neg\neg X$ and X in the light of (G1–G4). First, observe that *whenever it is true that it rains a lot it is certainly true that it does rain*; and indeed,

$$\vdash_{\mathscr{G}} X \to \neg\neg X \,. \tag{†}$$

However, *it may well rain without raining a lot* – precisification: it may well be that the measured precipitation is strictly greater than zero, but strictly less than $300\frac{mm}{day}$. So in case $\neg\neg X$ is true but X is not, *the implication $\neg\neg X \to X$ cannot be true*;[14] and indeed,

$$\nvdash_{\mathscr{G}} \neg\neg X \to X \,, \tag{‡}$$

where $\nvdash_{\mathscr{G}} \varphi$ means that $\vdash_{\mathscr{G}} \varphi$ does not hold, for $\varphi \in \text{FORM}$. Finally, let us consider the proposition

$$X \vee \neg X = \text{"Either it rains a lot, or it does not rain"} \,.$$

Since we already have precisified the truth conditions for the two disjuncts, we can just say, as in the classical case, that $X \vee \neg X$ is true if and only if either X is true, or $\neg X$ is true. Then $X \vee \neg X$ *is not a tautology*: for if the measured precipitation is strictly greater than zero, but strictly less than $300\frac{mm}{day}$ – as in the discussion of (‡) – *neither X nor $\neg X$ are true*. And sure enough,

$$\nvdash_{\mathscr{G}} X \vee \neg X \,.$$

The remaining logical relations needed to fully reconstruct the diagram in Figure 3.1 are dealt with similarly. This settles the claim. □

Remark 3.3 (The intrinsic logic of precisified events, II). The claim above shows that there is a way to precisify "It rains a lot", "It rains", and "It does not rain" so that the resulting propositions obey the laws of Gödel logic. However, there is something unsatisfactory about assumption (G4). Namely, although "It rains a lot" need not be true when "It rains" *simpliciter* is true, (G4) entails that whenever "It rains" is true, "It

---

[14] Observe, however, that the implication is *not false* either. Indeed, by (G4) the proposition "It rains a lot" is *neither true nor false* in case the measured precipitation lies in the open interval $(0, 300)\frac{mm}{day}$; whereas to conclude that $\neg\neg X \to X$ is false in such possible worlds we need to establish that the antecedent is true (which it is) and the consequent is false. It is important to note that (G4) does *not* warrant *any other* conclusion as to the truth value of "It rains a lot" other than the one just pointed out. In particular, statements such as "X is true to a degree" are *void of meaning* on the basis of (G1–G4).

rains a lot" *cannot be false* (without necessarily being true). It may well be argued that this does not adhere to the ordinary usage of the sentences "It rains a lot" and "It rains" by competent speakers. At any rate this is the best approximation to Gödel logic by vague sentences in natural language that I was able to isolate so far.

Remark 3.4 (Classical linguistic readings of non-classical connectives may be misleading). The claim above also shows that one admissible interpretation of $\neg\neg X$ in Gödel logic, given that $X :=$ "It rains a lot", is $\neg\neg X :=$ "It rains". This is hard to reconcile – to say the least – with the classical reading $\neg\varphi :=$ "It is not the case that φ".

Remark 3.5 (Gödel logic can do away with quantitative truth degrees). As the above indicates, it is possible to semantically define the one-variable tautologies of Gödel logic *without using magnitudes at all.* Indeed, it is well known that the semantics of Gödel logic in its entirety only depends on the order structure one imposes on truth degrees; the latter may be considered as elements of a linearly ordered set, with no further structure. Contemplation of conditions (1–5) in Section 3.1 should suffice to convince the reader of this. (More generally, this remark applies to synthetic truths relative to an arbitrary theory.)

3.3 The probability of one Gödelian event

Finally, since I am (provisionally) endorsing the analogies in the Prologue, I am obliged by the claim in the previous section to consider ($\star$) as linguistic descriptions of *Gödelian events* – non-classical events modelled by formulæ in Gödel logic. Question: *How are we to attach probability degrees to such events?* To tackle this, I will throughout make the blanket assumption that *probability degrees for Gödelian events are real numbers, as in the classical case.* While this might eventually turn out to be a false start, I will not discuss the assumption further in this paper. Let us then consider a function

$$P: \text{FORM}_1 \to \mathbb{R},$$

and let us set

$$P(X) = q,\ P(\neg\neg X) = p_1,\ P(\neg X) = p_2.$$

The question above becomes: *What conditions should be imposed on q, p_1, p_2 – more generally, on P – for these real numbers to be considered probability degrees of X, $\neg\neg X$, and $\neg X$, respectively?* A first answer might be: Just ask that the classical axioms (K0–K3) for probability assignments continue to hold. I will presently argue that this answer is not satisfactory. To see this, consider the next

Fact. With the notation above, the function P satisfies (K0–K3) if and only if the relations

(a) $q, p_1, p_2 \in [0, 1]$
(b) $q \leq p_1$
(c) $p_1 + p_2 = 1$

hold.   □

This is easy to prove directly from the definitions; for a general result along these lines, please see [1]. Here, (b) reflects (†), while (c) reflects (*′). But then, setting $p_1' = p_1 - q$, we conclude that $q, p_1, p_2 \in \mathbb{R}$ *satisfy* (a–c) *if and only if* $q, p_1', p_2 \in \mathbb{R}$ *satisfy* $q, p_1', p_2 \in [0, 1]$ *and* $q + p_1' + p_2 = 1$. In other words, the theory of probabilities of the three Gödelian events (*) would be equivalent to the theory of probabilities of the three *Boolean* events

"It rains a lot" , "It rains, but not a lot" , and "It does not rain" ,   (**)

precisified as follows. Call m the measured precipitation rate. The first event in (**) occurs if and only if $m \geq 300\frac{\text{mm}}{\text{day}}$, and does not occur otherwise; the second event occurs if and only if $m \in (0, 300)\frac{\text{mm}}{\text{day}}$, and does not occur otherwise; the third event occurs if and only if $m = 0\frac{\text{mm}}{\text{day}}$, and does not occur otherwise. In other words, q, p_1', and p_2 are just the probability degrees of three exhaustive, mutually exclusive classical events.

Unfortunately, this conclusion lends itself to a definitive objection. Since, as we have just seen, under (K0–K3) the theory of probabilities over Gödel logic boils down to the classical theory,[15] why use a non-classical logic at all in the first place?

At this point, we could just conclude that the theory of probabilities of Gödelian events is uninteresting. However, reflection shows that there is at least a second possibility. For here is something important that we have overlooked: in Gödel logic over the language $\text{VAR}_1 = \{X\}$, if $X :=$"It rains a lot", then *there is no formula that corresponds to the event* "It rains, but not a lot" – *one simply does not have the linguistic resources to utter a description of that event.* (The contents of Section 3.1 hopefully suffice to convince the reader of this.) Therefore, the conclusion that the

---

[15] In the more general case of a finite language, this follows from the results in [1].

theory of probabilities over the Gödelian event $X =$ "It rains" coincides with that of the three Boolean events in ($\star\star$) is palpably at variance with what we know about Gödel logic. It seems that something went wrong.

In the light of (G1–G4), the proposition "It rains a lot" is neither true nor false when the measured precipitation rate lies in $(0, 300) \frac{mm}{day}$. In order to show that the logic of the three propositions ($\star$) is Gödel logic, *we need assume no more*. But just like the transition from classical logic to the theory of probabilities calls for the additional assumption that truth values are additive – *cf.* Section 2.3 – here it may be reasonable, or even necessary, to assume more than is strictly needed to obtain the completeness theorem for Gödel logic. For the sake of fostering our would-be theory of probabilities I propose such an additional assumption.

(G5) The proposition "It rains a lot" is true *to some degree*[16] (but is neither absolutely true nor absolutely false) whenever the measured precipitation rate lies in the open interval $(0, 300) \frac{mm}{day}$.

It is clear that (G5) is compatible with (G1–G4). And albeit (G5) is not needed to establish the claim in the previous section that the logic of "It rains" is Gödel logic, it *does* have import when probabilities come into play, as I now argue. What constraints does a rational agent have in assigning a probability degree to ($\star$) under (G1–G5)? Suppose the rational agent assigns probability zero to the Gödelian event "It rains a lot". Can she then assign a non-zero probability degree to "It rains"? The answer is no: since, *whenever* it rains, it *does* rain a lot *to some degree* by (G5), banning altogether the possibility that it rains a lot also implies banning altogether the possibility that it rains at all. On the other hand, if under (G1–G5) the agent assigns a non-zero probability degree to "It rains a lot", then there is no constraint at all as to the remaining degrees, except (a–c) above which come as before from (G1–G4).

The discussion above may be summarised as a

Tentative definition. Assume (G1–G5). The numbers $q, p_1, p_2 \in \mathbb{R}$ are admissible degrees of probabilities for X, $\neg\neg X$, and $\neg X$ if and only if the relations (a–c) as in the above, along with the additional characteristic condition

(d) $q = 0$ implies $p_1 = 0$,

hold.   □

---

[16] Here one could elaborate on how to represent such a degree in quantitative terms, but I will not need to do so for my present purposes.

More generally, when the language VAR_n is finite this tentative definition can be formalised by an appropriate axiom[17] (K4) to be added to (K0–K3). Whether this can be regarded as an appropriate axiomatisation of probability assignments to formulæ in Gödel logic remains to be seen.

4 Epilogue

If mathematicians are analogy-makers, the duty befalls them to assess when their analogies are fruitful, and when barren. The former case may sometimes be recognised because it eventually leads to the proof of a few beautiful theorems; the latter case may often be recognised because it soon leads to the proof of very many uninteresting theorems.

To me, it is clear that we are not yet in a position to safely assess whether the analogies upon which a probability theory of many-valued events is based be fruitful or barren; further research is needed. I do hope to have convinced the reader, though, that this is research worth pursuing – independently of the final verdict.

ACKNOWLEDGEMENTS. I am indebted to S. Aguzzoli, C. Fermüller, M. Giaquinta, H. Hosni, F. Montagna, M. Mugnai, and D. Mundici for discussions and suggestions concerning the contents of this paper.

References

[1] S. Aguzzoli, B. Gerla and V. Marra, *De Finetti's no-Dutch-book criterion for Gödel logic*, Studia Logica **90** (2008), no. 1, 25–41.

[2] S. Aguzzoli, B. Gerla and V. Marra, *Gödel algebras free over finite distributive lattices*, Ann. Pure Appl. Logic **155** (2008), no. 3, 183–193.

[3] G. Boole, "An Investigation of The Laws of Thought", Walton and Maberly, London, 1854.

[4] C. C. Chang, *A new proof of the completeness of the Łukasiewicz axioms*, Trans. Amer. Math. Soc. **93** (1959), 74–80.

---

[17] For the sake of completeness, I spell out (K4). First, let $L(\Theta)$ denote the distributive lattice of equivalence classes of formulæ FORM_n, under logical equivalence relative to the theory Θ. Then (K4) reads: *Suppose each one of $x, y, z \in L(\Theta)$ are either the minimum of $L(\Theta)$ or a join-irreducible element. Suppose further that x is covered by y, and y is covered by z. Then whenever* $\mathrm{P}(x) = \mathrm{P}(y)$, *then* $\mathrm{P}(y) = \mathrm{P}(z)$. (I recall that a join-irreducible element of a distributive lattice is one that is not the minimum, and cannot be written non-trivially as the join of two elements; and that two elements x, y are such that x is covered by y if $x < y$ and there is no element properly between x and y.)

[5] R. L. O. CIGNOLI, ITALA M. L. D'OTTAVIANO and D. MUNDICI, "Algebraic Foundations of Many-Valued Reasoning", Trends in Logic—Studia Logica Library, Vol. 7, Kluwer Academic Publishers, Dordrecht, 2000.

[6] O. M. D'ANTONA and V. MARRA, *Computing coproducts of finitely presented Gödel algebras*, Ann. Pure Appl. Logic **142** (2006), no. 1-3, 202–211.

[7] A. DE MOIVRE, "The Doctrine of Chances", W. Pearson, London, 1718.

[8] D. P. ELLERMAN and G.-C. ROTA, *A measure-theoretic approach to logical quantification*, Rend. Sem. Mat. Univ. Padova **59** (1978), 227–246 (1979).

[9] EUCLID, "The thirteen books of Euclid's Elements translated from the text of Heiberg", Vol. I: Introduction and Books I, II. Vol. II: Books III–IX. Vol. III: Books X–XIII and Appendix, Dover Publications Inc., New York, 1956, Translated with introduction and commentary by Thomas L. Heath, 2nd ed.

[10] P. FERMAT, "Œuvres de Fermat", Vol. II, Gauthier-Villars, Paris, 1894, ed. by P. Tannery and C. Henry.

[11] G. FREGE, *Über Sinn und Bedeutung*, Zeitschrift für Philosophie und philosophische Kritik **100** (1892), 25–50, English translation as 'On Sense and Reference' by Max Black. In P.T. Geach and M. Black (eds.), "Translations form the Philosophical Writings of G. Frege", Blackwell, Oxford, 2nd ed., 1960.

[12] N. GALATOS, P. JIPSEN, T. KOWALSKI and H. ONO, "Residuated Lattices: an Algebraic Glimpse at Substructural Logics", Studies in Logic and the Foundations of Mathematics, Vol. 151, Elsevier B. V., Amsterdam, 2007.

[13] S. GIVANT and P. HALMOS, "Introduction to Boolean Algebras", Undergraduate Texts in Mathematics, Springer, New York, 2009.

[14] A. M. W. GLASS, "Partially Ordered Groups", Series in Algebra, Vol. 7, World Scientific Publishing Co. Inc., River Edge, NJ, 1999.

[15] P. HÁJEK, "Metamathematics of Fuzzy Logic", Trends in Logic—Studia Logica Library, Vol. 4, Kluwer Academic Publishers, Dordrecht, 1998.

[16] P. HALMOS and S. GIVANT, "Logic as Algebra", The Dolciani Mathematical Expositions, Vol. 21, Mathematical Association of America, Washington, DC, 1998.

[17] A. HORN, *Free L-algebras*, J. Symbolic Logic **34** (1969), 475–480.

[18] J. M. KEYNES, "A Treatise on Probability", Macmillan, London, 1921.

[19] D. A. KLAIN and G.-C. ROTA, "Introduction to Geometric Probability", Lezioni Lincee. [Lincei Lectures], Cambridge University Press, Cambridge, 1997.

[20] A. KOLMOGOROV, "Grundbegriffe der Wahrscheinlichkeitsrechnung", Springer-Verlag, Berlin, 1933.

[21] D. K. LEWIS, "On the Plurality of Worlds", Blackwell, Oxford, 1986.

[22] M. MCGRATH, *Propositions*, The Stanford Encyclopedia of Philosophy (Edward N. Zalta, ed.), Fall 2008 ed., 2008, Available on-line at http://plato.stanford.edu/archives/fall2008/entries/propositions.

[23] D. MUNDICI, *Interpretation of AF C^*-algebras in Łukasiewicz sentential calculus*, J. Funct. Anal. **65** (1986), no. 1, 15–63.

[24] J. B. PARIS, "The Uncertain Reasoner's Companion", Cambridge Tracts in Theoretical Computer Science, Vol. 39, Cambridge University Press, Cambridge, 1994.

[25] A. RÉNYI, *Foundations of probability*, Holden-Day Inc., San Francisco, Calif., 1970.

[26] G.-C. ROTA, "On the Combinatorics of the Euler Characteristic", Studies in Pure Mathematics (Presented to Richard Rado), Academic Press, London, 1971, pp. 221–233.

[27] G. SHAFER, "A Mathematical Theory of Evidence", Princeton University Press, Princeton, N.J., 1976.

[28] D. E. SMITH, "A Source Book in Mathematics", Dover Publications Inc., New York, 1959.

[29] N. J. J. SMITH, "Vagueness and Degrees of Truth", Oxford University Press, Oxford, 2008.

[30] A. S. TROELSTRA and D. VAN DALEN, "Constructivism in Mathematics", Vol. I, "Studies in Logic and the Foundations of Mathematics", Vol. 121, North-Holland Publishing Co., Amsterdam, 1988.

[31] S. M. ULAM, "Analogies between Analogies", Los Alamos Series in Basic and Applied Sciences, Vol. 10, University of California Press, Berkeley, CA, 1990.

[32] J. VON PLATO, "Creating Modern Probability", Cambridge Studies in Probability, Induction, and Decision Theory, Cambridge University Press, Cambridge, 1994.

3

UNCERTAINTY

On Giles style dialogue games and hypersequent systems

Christian G. Fermüller

1 Introduction

Modern formal logic deals with an almost unsurmountably vast realm of so-called non-classical logics. Even if we focus attention on propositional logics that deal with the usual logical connectives—conjunction, disjunction, (material) implication, and negation—it is hard even just to classify the many ways in which such logics can deviate from classical logic. There is intuitionistic logic with its well known reference to constructive reasoning; substructural logics, taking their departure from Gentzen's proof theoretic analysis of logical consequence; many-valued logics, seeking to generalize truth functionality from two to more 'truth values'; relevance logics; paraconsistent logics; and many more families of logics, that have been suggested as alternative models of classical reasoning. Here, we will be interested in certain many-valued logics. However, it is not so much this group of logics in itself that we are interested in, but rather a particular approach to formal models of logical reasoning that we want to highlight. This approach is based on a strategic game, where two antagonistic dialogue partners stepwise reduce arguments about logically complex statements to arguments about atomic formulas. The main point will be not only to explain how such a game leads to characterizations of certain logics, but also how a systematic search for winning strategies connects the game to so-called hypersequent calculi and thus to Gentzen type proof theory.

Most of the technical results mentioned below are not new, and those that are new amount to rather straightforward observations. It is not our aim to present a new mathematical concept or technical proofs. Rather, this paper can be seen as a kind of survey of results on the connection between Giles's game for Łukasiewicz logic and hypersequent based proof theory, as mainly presented in [14]. However we attempt to present those results by taking a somewhat different route: one, that emphasizes principles that go beyond the characterization of (just) Łukasiewicz logic and that might help to understand what roles can or should be played by dialogue games and hypersequent calculi, respectively. We hope that the emerging picture is attractive also to those, who don't share our interest in the proof theory of t-norm based fuzzy logics.

2   Syntactic presentations of logics: the good, the bad, and the ugly

Quite obviously, there is not only a plethora of non-classical logics, but also a number of different ways of presenting or specifying a logic. In fact, the question of appropriate presentations of logics can hardly be separated from the seemingly more fundamental question 'what is a logic?' For some purposes it is useful to call any set of propositional formulas that is closed under substitution and modus ponens a logic. In addition one might reasonably demand that it is effectively or even efficiently checkable whether a formula belongs to the logic in question. In such contexts a (Frege-)Hilbert style system, *i.e.*, a recursive list of axioms and simple inference rules, might be the obvious way to specify a logic.

But clearly, for many other purposes, we have good reasons to insist that a logic has to be specified *semantically* via an appropriate definition of validity or consequence. The question, whether a corresponding proof system of a particular type exists might then be secondary. A further alternative approach to logics, so-called 'proof theoretic semantics' [24], maintains that the meaning of logical connectives is specified by introduction and elimination rules of a system of natural deduction, or equivalently, the logical rules of a cut-free sequent system. Below, we will describe yet another approach to logics, that is sometimes traced back to antiquity, but in its modern form goes back to Paul Lorenzen's work, starting in the late 1950s: dialogue game semantics, where a formula F is considered valid if a proponent has a strategy to defend her assertion that F against all admissible attacks of an opponent according to rules that are determined by the logical connectives occurring in F.

Suppose I suggest to investigate the propositional logic that is given by the Hilbert style system with modus ponens and substitution as the only rules of inference, where the axioms are the following:

$$H1: ((A \supset B) \supset C) \supset (((B \supset A) \supset C) \supset C)$$
$$H2: (A \supset B) \supset ((B \supset C) \supset (A \supset C))$$
$$H3: \bot \supset A$$
$$H4: ((A \supset \bot) \supset \bot) \supset A$$
$$H5: (A \& B) \supset A$$
$$H6: (A \& B) \supset (B \& A)$$
$$H7: (A \& (A \supset B)) \supset (B \& (B \supset A))$$
$$H8: ((A \& B) \supset C) \supset (A \supset (B \supset C))$$
$$H9: (A \supset (B \supset C)) \supset ((A \& B) \supset C)$$

Although the resulting logic is indeed in the center of interest in the investigations below, it is clear that the suggested presentation of a logic is deeply unsatisfying. Of course, one might start to look for familiar schemes in the above list of axioms and relate the suggested system to similar ones, known from the literature. In particular, researchers familiar with t-norm based fuzzy logics as presented in the important monograph [22] will hardly have troubles to recognize that formulas H1–H3 and H5–H9 are actually the axioms that have been introduced by Petr Hájek for his *Basic Logic* **BL**. The remaining axiom H4 characterizes infinite-valued *Lukasiewicz Logic* $\mathbf{L}_\infty$ if added to **BL**. Therefore we could have presented the very same logic by a quite different list of axioms, also well known from the literature, namely

Ł1: $A \supset (B \supset A)$
Ł2: $(A \supset B) \supset ((B \supset C) \supset (A \supset C))$
Ł3: $(\neg A \supset \neg B) \supset (B \supset A)$
Ł4: $((A \supset B) \supset B) \supset ((B \supset A) \supset A)$.

Since Lukasiewicz's Ł1–Ł4 do not mention $\bot$ or & we should actually declare them defined symbols, *e.g.*, by stipulating $\bot =_{def} \neg(A \supset A)$ and $A \& B =_{def} \neg(A \supset \neg B)$. Moreover, since the system H1–H9 does not mention negation ($\neg$) we should add $\neg A =_{def} A \supset \bot$ there. The standard way to turn such a system for $\mathbf{L}_\infty$ into a system for n-valued Lukasiewicz logic $\mathbf{L}_n$ for $n \geq 2$ is to add a bunch of rather complex formulas as further axioms (see [20]). In any case, it should be clear that Hilbert style systems hardly amount to very informative presentations of logics if viewed in isolation.

Of course, Hájek's re-representation of Lukasiewicz logic as **BL** + $\neg\neg A \supset A$ is highly informative *relative* to the main result about **BL**: the set of formulas derivable in the mentioned system for **BL** coincides with the set of formulas that evaluate to 1 and under all assignments of values in [0, 1] to propositional variables if & is interpreted as any continuous t-norm and $\supset$ as its corresponding residuum [8]. Lukasiewicz logic $\mathbf{L}_\infty$ emerges as the logic based on the particular t-norm $x * y = \max(0, x + y - 1)$ and the corresponding residuum $x \Rightarrow_* y = \min(1, 1 - x + y)$ for all $x, y \in [0, 1]$. Most (mathematical) logicians will furthermore consider the following move an obvious one to make: instead of investigating the mentioned particular structure over the real closed unit interval [0, 1] one can isolate the properties needed to characterize $\mathbf{L}_\infty$, resulting in the important concept of an MV algebra. Analogously, **BL** is characterized by corresponding BL algebras.

The above considerations seem to suggest the following conclusion: a fully satisfying presentation of a propositional logic has to move away from *syntax*, *i.e.* from proof systems, and engage with *semantics*, which, in the standard approach, means *algebraic* characterizations. Indeed, soundness and completeness proofs that relate syntax and semantics in this sense are at the core of modern logic. However this view neglects a strong proof theoretic tradition in logic. Gentzen's remarkable and still central characterization of the relation between classical and intuionistic logic in terms of sequent and natural deduction systems [15] preceded Tarski style semantics. In any case it is completely independent of semantic characterizations in the (now) usual sense. Gentzen established the soundness and completeness of the sequent calculi **LK** and **LI** [1] relative to axiom systems for classical and intuitionistic logic that had been presented earlier by Hilbert and Heyting, respectively. For further reference, let us recall a suitable version of these two calculi, where the 'sequent arrow' ⊢ separates multi-sets of formulas. As usual we write A, Γ instead of $\{A\} \cup \Gamma$, etc.

Initial sequents of **LI** *and* **LK**:

$$\frac{}{A \vdash A} \ (ID) \qquad \frac{}{\bot \vdash} \ (\bot, l)$$

Structural rules of **LK**:

$$\frac{\Gamma \vdash \Delta}{A, \Gamma \vdash \Delta} \ (w, l) \qquad \frac{\Gamma \vdash \Delta}{\Gamma \vdash \Delta, A} \ (w, r)$$

$$\frac{A, A, \Gamma \vdash \Delta}{A, \Gamma \vdash \Delta} \ (c, l) \qquad \frac{\Gamma \vdash \Delta, A, A}{\Gamma \vdash \Delta, A} \ (c, r)$$

$$\frac{\Gamma_1 \vdash \Delta_1, A \qquad A, \Gamma_2 \vdash \Delta_2}{\Gamma_1, \Gamma_2 \vdash \Delta_1, \Delta_2} \ (cut)$$

Logical rules of **LK**:

$$\frac{\Gamma \vdash \Delta, A \qquad B, \Gamma \vdash \Delta}{A \supset B, \Gamma \vdash \Delta} \ (\supset, l) \qquad \frac{A, \Gamma \vdash \Delta, B}{\Gamma \vdash \Delta, A \supset B} \ (\supset, r)$$

$$\frac{A, \Gamma \vdash \Delta}{A \wedge B, \Gamma \vdash \Delta} \ (\wedge, l_1) \quad \frac{B, \Gamma \vdash \Delta}{A \wedge B, \Gamma \vdash \Delta} \ (\wedge, l_2) \quad \frac{\Gamma \vdash \Delta, A \qquad \Gamma \vdash \Delta, B}{\Gamma \vdash \Delta, A \wedge B} \ (\wedge, r)$$

$$\frac{A, \Gamma \vdash \Delta \qquad B, \Gamma \vdash \Delta}{A \vee B, \Gamma \vdash \Delta} \ (\vee, l) \quad \frac{\Gamma \vdash \Delta, A}{\Gamma \vdash \Delta, A \vee B} \ (\vee, r_1) \quad \frac{\Gamma \vdash \Delta, B}{\Gamma \vdash \Delta, A \vee B} \ (\vee, r_1)$$

---

[1] Gentzen's **LI** is often referred to as **LJ**. However this seem to rest on a typographic confusion: while the corresponding '**I**' in Gentzen's original paper indeed looks like a '**J**' in particular to eyes not familiar with certain typographic traditions in Germany, Gentzen's text clarifies that **LI** and **LK** are short for 'Logistischer **I**ntuitionistischer Kalkül' and 'Logistischer **K**lassischer Kalkül', respectively.

Rules for negation can be omitted if $\neg A$ is considered an abbreviation for $A \supset \perp$. As stated, the above rules are those of the classical calculus **LK**. The rules for **LI** are obtained from those of **LK** by constraining the right hand sides of all sequents to contain at most one formula. Thus rule (c, r) disappears, and Δ everywhere else has to be either empty or has to consist of at most one formula, accordingly.

Central to any Gentzen style characterization of a logic is the proof that applications of (cut) can be eliminated from all derivations. The resulting system is *analytic* in the sense that proofs only mention subformulas of the formulas occurring in the proven end-sequent. Gentzen actually considered his sequent systems auxiliary to natural deduction systems, where cuts correspond to redundancies (detours) that emerge when connectives are eliminated right after having been introduced in derivations. Intuitionistic natural deductions can be viewed as λ-terms according to the celebrated Curry-Howard isomorphism [21], which in turn is the basis for computational interpretations.

We do not have to join Jean-Yves Girard's well known polemic against the algebraic tradition in non-classical logic as merely replacing connectives by 'Broccoli' [19] to insist that it is not always clear whether we gain a better understanding of logical *reasoning* by providing algebraic semantics for a logical system. Indeed, the above mentioned computational interpretations of intuitionistic logic based on the Curry-Howard isomorphism, but also considerations associated with the concept of 'proof theoretic semantics' [24] show that certain forms of syntactic presentations of logics may be a key ingredient of formal models of logical reasoning.

Taking up the challenge to provide an analytic, Gentzen style proof system for Łukasiewicz logic $\mathbf{L}_\infty$ one quickly realizes that it seems impossible to turn the classical sequent system **LK** into a system for $\mathbf{L}_\infty$ as simply and elegantly as in the transition from **LK** to **LI**. Indeed, for many important non-classical logics it seems impossible to provide 'natural' analytic systems without deviating from Gentzen's original calculi in rather drastic ways. The state of the art concerning analytic systems for Łukasiewicz logic is nicely presented in the monograph [26] on the proof theory of fuzzy logics. The calculus for $\mathbf{L}_\infty$ that arguably comes closest to the spirit of Gentzen's presentation [2] of classical and intuitionistic

---

[2] In [26] one can also find a *sequent* system for $\mathbf{L}_\infty$. However, leaving aside the hypersequent format, the logical rules of that sequent system arguably deviate from Gentzen's rules even more drastically than those of **GL**. In particular, the central rule for implication does not respect the subformula principle. Thus, in contrast to **GL** one cannot reduce all compound formulas to less complex ones by applying the logical sequent rules backwards.

logic is the following *hypersequent system* **GL**.

Initial hypersequents of **GL**:

$$\overline{A \vdash A} \; (ID) \qquad \overline{\vdash} \; (EMP) \qquad \overline{\perp \vdash} \; (\perp, l)$$

Structural rules of **GL**:

$$\frac{\mathcal{H}}{\Gamma \vdash \Delta \mid \mathcal{H}} \; (EW) \qquad \frac{\Gamma \vdash \Delta \mid \Gamma \vdash \Delta \mid \mathcal{H}}{\Gamma \vdash \Delta \mid \mathcal{H}} \; (EC) \qquad \frac{\Gamma \vdash \Delta \mid \mathcal{H}}{A, \Gamma \vdash \Delta \mid \mathcal{H}} \; (IW)$$

$$\frac{\Gamma_1, \Gamma_2 \vdash \Delta_1, \Delta_2 \mid \mathcal{H}}{\Gamma_1 \vdash \Delta_2 \mid \Gamma_2 \vdash \Delta_1 \mid \mathcal{H}} \; (SPLIT) \qquad \frac{\Gamma_1 \vdash \Delta_1 \mid \mathcal{H} \qquad \Gamma_2 \vdash \Delta_2 \mid \mathcal{H}}{\Gamma_1, \Gamma_2 \vdash \Delta_1, \Delta_2 \mid \mathcal{H}} \; (MIX)$$

Logical rules of **GL**:

$$\frac{B, \Gamma \vdash \Delta, A \mid \mathcal{H}}{A \supset B, \Gamma \vdash \Delta \mid \mathcal{H}} \; (\supset, l) \qquad \frac{A, \Gamma \vdash \Delta, B \mid \mathcal{H} \qquad \Gamma \vdash \Delta \mid \mathcal{H}}{\Gamma \vdash \Delta, A \supset B \mid \mathcal{H}} \; (\supset, r)$$

$$\frac{A, \Gamma \vdash \Delta \mid \mathcal{H} \qquad \Gamma \vdash B, \Delta \mid \mathcal{H}}{A \vee B, \Gamma \vdash \Delta \mid \mathcal{H}} \; (\vee, l) \qquad \frac{\Gamma \vdash \Delta, A \mid \Gamma \vdash \Delta, B \mid \mathcal{H}}{\Gamma \vdash \Delta, A \vee B \mid \mathcal{H}} \; (\vee, r)$$

$$\frac{A, \Gamma \vdash \Delta \mid B, \Gamma \vdash \Delta \mid \mathcal{H}}{A \wedge B, \Gamma \vdash \Delta \mid \mathcal{H}} \; (\wedge, l) \qquad \frac{\Gamma \vdash \Delta, A \mid \mathcal{H} \qquad \Gamma \vdash \Delta, B \mid \mathcal{H}}{\Gamma \vdash \Delta, A \wedge B \mid \mathcal{H}} \; (\wedge, r)$$

$$\frac{A, B, \Gamma \vdash \Delta \mid \mathcal{H} \qquad \perp, \Gamma \vdash \Delta \mid \mathcal{H}}{A \& B, \Gamma \vdash \Delta \mid \mathcal{H}} \; (\&, l) \qquad \frac{\Gamma \vdash \Delta, A, B \mid \Gamma \vdash \Delta, \perp \mid \mathcal{H}}{\Gamma \vdash \Delta, A \& B \mid \mathcal{H}} \; (\&, r).$$

If, referring to the title of this section, Gentzen's **LK** and **LI** play the role of 'the good' and Hilbert style presentations serve as 'the bad', then **GL** seems to fit the role of 'the ugly'. Indeed, not only the hypersequent format, but also some of the logical rules will look strange to anyone familiar only with sequent systems. But at least some of the pressure to motivate the form of **GL** can be released right away.

The generalization of sequents to hypersequents goes back to [29] and [2] and has proved to be very useful to provide analytic systems for a wide range of non-classical logics (see, *e.g.*, [4,6,11,26]). Hypersequents are finite multisets of ordinary sequents that are interpreted disjunctively: the derivability of

$$\Gamma_1 \vdash \Delta_1 \mid \ldots \mid \Gamma_n \vdash \Delta_n$$

is intended to correspond to the claim that every interpretation validates at least one of the component sequents $\Gamma_i \vdash \Delta_i$, where $1 \le i \le n$. The symbol $\mathcal{H}$ in the above rules stands for a (possibly empty) multiset of further sequents, called the side-hypersequent of the rule. The logical rules of a hypersequent system do not depend on this context. For example, an elegant hypersequent characterization of Gödel-Dummett logic is given in [3], where the logical rules are exactly the same as in Gentzen's **LI**, except for the presence of side-hypersequents. The distinguishing feature of hypersequent systems are the (external) structural rules. Taking up the present example of **GL**, observe that external weakening (EW) and external contraction (EC) correspond to the fact that the symbol '|', separating the component sequents, is to be read as disjunction at the meta-level. Internal weakening (IW) is just ordinary left weakening at the sequent level. These rules alone do not add to the power of underlying sequent systems. However rules like ($SPLIT$) and (MIX) allow for the exchange of formulas between different components of hypersequents. They are essential in increasing the expressive power beyond standard sequent systems.

Note that we have not included a cut rule among the structural rules of **GL**. While we could have formulated such a rule, it is redundant, just as in **LK** and **LI**. The resulting system is indeed analytic: if we ignore the additional appearance of $\perp$ in the rules for strong conjunction (&) the rules respect the subformula property. ($\perp$ can be viewed as implicitly contained in strong conjunctions as subformula. In any case, it obviously does not spoil the fact that applying the logical rules of **GL** backwards always leads to atomic hypersequents in a finite number of steps.) The initial (hyper)sequents (ID) and ($\perp, l$) are already present in **LK** and in **LI**. In fact, the presentation of **GL** in [26] slightly differs from the above one in using $A \vdash A \mid \mathcal{H}$, $\vdash \mid \mathcal{H}$, and $\perp, \Gamma \vdash A \mid \mathcal{H}$ instead of our versions of (ID), (EMP), and ($\perp, l$). However, these simpler variants are easily shown to be equivalent to the original ones in the context of the other rules. Likewise we find it convenient to replace rule ($\supset, l$) by its variant

$$\frac{B, \Gamma \vdash \Delta, A \mid \mathcal{H}}{A \supset B, \Gamma \vdash \Delta \mid \Gamma \vdash \Delta \mid \mathcal{H}} \; (\supset, l)'.$$

Again, one can easily show that ($\supset, l$) and ($\supset, l$)' are equivalent in presence of the structural rules. We will still use **GL** to refer to the resulting system.

While the above remarks should help to understand the general format of the rules of **GL** one cannot deny that those rules still remain difficult to interpret in relation to Gentzen's sequent calculi. Note that the empty sequent '$\vdash$' invariably denotes inconsistency in standard sequent systems while **GL** declares it an initial sequent (axiom). Similarly, some of the logical rules of **GL** seem not to make sense if (component) sequents are interpreted analogously to classical or intuitionistic logic. Just pointing out that **GL** is sound and complete for $\mathbf{L}_\infty$, in the sense that $\vdash F$ is derivable iff F is valid in $\mathbf{L}_\infty$, is clearly not sufficient to maintain that the system represents logical reasoning according to first principles about fuzziness, vagueness, or partial truth. Yet our aim is to guide the reader to a satisfying interpretation of **GL**, based on a few rather general and basic principles from which its rules can be derived in a uniform manner. For this purpose we propose a detour through the presumably less familiar landscape of logical dialogue games.

3   Lorenzen's dialogue foundation of logic

Contemporary research usually draws a twofold picture of formal logics: on the one hand there is syntax (proof theory), on the other hand we have semantics (model theory). However, already in the late 1950s Paul Lorenzen [25] had introduced an analysis of logical reasoning that can be viewed as independent foundation of formal logic, as made clear in the following quote by Johan van Benthem in [23]:

> According to Lorenzen, valid arguments are those patterns from premises to conclusions in which the proponent of the conclusion has a winning strategy against any opponent granting the premises. Thus, there is a third independent pragmatic intuition of logical validity, based on viewing argumentation as a game. I have been converted to that view ever since, even though most of my professional life has been under camouflage as a model theorist, or occasionally a proof theorist.

Lorenzen [25] suggested to consider logical argumentation as a dialogue game where a proponent **P** tries to defend a logically complex sentence in face of attacks by an opponent **O** that refer to the logical connectives involved in the original statement. While Lorenzen included quantification, we will restrict attention to the propositional level here. The following table summarizes Lorenzen's rules for conjunction, disjunction, and implication. (If **X** is the proponent **P** then **Y** refers to the opponent **O**, and vice versa.)

Logical dialogue rules:

X:	attack by **Y**	defense by **X**
$A \wedge B$	l? or r? (**Y** chooses)	A or B, accordingly
$A \vee B$	?	A or B (**X** chooses)
$A \supset B$	A	B

Obviously, 'l?' and 'r?' refer to the left and right subformula of $A \wedge B$, respectively, while '?' simply denotes an attack on a disjunction. Note that attacking an implication involves the assertion of a sentence, which in turn can be attacked by the other player. In other words, the presence of implication entails that the roles of attacker and defender may switch. Negation is, again, defined as implication of *absurdum* ($\neg A =_{def} A \supset \bot$), where $\bot$ is an indefensible atomic statement.

A dialogue is a sequence of moves, which are either attacking or defending, in accordance with the presented logical rules. Each dialogue refers to a finite multiset of formulas that are initially granted by **O**, and to an initial formula to be defended by **P**. **P** wins the game when **O** has asserted *absurdum* or when **O** has already asserted the formula that **P** currently has to defend. The latter winning condition is aptly called *ipse dixisti* ('you said it yourself') by Lorenzen.

The logical rules and winning conditions alone do not yet determine a logic. We also need *structural rules* (*Rahmenregeln* in the diction of Lorenzen and his school) that regulate the succession of moves. Different versions of such rules can be found in the literature. Subtle changes can lead to different logics. Lorenzen aimed at a characterization of *constructive* reasoning, as opposed to classical logic or any other logic different from intuitionistic logic, which resulted in a number of attempts to justify a particular version of such rules from elementary pragmatic principles about reasoning. With hindsight, these attempts were not very successful. A more relaxed view on structural rules and a corresponding open-mindedness concerning alternative logics as candidates for formal models of reasoning allows to characterize and better understand the relation between a number of different logics as we will also see below, when looking at Łukasiewicz logic $\mathbf{L}_\infty$ and some related logics.

The following set of structural rules, due to [9, 10], turns out to be adequate for intuionistic logic.

Structural rules for E-dialogues:

    Start: The first move of the dialogue is carried out by **O** and consists in an attack on the initial formula asserted by **P**.

Alternate: Moves strictly alternate between **O** and **P**.

    Atom: Atomic formulas, including $\bot$, may be asserted by both players, but they can neither be attacked nor defended by **P**.

      E: Each (but the first) move of **O** reacts directly to the immediately preceding move by **P**. I.e., if **P** attacks a granted formula then **O**'s next move either defends this formula or attacks the formula used by **P** to launch this attack. If, on the other hand, **P**'s last move was a defending one then **O** has to attack immediately the formula stated by **P** in that defense move.

E-dialogues characterize intuitionistic logic in the following sense: **P** has a strategy to win a dialogue with her initial assertion B, where **O** initially grants sentences $A_1, \ldots, A_n$ iff B is an intuitionistic consequence of $A_1, \ldots, A_n$. In fact a version **LI′** of Gentzen's sequent calculus **LI** can defined that reveals a strong connection between sequent calculi and Lorenzen's game: any cut-free derivation of the sequent $A_1, \ldots, A_n \vdash B$ in **LI′** corresponds to a winning strategies for **P** with initial assertion B and initially granted sentences $A_1, \ldots, A_n$. (See, *e.g.*, [11] for details. That paper actually also describes how parallel versions of Lorenzen's game can be used to characterize some intermediate logics, *i.e.*, logics stronger that intuitionistic logic, but weaker than classical logic.)

4   Giles's game for Lukasiewicz logic

In the 1970s Robin Giles combined Lorenzen's logical dialogue rules with a different way to evaluate states at the resulting atomic level that was intended to capture realistic reasoning in the context of physical theories [16,17]. Later on, Giles presented his game explicitly as alternative semantics for fuzzy logic [18]. The game consists of two components that can be viewed as largely independent from each other:

(1) Betting for positive results of experiments.   Two players—following Giles, say *me* and *you*—agree to pay 1€ [3] to the opponent player for every false statement they assert. By $[p_1, \ldots, p_m \, \| \, q_1, \ldots, q_n]$ we denote an *elementary state* of the game, where I assert each q_i in the multiset $\{q_1, \ldots, q_n\}$ of atomic statements and you assert each atomic statement $p_i \in \{p_1, \ldots, p_m\}$.

Every propositional variable q refers to an experiment E_q with binary (yes/no) result. The statement q can be read as 'E_q yields a positive result'. This stipulation turns into a scenario for modelling vagueness or partial truth as the experiments may show dispersion; *i.e.*, the same experiment may yield different results when repeated. However, the results are not completely arbitrary: for every run of the game, a fixed *risk value* $\langle q \rangle \in [0, 1]$ is associated with q, denoting the proba-

---

[3] Giles had 1\$ (presumably Canadian dollar) instead of 1€. Matching the current background, we want to give the game a European twist. As will get clear below, we deviate from Giles in some important details. In particular we will include an analysis of strong conjunction, often considered the characteristic connective of Lukasiewicz logic, that is missing from Giles's account. (The corresponding rules can be found in [13] and in [14].)

bility that E_q yields a negative result.[4] For the special atomic formula $\perp$ (*falsum*) we define $\langle \perp \rangle = 1$. The risk associated with a multiset $\{p_1, \ldots, p_m\}$ of atomic formulas amounts to $\langle p_1, \ldots, p_m \rangle = \sum_{i=1}^{m} \langle p_i \rangle$. The risk $\langle \rangle$ associated with the empty multiset is 0. The risk associated with an elementary state $[p_1, \ldots, p_m \, \| \, q_1, \ldots, q_n]$ is calculated from my point of view as $\sum_{j=1}^{n} \langle q_j \rangle - \sum_{i=1}^{m} \langle p_i \rangle$. Therefore the condition $\langle p_1, \ldots, p_m \rangle \geq \langle q_1, \ldots, q_n \rangle$ expresses that I do not expect any loss of money (but possibly some gain) when betting on the truth of atomic statements according to the scheme explained above.

(2) A dialogue game for the reduction of compound formulas.   Giles refers to Lorenzen's logical dialogue rules, reviewed above. He states the rules for conjunction, disjunction, and implication as follows.

($R_\supset$) He who asserts $A \supset B$ agrees to assert B if his opponent will assert A.

($R_\vee$) He who asserts $A \vee B$ undertakes to assert either A or B at his own choice.

($R_\wedge$) He who asserts $A \wedge B$ undertakes to assert either A or B at his opponent's choice.

These formulations clearly match Lorenzen's rules, reviewed in Section 3 above. However, concerning conjunction, one may ask: why does an assertion of $A \wedge B$ not oblige to assert *both* A and B, if the opponent attacks this assertion? Giles justifies the format of the rule by reference to a 'principle of limited liability' that can best be seen at play in his rule for negation:

($R_\neg$) He who asserts $\neg A$ agrees to pay 1€ to his opponent if he (the opponent) will assert A.

Note that, given rule ($R_\supset$) and the fact that the risk value of $\perp$ is always 1, this is equivalent to stipulating that $\neg A$ abbreviates $A \supset \perp$. Thus 1€ is fixed as the maximal amount of loss associated with the assertion of a

---

[4] Giles insists that these probabilities are *subjective* and receive tangible meaning only through a corresponding betting scheme. However we want to quote verbatim a very interesting comment to an earlier version of this paper made by Hykel Hosni: "One interesting feature of this semantics is that it calls in probabilities. However, since they are meant to reflect the outcomes of physical experiments they really seem to have some 'objective' as opposed to 'subjective' character. In a certain sense, then, one would be tempted to say that degrees of probability do two distinct jobs in uncertain reasoning. They measure 'degrees of belief' if they are subjective, and they measure 'degrees of truth' if they are objective."

formula, whether compound or atomic. Applying this principle to conjunctions, it would indeed be problematic to require a defender of $A \wedge B$ to assert always both A and B, since this may result in a risk of paying 2€ in the worst case. Lorenzen's rule clearly circumvents this problem.

Nevertheless, one may argue, that a very natural alternative interpretation of conjunction involves an assertion of both conjuncts. But can this be maintained without violating the above principle of limited liability? We argue that it can—in fact in a very straightforward way, by using the following rule (formulated again in Giles's somewhat old-fashioned style):

($R_\&$) He who asserts $A \& B$ undertakes to assert either both, A and B, or else to pay 1€ to his opponent.

Of course, the choice involved in rule ($R_\&$) is the proponent's. It turns out that this version of conjunction ($\&$) behaves indeed differently from Giles's one ($\wedge$). While $\wedge$ corresponds to the 'weak' or 'lattice conjunction' algebraically represented by *infimum*, $\&$ turns out to be the 'strong conjunction' corresponding to the Łukasiewicz t-norm $x * y = \max(0, x + y - 1)$. This can most easily be seen by evaluating the risk $\langle p \& q \rangle$ associated with the assertion of $p \& q$. According to ($R_\&$) it is $\min(1, \langle p \rangle + \langle q \rangle)$, where $\langle p \rangle$ and $\langle q \rangle$ are the relevant risk values, *i.e.*, the probabilities that the experiments E_p and E_q associated with the atomic sentences p and q, respectively, yield a negative answer. If we stipulate that the truth value $v(p)$ is the inverse of the risk value $\langle p \rangle$ ($v(p) = 1 - \langle p \rangle$), we obtain the Łukasiewicz t-norm. In view of the approach to mathematical fuzzy logic initiated by Hájek [22], this augmentation of Giles's game, first published in [13], is rather central.

A distinctive feature of Giles's dialogue game, that provides a stark, positive contrast to Lorenzen's dialogue games for intuitionistic logic, is that no special regulations (*Rahmenregeln*) about the succession between moves of me and you—roughly corresponding to Lorenzen's players **P** and **O**—are imposed: it turns out that the order in which the players make their moves is immaterial. To bring this out more clearly, the notion of a *regulation* is defined in [14]. A regulation ρ assigns to each dialogue state $[A_1, \ldots, A_m \,\|\, B_1, \ldots, B_n]$ either the label **Y** to indicate that it is your turn to attack one of my (logically complex) assertions, or **I** to indicate that it is my turn to point to one of your assertions and oblige you to reply according to the logical rules. There is no restriction on regulations except the obvious one: the state must contain a corresponding logically complex assertion to attack; such regulations are called *consistent*.

Another characteristic of Giles's interpretation of Lorenzen's dialogue rules is that he stipulates that each assertion can be attacked at most once.

Again, Giles maintains that this is in accordance with the principle of limited liability. Technically, it is reflected in the removal of the attacked compound formula A from the dialogue state, *i.e.*, from the multiset of formulas currently asserted by a player, as soon as the other player has either attacked A according to the rules or has indicated that he will not attack F at all. In consequence, it is guaranteed that each run of the game ends in an elementary state $[p_1, \ldots, p_m \| q_1, \ldots, q_n]$ that can then be evaluated according to the presented scheme for betting on the results of elementary experiments.

There is yet another important feature of Giles's game that separates it from Lorenzen's game. Every run of the game refers to some fixed assignment of risk values to atomic formulas. Note that the resulting instances of Giles's dialogue *cum* betting scenario amount to a finite game with perfect information, only if the risk values are known to both players. Since risk value assignments correspond to fuzzy models, *i.e.*, to assignments of (inverted) values from $[0, 1]$ to propositional variables, the game should be classified as an *evaluation game*, where the aim is to determine the truth value of a formula in a given interpretation rather than to check its validity. In this respect, Giles's game is actually closer to Hintikka's evaluation game for classical predicate logic than to Lorenzen's game codifying intuitionistic validity. This is emphasized in the following summary of corresponding results, straightforwardly obtained as corollary from the proof of Theorem 4.2 in [14], which in turn—disregarding strong conjunction—is essentially due to Giles [16, 17]. To be able to state it concisely we recall the standard evaluation of formulas over the connectives $\supset$, $\wedge$, $\&$, $\vee$, $\neg$, and $\bot$, according to Łukasiewicz and Hájek.

Definition 4.1. Let v be an assignment of truth values in $[0, 1]$ to propositional variables. A corresponding (standard) *Łukasiewicz evaluation* is obtained by extending v to compound formulas as follows:

$$v(\bot) = 0 \qquad\qquad v(\neg A) = 1 - v(A)$$

$$v(A \wedge B) = \min(v(A), v(B)) \qquad v(A \vee B) = \max(v(A), v(B))$$

$$v(A \& B) = \max(0, v(A) + v(B) - 1) \quad v(A \supset B) = \min(1, 1 - v(A) + v(B))$$

A formula F is called *valid* in Łukasiewicz logic $\mathbf{L}_\infty$ if $v(F) = 1$ for all truth value assignments. If we restrict the set of truth values to $\{\frac{i}{n-1} : 0 \leq i < n\}$ we obtain n-valued Łukasiewicz logic $\mathbf{L}_n$ for all $n \geq 2$. (Note that $\mathbf{L}_2$ is classical logic.)

Theorem 4.2. *For every formula F, every risk value assignment $\langle \cdot \rangle$, and every consistent regulation ρ the following are equivalent:*

- *initially asserting F, I have a strategy for the game under regulation ρ to enforce a final elementary state, where my expected risk according to $\langle \cdot \rangle$ is not higher than $x \in$*
- *$v(F) = x$ according to Definition 4.1, where $v(p) = 1 - \langle p \rangle$ for all propositional variables p.*

It follows immediately from Theorem 4.2 that a formula F is valid in Łukasiewicz logic $\mathbf{L}_\infty$ ($\mathbf{L}_n$) iff if for every (properly restricted) risk value assignment I have a strategy to avoid expected loss in games that start with my initial assertion of F.

5   Giles's game stripped naked

To get a better view on the design choices that have been made by Giles in setting up his dialogue *cum* betting game for Łukasiewicz logic, let us investigate the game from a game theoretic point of view. In standard game theoretic terminology (see, *e.g.*, [28]) we obtain a concrete *extensive game form* if we fix a consistent regulation and a concrete initial state, in our case these are the formulas that the players, me and you, initially assert. This game form is just the tree of states that records the possible moves that can be made according to the logical rules by either me or you, depending on whether the state is labeled by $\mathbf{I}$ or by $\mathbf{Y}$ by the regulation. In a state, say, $[A_1, \ldots, A_m \,\|\, B_1, \ldots, B_n]^{\mathbf{Y}}$ you first have to choose one of the logically complex formulas among my assertions $B_1, \ldots, B_n$ for attack or to announce that you will not attack it at all. We will consider such a choice as a move in its own right and denote the resulting successor state by underlining the formula that has been chosen. The game turns into an *extensive game* if we fix an assignment of risk values to all propositional variables occurring in the initially asserted formulas, because such an assignment determines a concrete *payoff* in form of the expected amount of money that I have to pay to you or vice versa. Clearly, the resulting game is *finite*, *zero sum*, and with *perfect information*. Indeed, the proof of Theorem 4.2 follows the standard backward induction that can be applied to such determined games. A *run* of the game is just a branch of the tree. A *strategy for me* (we will not care about strategies for you here) is obtained from the full game tree by removing for each state labeled by $\mathbf{I}$ all but one successor states. This, of course, is equivalent to the view of a strategy for a player $\mathbf{X}$ as a *function* that assigns an admissible successor state to every state where it is $\mathbf{X}$'s turn to move.

Example 5.1. The following tree is a strategy for $(\neg p \supset \neg q) \supset (q \supset r)$ for me, where the regulation is indicated by the superscripts. Recall that $\neg A =_{def} A \supset \bot$.

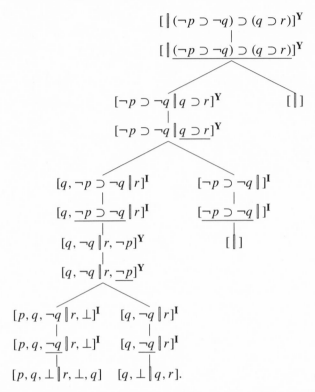

In this particular case, there is always only one currently asserted compound formula that can be singled out for attack by the opponent player. All these formulas are implications and therefore the attacking player can choose to either attack by asserting the left subformula or to declare that the formula will not be attacked. Since this is a strategy for me, your corresponding choices result in a branching of the tree at corresponding Y-labeled nodes, while the strategy determines that I choose to attack $\neg q$ to oblige you to assert $\bot$ in reply to my assertion of q as can be seen in the two left most final states. Clearly, I do not have to expect to lose money at the elementary states $[\,\|\,]$ and $[q, \bot \| q, r]$, independently of the risk values assigned to q and r. However, for $[p, q, \bot \| r, \bot, q]$ my expected loss of money is $(\langle r \rangle - \langle p \rangle)\varepsilon$, which is only non-positive if $\langle r \rangle \leq \langle p \rangle$.

Note that Giles's elaborated story in [17] (which we have only very cursorily reviewed here) about why and how to evaluate atomic formulas by betting on the results of associated dispersive experiments for which success probabilities are known, boils down to just some assignment of concrete payoff values to final states in a game. In particular, the reference to probabilities completely disappears: only the *expected amount of money* to be paid or received is relevant. From the game theoretic point of view, we only need a real number as payoff value for each final state.

To formulate an especially transparent variant of Giles's game we suggest a few simplifications. We want to maintain the principle that the payoff value for an elementary state $[p_1, \ldots, p_m \| q_1, \ldots, q_n]$ results from summing up values associated with the individual atomic formulas. However, we don't care what these values stand for; consequently *any* real number $v(p)$ can be assigned to a propositional variable now. To bring the definition of corresponding payoff closer to standard terminology we speak of *gain* instead of *risk* associated with my statements. Since we allow also negative values, this gain can also be negative. The principle of symmetric evaluation is maintained: my gain is your loss, and vice versa. We thus define my payoff in a final (*i.e.*, elementary) state by

$$v([p_1, \ldots, p_m \| q_1, \ldots, q_n]) =_{def} 1 - \sum_{1 \leq i \leq n} (1 - v(q_i)) + \sum_{1 \leq j \leq m} (1 - v(p_j)),$$

where for *absurdum* we define $v(\bot) = 0$.

Another simplification concerns the possibility, in Giles's game, to declare that an occurrence of $A \supset B$ will not be attacked at all, instead of attacking it by asserting A to oblige the opposing player to assert B. We will drop this option: all compound formulas are to be reduced to their subformulas in succeeding states. In fact, we drop the principle of limited liability altogether. I.e., asserting a strong conjunction $A \& B$ obliges one to assert both A and B, without the option to assert $\bot$ instead. However we stick to the principle that an occurrence of a formula is removed from the state upon attacking it.

The logical rules of the resulting game, seen from my point of view, are summarized in Table 1. Note the perfect symmetry with respect to the left and right hand side of states in these rules, which is in stark contrast to the asymmetry of the logical rules of Gentzen's sequent calculi.

As a consequence of our simplifications only your attacks to weak conjunctions asserted by me, as well as my attacks to your assertions of disjunctions, require branching in corresponding strategies for me. Where such formulas do not occur, the runs of the game are now fully deterministic: my strategies lead straight to elementary states without involving possible choices by you. Returning to Example 5.1 this means that the exhibited strategy for $(\neg p \supset \neg q) \supset (q \supset r)$ reduces to its left most branch.

Of course, the simplified game is no longer an evaluation game for Łukasiewicz logic. However it turns out to be adequate for a further

$$\frac{[\Pi \,\|\, \Sigma, \underline{A \supset B}]^{\mathbf{Y}}}{[A, \Pi \,\|\, \Sigma, B]} \qquad\qquad \frac{[\underline{A \supset B}, \Pi \,\|\, \Sigma]^{\mathbf{I}}}{[B, \Pi \,\|\, \Sigma, A]}$$

$$\frac{[\Pi \,\|\, \Sigma, \underline{A \vee B}]^{\mathbf{I}}}{[\Pi \,\|\, \Sigma, A]} \qquad \frac{[\Pi \,\|\, \Sigma, \underline{A \vee B}]^{\mathbf{I}}}{[\Pi \,\|\, \Sigma, B]}$$

$$\frac{[\underline{A \vee B}, \Pi \,\|\, \Sigma]^{\mathbf{Y}}}{[A, \Pi \,\|\, \Sigma] \quad [B, \Pi \,\|\, \Sigma]}$$

$$\frac{[\Pi \,\|\, \Sigma, \underline{A \wedge B}]^{\mathbf{Y}}}{[\Pi \,\|\, \Sigma, A] \quad [\Pi \,\|\, \Sigma, B]} \qquad\qquad \frac{[\underline{A \wedge B}, \Pi \,\|\, \Sigma]^{\mathbf{I}}}{[A, \Pi \,\|\, \Sigma]}$$

$$\frac{[\underline{A \wedge B}, \Pi \,\|\, \Sigma]^{\mathbf{I}}}{[B, \Pi \,\|\, \Sigma]}$$

$$\frac{[\Pi \,\|\, \Sigma, \underline{A \,\&\, B}]^{\mathbf{I}}}{[\Pi \,\|\, \Sigma, A, B]} \qquad\qquad \frac{[\underline{A \,\&\, B}, \Pi \,\|\, \Sigma]^{\mathbf{Y}}}{[A, B, \Pi \,\|\, \Sigma]}$$

Table 1. Simplified rules for $(R_\supset)$, $(R_\vee)$, $(R_\wedge)$, and $(R_\&)$.

interesting logic: Abelian logic **A**, that has been introduced and motivated as relevance logic by Meyer and Slaney in [27], and independently by Casari [5] as a 'comparative logic'. While **A** is rather exotic in counting formulas like $((A \supset B) \supset B) \supset A$ as valid that are not valid in classical logic, it has an elegant algebraic interpretation as the logic of lattice-ordered abelian groups, or for short, abelian ℓ-groups, as shown in [27]. We do not want to review the definition of ℓ-groups here, but rather point out the well known fact that the corresponding variety is generated by the structure $\mathcal{R} = \langle \mathbb{R}, +, \max, -, 0 \rangle$. In our context this means that evaluations for **A** are obtained by extending assignments of arbitrary real numbers from propostional variables to compound formulas as follows:

$$v(\bot) = 0 \qquad\qquad v(\neg A) = -v(A)$$
$$v(A \wedge B) = \min(v(A), v(B)) \quad v(A \vee B) = \max(v(A), v(B))$$
$$v(A \,\&\, B) = v(A) + v(B) \qquad v(A \supset B) = v(B) - v(A))$$

It is not difficult to see that, given the above, new definition of payoff, the reductions codified in the rules of Table 1 indeed correspond to such an evaluation.

The above analysis might be summarized as follows: if we relax the assignment of values in [0, 1] to propositional variables to assignments of arbitrary real numbers and if we moreover strip the principle of limited liability from Giles's game then Abelian logic **A** appears beneath this clothing of Łukasiewicz logic $\mathbf{L}_\infty$. Taking our simplified game (for **A**) as a starting point, we may re-introduce the principle of limited liability in a more systematic manner than done by Giles. Consider the following:

Attack principle of limited liability: Every player can, instead of attacking a compound formula F asserted by the opponent player, declare that he will not attack F at all.

Defense principle of limited liability: Every player can, instead of defending a compound formula F asserted by him according to the rules in Table 1, alternatively assert $\bot$.

Taking into account the attack principle of limited liability the simplified implication rules of Table 1 turn into the original ones of Giles. They can be denoted as follows:

$$[\Pi \,\|\, \Sigma, \underline{A \supset B}]^{\mathbf{Y}}$$
$$[A, \Pi \,\|\, \Sigma, B] \quad [\Pi \,\|\, \Sigma]$$

$$[\underline{A \supset B}, \Pi \,\|\, \Sigma]^{\mathbf{I}}$$
$$[B, \Pi \,\|\, \Sigma, A]$$

$$[\underline{A \supset B}, \Pi \,\|\, \Sigma]^{\mathbf{I}}$$
$$[\Pi \,\|\, \Sigma]$$

The defense principle of limited liability applied to the rules for strong conjunction leads to the following rules:

$$[\Pi \,\|\, \Sigma, \underline{A \,\&\, B}]^{\mathbf{I}}$$
$$[\Pi \,\|\, \Sigma, A, B]$$

$$[\Pi \,\|\, \Sigma, \underline{A \,\&\, B}]^{\mathbf{I}}$$
$$[\Pi \,\|\, \Sigma, \bot]$$

$$[\underline{A \,\&\, B}, \Pi \,\|\, \Sigma]^{\mathbf{Y}}$$
$$[A, B, \Pi \,\|\, \Sigma] \quad [\bot, \Pi \,\|\, \Sigma]$$

We have thus re-established Giles's original game (augmented by strong conjunction). But why is this twofold principle of limited liability not applied uniformly to *all* connectives? The answer is: because it would only complicate the game without affecting its expressive power. In other words, while we could insist on versions of the logical rules that respect both forms of the principle of limited liability, the resulting version of Giles's game will still just characterize Łukasiewicz logic $\mathbf{L}_\infty$ in the sense of Theorem 4.2.

6   Playing in ignorance of risk

While Giles could hardly have expected a tight relation between his game for $\mathbf{L}_\infty$ and a hypersequent calculus like **GL** (which was discovered only much later), he himself in a paper with Adamson [1] attempted to connect the game with Gentzen style proof theory. Indeed, it seems obvious that a game state $[A_1, \ldots, A_m \parallel B_1, \ldots, B_n]$ can be viewed as a sequent $A_1, \ldots, A_m \vdash B_1, \ldots, B_n$. However the attempts in [1] to establish a sequent calculus for $\mathbf{L}_\infty$ using logical rules that directly correspond to the dialogue rules $(R_\supset)$, $(R_\vee)$, $(R_\wedge)$, and $(R_\neg)$ are spoilt by the fact that an additional rule of the form

$$\frac{A, \Gamma \vdash \Delta, A}{\Gamma \vdash \Delta} \ (cancel)$$

has to be added to the sequent system to achieve completeness. While the soundness of $(cancel)$ with respect to the intended semantics is straightforward, the problem with this rule is that it is not analytic: applying it backwards we have to guess a formula that does not occur in the lower sequent. Indeed, it can be shown that $(cancel)$ is equivalent to the usual cut rule. As a consequence, in contrast to Giles's game, the sequent calculus presented in [1] cannot be interpreted as a tool for the stepwise reduction of complex arguments to logically simpler ones. In this respect, while hardly 'ugly', it is as 'bad' as the Hilbert-style systems reviewed in Section 2.

To obtain a game based characterization of *validity* in Łukasiewicz logic and not just a game based evaluation with respect to a given assignment of truth values (or risk values) to propositional variable, we obviously have to take into account all possible assignments. At a first glimpse this might seem unfeasible to manage within a single finite formalism, since there are uncountably many different assignments of values in $[0, 1]$ to the propositional variables occurring in a given formula. However, since the *game form* is finite and moreover is identical for all evaluation games pertaining to the same initial state, we can define a finite mechanism that keeps track of all possible runs of the game and thus allows to represent strategies at a level that abstracts away from concrete truth value assignments. The key to such a bookkeeping mechanism is the notion of a *state disjunction*, which is completely independent of the concrete dialogue game at hand: it can be applied to any family of games that share a common game form.[5]

Let us use $D = S_1 \bigvee \ldots \bigvee S_n$ to denote a *state disjunction*. Since the order of its *component states* (ordinary game states) $S_1, \ldots, S_n$ is irrelevant, a state disjunction may be viewed as a multiset of states. A *disjunctive strategy* for D respecting a regulation ρ, *i.e.*, an a assignment of players to component states, is a tree of state disjunctions with root

---

[5] The following passage borrows directly from [14].

D where the successor nodes are in principle determined in the same way as for ordinary strategies. However, we also allow for the possibility of *duplicating* a component of a state in order to let disjunctive strategies for a player P record different ways of proceeding for P in identical component states. More precisely, there are two kinds of non-leaf nodes $D = S_1 \bigvee \ldots \bigvee S_n$ in a disjunctive strategy for P:

1. *Playing nodes*, focused on some component S_i $(1 \le i \le n)$ of D. The successor nodes are like those for S_i in ordinary strategies, except for the presence of additional components (that remain unchanged). If, according to ρ, it is P's turn to play at S_i, then there is a single successor state disjunction where the component S_i of D is replaced by some S_i' corresponding to some move of P. If the opposing player Q is to move at S_i, then all possible moves of Q determine the successor nodes of D where S_i is replaced by a state obtained by the corresponding move.
2. *Duplicating nodes*, where the single successor node is obtained by duplicating one of the components in D.

A disjunctive *winning* strategy for P is one where at least one component of all leaf nodes (final state disjunctions) of the tree is a winning state for P in the underlying game.

There are quite different ways to interpret a tree of state disjunctions. It may be considered as a mere bookkeeping device that records possible runs of different individual games within a single formal structure. An alternative, maybe more interesting, interpretation takes state disjunctions as states of a new type of game, namely one that arises from playing the underlying individual games *in parallel*. In the latter case one should augment the regulation ρ, that specifies who's turn it is to move in the individual components, by similar information at the disjunctive level: who's turn it is to pick a component to focus on it or to duplicate a component. However, such information turns out to be irrelevant for our purpose. Different regulations, also on the disjunctive level, lead to the same final states in the case of Giles's game. In any case, an attractive interpretation of the disjunctive version of the game emerges: since we can explore all possible moves by duplicating nodes whenever different options arise, we do not have to know risk value for propositional variables in advance. In other words, we may speak of 'playing in ignorance of risk'.

Example 6.1. To illustrate the notion of a disjunctive strategy, consider a game form where player P has two possible moves at state S_0 leading to successor states S_1 and S_2, respectively. In each of these states, the other player Q has two possible moves, resulting in terminal states S_3, S_4, and S_5, S_6, respectively. This gives P two different (ordinary) strategies:

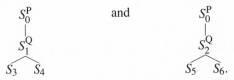

Such ordinary strategies are disjunctive strategies by definition. However, more interesting is the following combination of the two strategies for P into a single disjunctive strategy for P, where the root is a duplicating node:

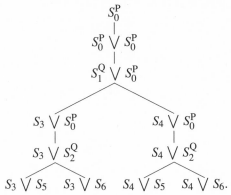

Let us consider a collection of games $\mathcal{W}$ with the same game form. (As already mentioned, for Giles's game we obtain such a collection by adding risk assignments to a game form determined by the logical rules, the initial state, and some consistent regulation). A disjunctive strategy Σ for $\mathcal{W}$ for player P is a *disjunctive winning strategy* for $\mathcal{W}$ if for every game in $\mathcal{W}$, at least one component of each leaf node of Σ is a winning state for P in that game. In particular, a disjunctive winning strategy (for me) for the family of all instances of Giles's game starting in an initial state $[\Gamma \parallel \Delta]$ and respecting a consistent regulation ρ is a disjunctive strategy where in each leaf node for *every* risk assignment $\langle \cdot \rangle$ there is at least one component, *i.e.*, elementary d-state, $[p_1, \ldots, p_m \parallel q_1, \ldots, q_n]$ such that $\langle p_1, \ldots, p_m \rangle \geq \langle q_1, \ldots, q_n \rangle$.

Proposition 6.2 ([14]). *Let $\mathcal{W}$ be a collection of games based on the same finite game form. Then there exists a disjunctive winning strategy for $\mathcal{W}$ for player P iff there is an ordinary winning strategy for P for every game in $\mathcal{W}$.*

Applied to Giles's game we obtain the following consequence of Theorem 4.2.

Corollary 6.3. *A formula F is valid in* $\mathbf{L}_\infty$ *iff I have a disjunctive winning strategy for every instance of Giles's game with initial state* $[\,\|\,F]$. *This also holds for* $\mathbf{L}_n$ *for* $n \geq 2$ *if we restrict the possible risk values to* $\{\frac{i}{n-1} : 0 \leq i < n\}$.

It is interesting to observe that for every instance F of the Łukasiewicz axiom schema (L1)-(L3), there exists a single ordinary winning strategy for *any* game with initial state $[\,\|\,F]$, *i.e.*, a strategy for me that is winning for all risk assignments. For (L4), there exists an ordinary winning strategy for certain regulations, but for others, a disjunctive winning strategy with duplication is needed. More generally, there are other valid formulas of the logic that require duplication for any regulation.

Example 6.4. It is easy to see that for the initial state $[\,\|\,(p \supset q) \vee (q \supset p)]$ there is an ordinary strategy for me that is winning for any assignment $\langle \cdot \rangle$ such that $\langle p \rangle \leq \langle q \rangle$ and another ordinary winning strategy for any assignment where $\langle q \rangle \leq \langle p \rangle$. However, no single strategy is winning for *all* assignments. Instead, we can combine ordinary strategies to obtain the disjunctive winning strategy exhabited in Table 2.

$$[\,\|\,(p \supset q) \vee (q \supset p)]^{\mathbf{Y}}$$
$$|$$
$$[\,\|\,(p \supset q) \vee (q \supset p)]^{\mathbf{Y}} \bigvee [\,\|\,(p \supset q) \vee (q \supset p)]^{\mathbf{Y}}$$
$$|$$
$$[\,\|\,\underline{(p \supset q) \vee (q \supset p)}]^{\mathbf{I}} \bigvee [\,\|\,(p \supset q) \vee (q \supset p)]^{\mathbf{Y}}$$
$$|$$
$$[\,\|\,p \supset q]^{\mathbf{Y}} \bigvee [\,\|\,(p \supset q) \vee (q \supset p)]^{\mathbf{Y}}$$
$$|$$
$$[\,\|\,p \supset q]^{\mathbf{Y}} \bigvee [\,\|\,\underline{(p \supset q) \vee (q \supset p)}]^{\mathbf{I}}$$
$$|$$
$$[\,\|\,p \supset q]^{\mathbf{Y}} \bigvee [\,\|\,q \supset p]^{\mathbf{Y}}$$
$$|$$
$$[\,\|\,\underline{p \supset q}]^{\mathbf{Y}} \bigvee [\,\|\,q \supset p]^{\mathbf{Y}}$$

$$[p\,\|\,q] \bigvee [\,\|\,\underline{q \supset p}]^{\mathbf{Y}} \qquad\qquad [\,\|\,] \bigvee [\,\|\,\underline{q \supset p}]^{\mathbf{Y}}$$
$$| \qquad\qquad\qquad\qquad\qquad |$$
$$[p\,\|\,q] \bigvee [\,\|\,\underline{q \supset p}]^{\mathbf{Y}} \qquad\qquad [\,\|\,] \bigvee [\,\|\,\underline{q \supset p}]^{\mathbf{Y}}$$

$$[p\,\|\,q] \bigvee [q\,\|\,p] \quad [p\,\|\,q] \bigvee [\,\|\,] \qquad [\,\|\,] \bigvee [q\,\|\,p] \quad [\,\|\,] \bigvee [\,\|\,].$$

Table 2. Disjunctive winning strategy for $(p \supset q) \vee (q \supset p)$.

7 System GL rediscovered: hypersequents as state disjunctions

With hindsight, we can see that Adamson and Giles [1] had looked for a relation between the dialogue game and analytic calculi at the wrong level of generality. Once we accept that every concrete run of the game refers to a concrete risk value assignment—and therefore amounts to playing an *evaluation game*—it is natural to ascend to the disjunctive level, as explained in the last section, to obtain a *game for validity*.

We have finally arrived at the intended interpretation of system **GL** already announced in the introduction: hypersequents can be viewed as state disjunctions for Giles's game. The dialogue game rules, if formulated at the disjunctive level, directly correspond to the logical rules of **GL**. For example, in case of weak conjunction ($\wedge$) the dialogue rules at the conjunctive level can be written as follows, where $\mathcal{D}$ denotes the remaining part of the state disjunction.

$$[\Pi \| \Sigma, \underline{A \wedge B}]^{\mathbf{Y}} \bigvee \mathcal{D}$$

$$[\Pi \| \Sigma, \underline{A}] \bigvee \mathcal{D} \quad [\Pi \| \Sigma, \underline{B}] \bigvee \mathcal{D}$$

$$[\underline{A \wedge B}, \Pi \| \Sigma]^{\mathbf{I}} \bigvee \mathcal{D}$$
$$[A, \Pi \| \Sigma] \bigvee [B, \Pi \| \Sigma] \bigvee \mathcal{D}$$

This obviously is only a notational variant of the corresponding **GL**-rules:

$$\frac{\Gamma \vdash \Delta, A \mid \mathcal{H} \quad \Gamma \vdash \Delta, B \mid \mathcal{H}}{\Gamma \vdash \Delta, A \wedge B \mid \mathcal{H}} \ (\wedge, r) \qquad \frac{A, \Gamma \vdash \Delta \mid B, \Gamma \vdash \Delta \mid \mathcal{H}}{A \wedge B, \Gamma \vdash \Delta \mid \mathcal{H}} \ (\wedge, l).$$

The case for disjunction is completely analogous.

For implication we have to take into account the attack principle of limited liability, as explained at the end of Section 5, and obtain

$$[\Pi \| \Sigma, \underline{A \supset B}]^{\mathbf{Y}} \bigvee \mathcal{D}$$

$$[A, \Pi \| \Sigma, B] \bigvee \mathcal{D} \quad [\Pi \| \Sigma] \bigvee \mathcal{D}$$

$$[\underline{A \supset B}, \Pi \| \Sigma]^{\mathbf{I}} \bigvee \mathcal{D}$$
$$[A, \Pi \| \Sigma, B] \bigvee [\Pi \| \Sigma] \bigvee \mathcal{D}$$

which is a notational variant of the **GL**-rules

$$\frac{A, \Gamma \vdash \Delta, B \mid \mathcal{H} \quad \Gamma \vdash \Delta \mid \mathcal{H}}{\Gamma \vdash \Delta, A \supset B \mid \mathcal{H}} \ (\supset, r) \qquad \frac{B, \Gamma \vdash \Delta, A \mid \mathcal{H}}{A \supset B, \Gamma \vdash \Delta \mid \Gamma \vdash \Delta \mid \mathcal{H}} \ (\supset, l)'.$$

Finally, for strong conjunction the defense principle of limited liability gives

$$[\Pi \, \| \, \Sigma, \underline{A \,\&\, B}]^{\mathbf{Y}} \bigvee \mathcal{D}$$
$$[\Pi \, \| \, \Sigma, A, B] \bigvee [\Pi \, \| \, \Sigma, \bot] \bigvee \mathcal{D}$$

$$[\underline{A \,\&\, B}, \Pi \, \| \, \Sigma]^{\mathbf{I}} \bigvee \mathcal{D}$$
$$[A, B, \Pi \, \| \, \Sigma] \bigvee \mathcal{D} \quad [\bot, \Pi \, \| \, \Sigma] \bigvee \mathcal{D}$$

which directly corresponds to

$$\frac{\Gamma \vdash \Delta, A, B \, | \, \Gamma \vdash \Delta, \bot \, | \, \mathcal{H}}{\Gamma \vdash \Delta, A \& B \, | \, \mathcal{H}} \ (\&, r) \qquad \frac{A, B, \Gamma \vdash \Delta \, | \, \mathcal{H} \quad \bot, \Gamma \vdash \Delta \, | \, \mathcal{H}}{A \& B, \Gamma \vdash \Delta \, | \, \mathcal{H}} \ (\&, l).$$

The logical rules of the hypersequent system **GL** have thus been interpreted as Giles's dialogue game rules viewed at the disjunctive level. But what about the structural rules and the initial hypersequents of **GL**? As shown in [14], the structural rules (EW), (EC), (IW), $(SPLIT)$, and (MIX) are only needed at the atomic level. Applied backwards they allow one to reduce every valid hypersequent that consists only of atomic formulas to sequents of the form $p \vdash p$, or $\bot \vdash A$, or the empty sequent $\vdash$. For example, the final (elementary) state disjunction

$$[p \, \| \, q] \bigvee [q \, \| \, p] \bigvee [q \, \| \, \bot]$$

is a winning state in the sense of Section 6 although neither $p \vdash q$ nor $q \vdash p$ nor $q \vdash \bot$ are initial sequents of **GL**: for every assignment of risk values at least one of the three component states carries no positive expected risk for me. The following derivation reduces the corresponding hypersequent to initial sequents.

$$\frac{\dfrac{p \vdash p \quad q \vdash q}{p, q \vdash q, p} \ (MIX)}{\dfrac{p \vdash q \, | \, q \vdash p}{p \vdash q \, | \, q \vdash p \, | \, q \vdash \bot} \ (SPLIT)} \ (EW)$$

One may ask whether the structural rules are needed at all. Couldn't we just declare all valid *atomic* hypersequents to be initial hypersequents? This is indeed a reasonable option if we can decide efficiently whether a given atomic hypersequent is valid with respect to Łukasiewicz logic. As shown in [7] one can reduce the validity problem for atomic hypersequents of $\mathbf{GL}$[6] to linear programming. Thus the problem is in PTIME, implying that there is a feasible test for accepting given atomic hypersequents as initial. An additional advantage of this version of the hypersequent system over that with structural rules is that it is also complete for the finite valued Łukasiewicz logics $\mathbf{L}_n$, where $n \geq 2$. We simply have to restrict the possible truth or risk values to $\{\frac{i}{n-1} : 0 \leq i < n\}$.

8 Conclusion

We have motivated our investigation by pointing out that many non-classical logics are frequently presented by Hilbert style proof systems and corresponding algebraic semantics. This is in particular true for deductive fuzzy logics. In this context the challenge to find sound and complete analytic Gentzen type proof systems is often viewed as a side issue, important only to those interested in automated proof search or in certain traditions in proof theory. However we hope to have shown here, at least for the case of Łukasiewicz logics $\mathbf{L}_\infty$, $\mathbf{L}_n$ where $n \geq 2$ (which includes classical logic), and Abelian logic $\mathbf{A}$, that there is a close connection between a certain variant of hypersequent systems and dialogue games of the type introduced by Robin Giles for Łukasiewicz logic. This might also shed new light on the role that Gentzen style systems may play in investigating models of reasoning based on particular fuzzy logics and related logics.

ACKNOWLEDGEMENTS. This work has been partially supported by Eurocores-ESF/FWF grant I143-G15 (LogICCC-LoMoReVI).
Moreover, we wish to thank George Metcalfe, who has been centrally involved in some of the developments reported in this paper and Hykel Hosni for interesting comments suggesting directions for further work related to Giles's game.

References

[1] A. ADAMSON and R. GILES, *A game-based formal system for L_∞*, Studia Logica, **1** (1979), 49–73.

---

[6] Actually the corresponding proof in [7] refers to a slightly more general version of hypersequents, where there are two types of 'sequent arrows', one corresponding semantically to '$\leq$' and one corresponding to '$<$'. The second type of sequent is not needed here, but allows to treat also Gödel logic $\mathbf{G}$ and Product logic $\mathbf{P}$ in the same style as Łukasiewicz logic, see [7, 12].

[2] A. AVRON, *A constructive analysis of RM*, Journal of Symbolic Logic **52** (1987), 939–951.

[3] A. AVRON, *Hypersequents, logical consequence and intermediate logics for concurrency*, Annals of Mathematics and Artificial Intelligence **4** (1991), 225–248.

[4] M. BAAZ, A. CIABATTONI, C. G. FERMÜLLER and H. VEITH, *Proof theory of fuzzy logics: Urquhart's C and related logics*, In: L. Brim, J. Gruska and J. Zlatuska (eds.), "Proc. 23rd International Symposium Mathematical Foundations of Computer Science", Brno, Czech Republic, Vol. 1450 of *LNCS*, pages 203–212, 1998.

[5] E. CASARI, *Comparative logics and abelian ℓ-groups*, In: C. Bonotto, R. Ferro, S. Valentini and A. Zanardo (eds.), "Logic Colloquium '88", pages 161–190, Elsevier, 1989.

[6] A. CIABATTONI and C. G. FERMÜLLER, *Hypersequents as a uniform framework for Urquhart's C, MTL and related logics*, In: "Proceedings of the 31st IEEE International Symposium on Multiple-Valued Logic", Warsaw, Poland, pages 227–232. IEEE Computer Society Press, May 2001.

[7] A. CIABATTONI, C. G. FERMÜLLER and G. METCALFE, *Uniform rules and dialogue games for fuzzy logics*, In: "Proceedings of LPAR 2004", Vol. 3452 of *LNAI*, pages 496–510. Springer, 2005.

[8] R. CIGNOLI, F. ESTEVA, L. GODO and A. TORRENS, *Basic fuzzy logic is the logic of continuous t-norms and their residua*, Soft Computing **4** (2000), 106–112.

[9] W. FELSCHER, *Dialogues, strategies, and intuitionistic provability*, Annals of Pure and Applied Logic **28** (1985), 217–254.

[10] W. FELSCHER, *Dialogues as foundation for intuitionistic logic*, In: D. Gabbay and F. Günther (eds.), "Handbook of Philosophical Logic", Vol. III, pages 341–372, Reidel, 1986.

[11] C. G. FERMÜLLER, *Parallel dialogue games and hypersequents for intermediate logics*, In: M. Cialdea Mayer and F. Pirri (eds.), "Proceedings of TABLEAUX 2003", Vol. 2796 of *LNCS*, pages 48–64, Springer, 2003.

[12] C. G. FERMÜLLER, *Revisiting Giles - connecting bets, dialogue games, and fuzzy logics*, In: O. Majer, A. V. Pietarinen and T. Tulenheimo (eds.), "Games: Unifying Logic, Language, and Philosophy", pages 209–227, Springer, 2009.

[13] C. G. FERMÜLLER and R. KOSIK, *Combining supervaluation and degree based reasoning under vagueness*, In: "Proceedings of LPAR 2006", Vol. 4246 of *LNAI*, pages 212–226, Springer, 2006.

[14] C. G. FERMÜLLER and G. METCALFE, *Giles's game and the proof theory of łukasiewicz logic*, Studia Logica **92** (2009), 27–61.

[15] G. GENTZEN, *Untersuchungen über das Logische Schliessen*, Math. Zeitschrift **39** (1935), 176–210, 405–431.

[16] R. GILES, *A non-classical logic for physics*, Studia Logica **4** (1974), 399–417.

[17] R. GILES, *A non-classical logic for physics*, In: R. Wojcicki and G. Malinkowski (eds.), "Selected Papers on Łukasiewicz Sentential Calculi", pages 13–51, Polish Academy of Sciences, 1977.

[18] R. GILES, *Semantics for fuzzy reasoning*, International Journal of Man-Machine Studies **17** (1982), 401–415.

[19] J.-Y. GIRARD, *On the meaning of logical rules I: syntax vs. semantics*, In: "Computational Logic", pages 215–272, Springer, 1999.

[20] R. GRIGOLIA, *Algebraic analysis of Łukasiewicz-Tarski n-valued logical systems*, In: R. Wòjciki and G. Malinowski (eds.), "Selected Papers on Łukasiewicz Sentential Calculi", pages 81–91, Polish Acad. of Sciences, 1977.

[21] PH. DE GROOTE (ed.), "The Curry-Howard Isomorphism", Vol. 8 of *Cahiers du Centre de Logique (Universitè catholique de Louvain)*, Academia-Bruylant, 1995.

[22] P. HÁJEK, "Metamathematics of Fuzzy Logic", Kluwer, Dordrecht, 1998.

[23] P. G. HANSEN and V. F. HENDRICK (eds.), "Game Theory: 5 Questions", Automatic Press / VIP, 2007.

[24] R. KAHLE and P. SCHROEDER-HEISTER (eds.), "Special issue on Proof Theoretic Semantics", Vol. 148 of *Synthese*. Springer, 2006.

[25] P. LORENZEN, *Logik und Agon*, In: "Atti Congr. Internaz. di Filosofia", pages 187–194, Sansoni, 1960.

[26] G. METCALFE, N. OLIVETTI and D. GABBAY, "Proof Theory for Fuzzy Logics", Vol. 36 of *Applied Logic*, Springer, 2008.

[27] R. K. MEYER and J. K. SLANEY, *Abelian logic from A to Z*, In: G. Priest *et al.* (eds.), "Paraconsistent Logic: Essays on the Inconsistent", pages 245–288, Philosophia Verlag, 1989.

[28] M. J. OSBORNE and A. RUBINSTEIN, "A Course in Game Theory", MIT Press, 1994.

[29] G. POTTINGER, *Uniform, cut-free formulations of T, S4 and S5* (abstract). Journal of Symbolic Logic **48** (1983), 900.

Poset representation for free RDP-algebras

Diego Valota

1 Introduction

Many-Valued Logics deal with sets of truth values that extend the classical set $\{0, 1\}$. In this paper we consider only Many-Valued Logics with the unitary real interval as set of truth values.

A minimal set of properties for a conjunction connective on the set $[0, 1]$ leads to the notion of *t-norm*, that is a binary, commutative, associative and monotonically non-decreasing operation that has 1 as unit element. In [8], Hájek has introduced a class of logics deeply studied in literature, namely truth-functional propositional Many-Valued Logics based on continuous *t-norms*. In these logical systems *t*-norms model conjunction connectives. Implication connectives are modelled by the *residuum* of the *t*-norms, that is $x \Rightarrow y = \max\{z \mid x \odot z \leq y\}$ for all $x, y, z \in [0, 1]$, where $\odot$ is a *t*-norm. It has been shown that the necessary and sufficient condition for a *t*-norm to have a residuum is the left-continuity. Hence, in order to obtain the logic of all left-continuous *t*-norms and their residua, in [6] the authors have introduced the *Monoidal T-norm based Logic*. Jenei and Montagna ([12]) have shown that MTL logic is complete with respect to the class of *MTL algebras*.

An algebra $A = \langle A, \vee, \wedge, \odot, \rightarrow, \perp, \top \rangle$ is an MTL algebra if $\langle A, \vee, \wedge, \perp, \top \rangle$ is a bounded lattice, $\langle A, \odot, \top \rangle$ is a commutative monoid, if the *residuation* condition holds, that is $x \odot y \leq z$ if and only if $x \leq y \rightarrow x$, and the *prelinearity* equation $(x \rightarrow y) \vee (y \rightarrow x) = \top$ holds in A. An MTL algebra that satisfies also the *weak nilpotent minimum* equation

$$\neg(x \odot y) \vee ((x \wedge y) \rightarrow (x \odot y)) = \top,$$

is called a *WNM algebra*. If A is totally ordered we call it a *chain*.

MTL algebras form an algebraic variety $\mathbb{V}(MTL)$ whose subdirectly irreducible members are chains. Each schematic extension $\mathcal{L}$ of MTL logic determines a subvariety $\mathbb{V}(\mathcal{L})$ of $\mathbb{V}(MTL)$. Given a logic $\mathcal{L}$, two formulas φ and ψ are logical equivalent, in symbols $\varphi \equiv_{\mathcal{L}} \psi$, if and only if $\varphi \leftrightarrow \psi$ is a tautology. The *Lindenbaum Algebra* $\mathbf{L}(\mathcal{L})$ of $\mathcal{L}$ is the algebra whose elements are the equivalence classes of formulas of $\mathcal{L}$, with respect to $\equiv_{\mathcal{L}}$.

An algebra F in a variety $\mathbb{V}(\mathcal{L})$, is said *free* over the set of *generators* X, if given an algebra $B \in \mathbb{V}(\mathcal{L})$, each map $X \to B$ uniquely extends to an homomorphism $F \to B$. By universal algebraic facts ([4]), free algebra exists in any variety. We wrote F_n to denote the free algebra over X, where $n = |X|$.

The free n-generated algebra F_n in $\mathbb{V}(\mathcal{L})$ is the Lindenbaum algebra $\mathbf{L}(\mathcal{L})$ of the formulas φ over the first n variables. This relation shows the importance of concrete representations of the free algebra in order to do real computations or to define normal forms for $\mathcal{L}$. More generally, a characterization of the free algebra is useful for a better understanding of its underlying logic.

When the variety $\mathbb{V}(\mathcal{L})$ is *locally finite*, that is the finitely generated free algebras in $\mathbb{V}(\mathcal{L})$ are finite, one can hope to have a discrete combinatorial representation of the free algebra F_n. This has been shown to be the case for the t-norm based many-valued logics of *Gödel* and *Nilpotent Minimum* in [1]. In this paper, we adapt the methodology of [1] to the case of *Revised Drastic Product* (RDP) logic ([14]). All these logics are schematic extension of *Weak Nilpotent Minimum* logic, whose corresponding algebraic variety $\mathbb{V}(WNM)$ is locally finite. Hence, RDP logic together with Gödel, NM and NMG logics ([1]) is a useful case study towards the analysis of WNM logic.

The left-continuous RDP t-norm is obtained starting from the non-continuous *drastic product* t-norm (that is the smallest t-norm in the pointwise order) and applying the construction detailed in [11]. It was introduced by Jenei ([11]) as the first example of a t-norm which is an ordinal sum of t-subnorms [1]. Indeed, the RDP t-norm arises as the ordinal sum of the t-subnorm 0 (that is, the constant function 0) and the Gödel t-norm. Details on the ordinal sum construction could be found in [11]. This fact allows to adapt some techniques used to characterize free n-generated algebras corresponding to Gödel logic to the case of RDP logic.

Throughout the paper we will use the same symbols for logic's connectives and their algebraic interpretations.

---

[1] A t-subnorm S on the set $[0, 1]$, is a commutative, associative and monotonically non-decreasing operation such that $S(x, y) \leq \min(x, y)$.

2 Background

The propositional language of RDP logic contains the constant $\bar{0}$ and the binary connectives $\odot, \rightarrow, \wedge$. Negation connective is derived from implication: $\neg x = x \rightarrow \bar{0}$ and $\bar{1} = \neg \bar{0}$. The only inference rule is *modus ponens*. RDP logic is axiomatised by the axioms of WNM plus the *revised drastic product* axiom:

$$\neg\neg\varphi \vee (\varphi \rightarrow \neg\varphi). \tag{RDP}$$

Since RDP logic is a schematic extension of MTL, RDP algebras form a subvariety $\mathbb{V}(RDP)$ of $\mathbb{V}(MTL)$. Standard RDP chains (that is, RDP chains having $[0,1]$ as domain) have the form $[0,1]_e = \langle [0,1], \odot, \rightarrow, \wedge, 1 \rangle$, where $e \in (0,1)$ and for every $x, y \in [0, 1]$ t-norm and residuum are defined as:

$$x \odot y = \begin{cases} 0 & x, y \leq e, \\ \min(x, y) & \text{otherwise.} \end{cases} \tag{2.1}$$

$$x \rightarrow y = \begin{cases} 1 & x \leq y, \\ e & y < x \leq e, \\ y & e < x, y < x. \end{cases} \tag{2.2}$$

Moreover, for every $x \in [0, 1]$ the negation is defined as:

$$\neg x = \begin{cases} 1 & x = 0, \\ e & 0 < x \leq e, \\ 0 & e < x. \end{cases} \tag{2.3}$$

Note that $\neg\neg x = 1$ if and only if $e < x$, $\neg\neg x = e$ if and only if $0 < x \leq e$ and $\neg\neg x = 0$ if and only if $x = 0$.

Every standard RDP chain is isomorphic to the standard RDP chain $[0, 1]_{\frac{1}{2}}$ ([14]). Hence, throughout the paper we shall refer to this chain as "the" standard RDP chain.

Let φ be a logical formula with variables among $x_1, ..., x_n$ and A be a RDP algebra with $a_1, ..., a_n \in A$. We denote by $\varphi^A(a_1, ..., a_n)$ the element of A obtained by the evaluation of φ in A interpreting every x_i with the corresponding a_i, in particular $x_i^A = a_i$. With this notation a formula φ is a tautology of RDP logic if and only if for every algebra $A \in \mathbb{V}(RDP)$ and for every $a_1, ..., a_n \in A$, $\varphi^A(a_1, ..., a_n) = \bar{1}^A$.

The algebra $[0, 1]_{\frac{1}{2}}$ generates the variety $\mathbb{V}(RDP)$ ([14]). Hence, by well known universal algebraic facts ([4]) the free n-generated algebra $F_n(RDP)$ is the subalgebra of $[0, 1]_{\frac{1}{2}}^{[0,1]_{\frac{1}{2}}^n}$ generated by the projections

$x_i(t_1, ..., t_n) \mapsto t_i$. The elements of $F_n(RDP)$ are the classes of equivalent formulas in RDP logic, we call these elements *truth functions*.

Let $C = \langle C, \odot, \rightarrow, \wedge, \perp, \top \rangle$ be a WNM chain. Operations in C are completely determined by its negation $\neg$ ([13]):

$$x \odot y = \begin{cases} \perp & x \leq \neg y \\ \min(x, y) & \text{otherwise} \end{cases} \tag{2.4}$$

$$x \rightarrow y = \begin{cases} \top & x \leq y, \\ \neg x & y < x \leq \neg x, \\ y & \neg x < x, y < x. \end{cases} \tag{2.5}$$

Gödel logic is the schematic extension of RDP logic (and WNM logic too) obtained by adding the *idempotency* axiom, $\varphi \rightarrow (\varphi \odot \varphi)$. Operations on Gödel chains are defined as:

$$x \odot y = \min(x, y) \tag{2.6}$$

$$x \rightarrow y = \begin{cases} \top & x \leq y, \\ y & \text{otherwise.} \end{cases} \tag{2.7}$$

Note that $\mathbb{V}(RDP)$ is locally finite, since it is a subvariety of $\mathbb{V}(WNM)$. Let $C = \langle C, \odot, \rightarrow, \wedge, \perp, \top \rangle$ be a RDP chain. By equations (2.4), (2.5) and the axiom (RDP), for all $x, y \in C$:

$$x \odot y = \begin{cases} \perp & x, y \leq \neg x, \neg y, \\ \min(x, y) & \text{otherwise.} \end{cases} \tag{2.8}$$

$$x \rightarrow y = \begin{cases} \top & x \leq y, \\ \neg x & y < x \leq \neg x, \neg y, \\ y & \neg x < x, y < x. \end{cases} \tag{2.9}$$

Given an RDP chain C generated by $g_1, ..., g_n$, let $\varphi(x_1, ..., x_n)$ be a formula. Then, $\varphi^C(g_1, ..., g_n) = c$ for some $c \in C$. By equations (2.8), (2.9) and the fact that $\neg\neg\neg x = \neg x$ holds in MTL, it follows that $\varphi^C(g_1, ..., g_n) = \psi^C(g_1, ..., g_n)$, for some ψ in the set

$$\mathcal{F}_n = \{\bar{0}, x_1, ...x_n, \neg x_1, ..., \neg x_n, \neg\neg x_1, ..., \neg\neg x_n, \bar{1}\}.$$

For every element $c_i \in C$ we call W_i the set of all formulas in $\mathcal{F}_n$ that are interpreted in c_i. That is $W_i = \{\varphi \in \mathcal{F}_n \mid \varphi^C(g_1, ..., g_n) = c_i\}$. Note that $\bar{0} \in W_0$ and $\bar{1} \in W_k$. We denote $\mathcal{W}$ the chain $W_0 < W_2 < ... < W_k$. The RDP chain $\mathcal{W}$ is isomorphic to C via the map that sends every c_i to W_i.

A *partition* of a set S is a family of non-empty subsets (called *blocks*), which are pairwise disjoint and their union is S. An *ordered partition* of S is a partition of S equipped with a total order among its blocks. Clearly, every $\mathcal{W}$ is a partition of $\mathcal{F}_n$ with a total order inherited by the RDP chain C. Hence, we call $\mathcal{W}$ an *ordered partition* of $\mathcal{F}_n$, and each W_i is called *block* of $\mathcal{W}$. We shall tacitly use this identification throughout the paper. As a consequence, when we consider the set of all n-generated RDP chains, we identify it with the set of all ordered partitions of $\mathcal{F}_n$ equipped with a structure of RDP algebra. The block W_i of a chain C that contains the element x_i will be denoted x_i^C.

2.1 Preliminaries on posets and normal forms

In this section we give concepts and results (presented in [1]) that will be useful in the rest of the paper, for additional background on *partially ordered sets* see [5].

Given a *poset* $(P, \leq)$, its *dual* $(P^\partial, \leq^\partial)$ is obtained reversing the original order relation: $x \leq^\partial y$ if and only if $y \leq x$, and $P^\partial = P$.

Given two disjoint posets A and B, their *horizontal sum* $A \sqcup B$ is the poset over $A \cup B$ formed by defining $x \leq y$ if and only if $x, y \in A$ and $x \leq y$ in A or $x, y \in B$ and $x \leq y$ in B. With nA we denote the poset obtained by applying to A horizontal sum for n times, that is $A \sqcup A \sqcup ... \sqcup A$. Another construction is the *vertical sum*. Let A and B be two disjoint poset, their vertical sum $A \oplus B$ is the poset over $A \cup B$ obtained by taking the order relation defined in the following way: let x and y be two elements that belong to $A \cup B$, then $x \leq y$ if the pair (x, y) fall in one of the following three mutually disjoint cases; $x \leq y$ if and only if $x, y \in A$ and $x \leq y$ in A, second $x, y \in B$ and $x \leq y$ in B and finally $x \in A$ and $y \in B$. A special case of vertical sum is the *lifting*, $A_\perp := \{\perp\} \oplus A$, with $\perp \notin A$.

We denote with $\mathbf{1}$ the poset containing only one element.

The poset obtained by $\mathbf{1} \oplus ... \oplus \mathbf{1}$ applied n times, is isomorphic to a chain of n elements, which we denote by $\mathbf{n}$.

If we replace every element of a poset $(P, \leq)$ with a copy of $\mathbf{1}$ we obtain a poset $(o(P), \leq)$ that is order isomorphic to $(P, \leq)$. We call $(o(P), \leq)$ the *type* of $(P, \leq)$. The type provides only the order-theoretic structure of a poset.

Definition 2.1. $(P, \leq)$ is *nice* if its type $(o(P), \leq)$ could be constructed from finitely many copies of $\mathbf{1}$ and using only operations $\oplus$ e $\sqcup$. A maximal chain in $(P, \leq)$ is called a *branch*. The set of branches of $(P, \leq)$ is denoted by $\mathcal{B}(A)$. A maximal antichain over $(P, \leq)$ is called a *section*. We denote with $S(P)$ the set of all sections over $(P, \leq)$.

We denote with $[p_B]_{\mathcal{B}(P)}$ a section over $(P, \leq)$, where $p_B \in B$ for every branch B in $\mathcal{B}(P)$. The set $S(P)$ can be equipped with an order structure, for any two section $[p_B]_{\mathcal{B}(P)}$ and $[q_B]_{\mathcal{B}(P)}$, we define:

$$[p_B]_{\mathcal{B}(P)} \leq [q_B]_{\mathcal{B}(P)},$$

if and only if $p_B \leq q_B$, for every $B \in \mathcal{B}(P)$.

Given a poset $(P, \leq)$, if $P = P_1 \sqcup P_2$ then $S(P) = S(P_1) \times S(P_2)$ and if $P = P_1 \oplus P_2$ then $S(P) = S(P_1) \oplus S(P_2)$. Hence we can compute the number of elements of a nice poset and of the set of its sections, starting from its type:

$$\begin{array}{ll}
|\mathbf{1}| = 1 & |S(\mathbf{1})| = 1 \\
|P_1 \sqcup P_2| = |P_1| + |P_2| & |P_1 \oplus P_2| = |P_1| + |P_2| \\
|S(P_1 \sqcup P_2)| = |S(P_1)| \cdot |S(P_2)| & |S(P_1 \oplus P_2)| = |S(P_1)| + |S(P_2)|.
\end{array}$$

Definition 2.2. Let C_1 and C_2 be two chains, we call *common prefix* of C_1 and C_2, the chain P such that $C_1 = P \oplus C_1'$ and $C_2 = P \oplus C_2'$. If C_1' and C_2' have the empty set as the only common prefix, then P is *longest common prefix* of C_1 and C_2.

Definition 2.3. Let B a branch of a finite poset $(P, \leq)$. We call *semantical maxterm* for $p_B \in B$, the section τ_{p_B} composed by p_B and the top elements of every branch B' in $(P, \leq)$ such that $B' \neq B$.

Remark 2.4. In [1], semantical maxterms are defined slightly more generally. But, for our purposes we are satisfied with this level of generality.

Theorem 2.5. *Let $[p_B]_{\mathcal{B}(P)}$ be an element of $S(P)$, then:*

$$[p_B]_{\mathcal{B}(P)} = \bigwedge_{B \in \mathcal{B}(P)} \tau_{p_B}.$$

Notation. Let B be a branch of the poset $(P, \leq)$ and x an element in P. We write $x \in\in B$ to mean that there is a block p in B that contains x. If $x \in p$, then we denote with τ_{x_B}, the semantical maxterm τ_{p_B}.

For any axiomatic extension L of WNM, we define $\mathcal{C}_n^L$ the set of all chains in the variety $\mathbb{V}(L)$ generated by $\{x_1, ..., x_n\}$. Every chain $C = \{c_1 < c_2 < ... < c_k\}$ in $\mathcal{C}_n^L$ could be represented as an ordered partition $\mathcal{W}$ of $\mathcal{F}_n$. Note that the set of blocks $\{x_i^C | 1 \leq i \leq n\}$ generates the chain C.

Taking the quotient C' of C by a congruence R, we obtain a chain $C' = C/R$ whose elements are the equivalence classes $[c_i]_R$, that is, subsets of $\{c_1, ..., c_k\}$. Identifying each block W_i' of C', as the union of all blocks containing R-equivalent elements of C, we could represent also C' as an ordered partition $\mathcal{W}'$ of $\mathcal{F}_n$. Under this identification, we denote with $\mathcal{K}_n^L$ the set of all chains generated by $\{x_1, ..., x_n\}$ that are not proper quotients of elements of $\mathcal{C}_n^L$. Note that, given $D \in \mathcal{K}_n^L$, the homomorphic images of D that are not proper quotients of D (but only isomorphic to proper quotients), belong to $\mathcal{K}_n^L$. For instance, in the following example C' is a proper quotient of C.

Example 2.6. Let $C' = \{\{\bar{0}, \neg x_2\} < \{x_1\} < \{\neg x_1, \neg\neg x_1\} < \{\bar{1}, x_2, \neg\neg x_2\}\}$. Clearly, C' is obtained as a quotient of $C = \{\{\bar{0}, \neg x_2\} < \{x_1\} < \{\neg x_1, \neg\neg x_1\} < \{x_2\} < \{\bar{1}, \neg\neg x_2\}\}$, by the congruence R such that the only proper equivalences are $x_2 R \bar{1}$ and $x_2 R \neg\neg x_2$.

Lemma 2.7 ([1]). *Let* $C = \{c_0 < ... < c_k\}$ *be a chain in* $\mathbb{V}(L)$*, where* L *is a schematic extension of WNM. Then, for any congruence* R *on* C*, if* $c_i \leq c_j$ *and* $c_i R c_k$ *then* $c_j R c_k$*. Moreover, if the equivalence class of* c_i *under* R *is not equal to the singleton* $\{c_i\}$*, then* $c_i R c_k$ *or* $\neg c_i R c_k$*.*

Lemma 2.8 ([1]). *The map*

$$\varphi \in F_n(L) \to (\varphi^C)_{C \in \mathcal{K}_n^L} \in \prod_{C \in \mathcal{K}_n^L} C$$

is a monomorphism.

Let L_n be a finite poset. Suppose to have proved that, for a logic L, the free algebra $F_n(L)$ can be represented as an algebra $\mathbf{S}(L_n)$ whose lattice reduct is $S(L_n)$. Then,

Definition 2.9. If B is a branch of $S(L_n)$ and $p \in B$, a *syntactical maxterm* is a formula v_{pB} such that the section $v_{pB}^{S(L_n)} = [v_{pB}^{B'}]_{B' \in \mathcal{B}(L_n)}$ is a semantical maxterm for p_B.

As a consequence of Theorem 2.5, we have:

Corollary 2.10. *If* $F_n(L)$ *is isomorphic to* $\mathbf{S}(L_n)$ *and* φ *is a formula of* L*, then:*

$$\varphi \equiv_L \bigwedge_{B \in \mathcal{B}(L_n)} v_{\varphi_B^B}$$

With $\equiv_L$ we mean the logical equivalence in the logic L. Since we have defined $\bar{1}$ in our language, we can define normal forms:

Definition 2.11. Let $F_n(L)$ be isomorphic to $\mathbf{S}(L_n)$ and φ a formula of L, then:

- if $\varphi \equiv_L \bar{1}$ then $NF(\varphi) := \bar{1}$

- otherwise:

$$NF(\varphi) := \bigwedge_{B \in \mathcal{B}(L_n), \varphi^B \neq \bar{1}^B} \nu_{\varphi_B^B}.$$

It follows that:

Theorem 2.12. $\varphi \equiv_L NF(\varphi)$.

2.2 Gödel logic

In this section we recall the combinatorial characterization of free n-generated Gödel algebras found in [1]. Since in each Gödel chain represented as an ordered partitions of $\mathcal{F}_n$, the block to which $\neg x_i$ belongs is determined by the block that contains x_i, negation of elements do not brings any information. Hence we can eliminate the elements $\neg x_i$ in every ordered partition $\mathcal{W}$ that represents a Gödel chain. Throughout this section, by ordered partition we mean an ordered partition of the set $\{\bar{0}, x_1, ..., x_n, \bar{1}\}$.

Lemma 2.13 ([1]). *A Gödel chain C represented as an ordered partition $\{W_0 < ... < W_k\}$, belongs to $\mathcal{K}_n^G$ if and only if $W_k = \{1\}$.*

Lemma 2.14 ([1]). *For any formula $\varphi(x_1, ..., x_n), (\varphi^C)_{C \in \mathcal{K}_n^G}$ is an element of $\prod_{C \in \mathcal{K}_n^G} C$ that satisfies the* prefix property*:*

let C, C' be chains in $\mathcal{K}_n^G$ with a common prefix $W_0 < ... < W_h$, and $\varphi^C = W_i$ for some $0 \leq i \leq h$, then $\varphi^{C'} = W_i$.

Given a set of Gödel chains $\mathcal{C}$, we construct a poset $\Gamma(\mathcal{C})$. Let C_1 and C_2 be two chains in $\mathcal{C}$, with P their longest common prefix. Then, the poset $P \oplus (C_1' \sqcup C_2')$ belongs to $\Gamma(\mathcal{C})$. For each chain C in $\mathcal{C}$ there is a unique branch of $\Gamma(\mathcal{C})$ that is a unique copy of C and every branch of $\Gamma(\mathcal{C})$ is a copy of a unique chain in $\mathcal{C}$. We call G_n the poset $\Gamma(\mathcal{K}_n^G)$ and $S(G_n)$ the poset of sections over G_n. Note that $\mathbf{S}(G_n) = \langle S(G_n), \wedge, \rightarrow, \perp \rangle$ is a Gödel algebra, where $\perp$ is the least section in $S(G_n)$ and the operations are defined branch-wise. That is, $[p_B]_{\mathcal{B}} \wedge [q_B]_{\mathcal{B}} = [p_B \wedge q_B]_{\mathcal{B}}, [p_B]_{\mathcal{B}} \rightarrow [q_B]_{\mathcal{B}} = [p_B \rightarrow q_B]_{\mathcal{B}}$, where $\mathcal{B} = \mathcal{B}(S(G_n))$.

2.2.1 Normal forms   In order to describe Gödel maxterms consider branches of the form $B = \{W_0 < ... < W_w < W_{w+1} = \{1\}\}$ and define formulas for $i \in \{1, ..., w\}$:

$$\pi_i := \bigvee_{y \in W_i, y \neq z_i} (z_i \leftrightarrow y) \rightarrow z_i$$

The behaviour of π_i is such that $\pi_i = z_i$ over the Gödel chain B, since every y belongs to W_i. For every other Gödel chain $B' \neq B$, $\pi_i^{B'} = \bar{1}$.

Let B be the same Gödel chain as above, we define terms for each $i \in \{0, ..., w - 1\}$ and each $j \in \{0, ..., w\}$, $k \in \{j + 1, ..., w + 1\}$:

$$\rho_i := z_{i+1} \to z_i \qquad \rho_j^k := \bigvee_{y \in W_k} (y \to z_j).$$

The behaviour of the above defined terms are such that $\rho_i = z_i$ and $\rho_i^k = z_j$ when evaluated over B. Otherwise $\rho_i = \bar{1}$ if $z_{i+1} \leq z_i$ and $\rho_i^k = \bar{1}$ if $z_j \geq y$ for some y. This could be the case for some Gödel chain $B' \neq B$.

Note that $\rho_j^{w+1} \equiv_G z_j$, for each element $p \in\in B$, with $p \in W_j$ for some $j \in \{0, ..., w\}$ we set:

$$v_{p_B} := p_B \vee \bigvee_{i=0}^{j} \pi_i \vee \bigvee_{i=0}^{j-1} \rho_i \vee \bigvee_{i=j+1}^{w} \rho_j^i.$$

Theorem 2.15 ([1]). *For each branch $B = \{W_0 < ... < W_w < W_{w+1} = \{1\}\} \in \mathcal{B}(G_n)$ and each element $p \in\in B$, the formula v_{p_B} is a syntactical maxterm for G_n.*

Let $NF(\varphi)$ defined as in Definition 2.11, from Corollary 2.10 and Theorem 2.12 we conclude:

$$\varphi \equiv_G NF(\varphi).$$

2.2.2 Combinatorial description   The construction of the previous part leads to the following representation theorem.

Theorem 2.16 ([1]). *The free algebra $F_n(G)$ is isomorphic to the algebra of sections $\mathbf{S}(G_n)$.*

Theorem 2.17 ([1]). *The poset G_n is nice and its type is $o(G_n) = H_n \sqcup (H_n)_\perp$, with $H_0 = 1$ and*

$$H_n = \bigsqcup_{i=0}^{n-1} \binom{n}{i} (H_i)_\perp$$

Thanks to the two previous theorems, it is possible to reobtain recurrence formula for computing the cardinality of the free n-generated Gödel algebras.

Corollary 2.18 ([9]). $|F_n(G)| = |S(H_n)|^2 + |S(H_n)|$, *where*

$$|S(H_n)| = \prod_{i=0}^{n-1} (|S(H_i)| + 1)^{\binom{n}{i}}.$$

3   Free RDP-algebras

We represent RDP chains generated by $\{x_1, ..., x_n\}$ as ordered partitions $\mathcal{W}$ of the set $\mathcal{F}_n$. Before introducing a characterization of RDP chains, we give an example of how negations and double negations behave in RDP chains.

Example 3.1. Given a RDP chain $C = \{W_0 < W_1 < W_2 < W_3 < W_4\}$ generated by $\{x_1, x_2, x_3\}$. Then, $\bar{0} \in W_0$ and $\bar{1} \in W_4$. Let $x_1 \in W_1$, $x_2 \in W_2$, $x_3 \in W_3$ and $x_2 = \neg x_2$. Then, by equation (2.3), $x_1 < \neg x_1$ and $\neg x_3 = \bar{0}$. Hence, $W_1 = \{\bar{0}, \neg x_3\}$, $W_1 = \{x_1\}$, $W_2 = \{x_2, \neg x_2, \neg\neg x_2, \neg x_1, \neg\neg x_1\}$, $W_3 = \{x_3\}$ and $W_4 = \{\bar{1}, \neg\neg x_3\}$.

Every ordered partition $\mathcal{W}$ will be of the form $\{W_0 < ... < W_M < ... < W_k\}$ where:

$$x_i \in W_0 \text{ if and only if } \neg x_i \in W_k \text{ if and only if } \neg\neg x_i \in W_0;$$
$$\text{if } x_i > \neg x_i \text{ then } \neg x_i \in W_0, \neg\neg x_i \in W_k \text{ and } x_i \in W_j \text{ for some } j > M; \quad (3.1)$$
$$\text{if } \bar{0} < x_i \leq \neg x_i \text{ then } \neg x_i, \neg\neg x_i \in W_M.$$

From these three conditions, it follows that the set W_M is given by $\{x_i \mid x_i = \neg x_i\} \cup \{\neg x_i, \neg\neg x_i \mid \bar{0} < x_i \leq \neg x_i\}$. If $W_M = \emptyset$ then every x_i belongs to some W_j such that $M < j$ or $j = 0$. Moreover, for any generator x_i, the block which $\neg\neg x_i$ belongs to is uniquely determinated by the blocks that contain x_i and $\neg x_i$. Therefore we can remove double negations from $\mathcal{F}_n$. Throughout the rest of the paper, by ordered partition we mean an ordered partition of the set $\{\bar{0}, x_1, ...x_n, \neg x_1, ..., \neg x_n, \bar{1}\}$. We let $\mathcal{C}_n^{RDP}$ the set of all RDP chains generated by $\{x_1, ..., x_n\}$.

Remark 3.2. If x is a variable, then we write x^{-1} to denote its negation $\neg x$ and x^1 to denote the variable itself. Given a chain $C = \{W_0 < ... < W_M < ... < W_k\}$ in $\mathcal{C}_n^{RDP}$, let $\mu_i \in \{1, -1\}$. Then, by (3.1) the sets $W_0 \backslash \{\bar{0}\}$, W_M and $W_k \backslash \{\bar{1}\}$ are subsets of $\{x_1^{\mu_1}, ..., x_n^{\mu_n}\}$, where a pair of the form $\{x_i^1, x_i^{-1}\}$ may belong only to W_M. The blocks W_l with $0 < l < M$ or $M < l < k$, are subsets of $\{x_1^1, ..., x_n^1\}$.

Lemma 3.3. $C = \{W_0 < ... < W_M < ... < W_k\} \in \mathcal{K}_n^{RDP}$ if and only if $\{x_1, ..., x_n\} \cap W_k = \emptyset$.

Proof. Given $C = \{W_0 < ... < W_M < ... < W_k\}$, suppose that $x_i \in W_k$ then by Lemma 2.7, there exists a chain C' of the form $\{W_0 < ... < \{x_i\} \cup S < W_k\}$ for some S, and a congruence R on C' such that $C = C'/R$ and $x_i R \bar{1}$. Hence $C \notin \mathcal{K}_n^{RDP}$.

Let $C = \{W_0 < ... < W_M < ... < W_k\}$ be a RDP chain in $\mathcal{C}_n^{RDP}$ with $\{x_1, ..., x_n\} \cap W_k = \emptyset$. If C is a quotient of some RDP chain C' by a congruence R, then by Lemma 2.7 $C = C'$. Hence $C \in \mathcal{K}_n^{RDP}$. $\qquad \square$

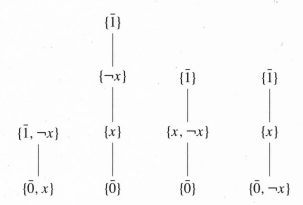

Figure 3.1. The four chains in $\mathcal{K}_1^{RDP}$. Note that W_M is empty in the first chain and in the last chain, W_M is equal to $\{\neg x\}$ and $\{x, \neg x\}$ in the second and in the third chain respectively.

Every chain C in $\mathcal{C}_n^{RDP}$ has the form $C = W_0 \oplus C_\downarrow \oplus W_M \oplus C_\uparrow$, where W_M may be empty. Further, consider the subchain $W_0 \oplus C_\uparrow$. Then, $W_0 \oplus C_\uparrow$ is a Gödel subalgebra of C generated by $\{x_1, ..., x_n\} \cap \bigcup W_l$, with $l = 0$ or $l > M$.

Lemma 3.4. *For any formula* $\varphi(x_1, ..., x_n)$, $(\varphi^C)_{C \in \mathcal{K}_n}$ *is an element of* $\prod_{C \in \mathcal{K}_n} C$ *satisfying the following form of the* prefix *property:*
  if U, V are chains in $\mathcal{K}_n^{RDP}$ with a common prefix $W_0 < ... < W_h$ for some $h \geq M$ (hence $U_\downarrow = V_\downarrow$), and $\varphi^U = W_i$ for some $i \leq h$, then $\varphi^V = W_i$.

Proof. Recall that with φ^C we denote the block in C containing the value of φ. The proof is by induction on the structure of φ. The base case is $\varphi = x_i$. Let $\varphi^U = W_l$, hence W_l is the unique block in U that contains x_i. If $l \leq h$, then W_l is the unique block of V that contains x_i, hence $\varphi^V = W_l$.
  We suppose that the prefix property holds for ψ_1, ψ_2.

$\underline{\varphi = \psi_1 \wedge \psi_2}$. Let $\psi_1^U = W_i$ and $\psi_2^U = W_j$. Without loss of generality we suppose that $W_i \leq W_j$, then $\varphi^U = W_i$. If $\psi_1^U \leq \psi_2^U \leq W_h$, then by induction $\psi_1^V \leq \psi_2^V \leq W_h$. We conclude that $\varphi^V = W_i$. If $\psi_1^U \leq W_h < \psi_2^U$, then by induction $W_i = \psi_1^V < \psi_2^V = W_j$, for some $j > h$. We deduce that $\varphi^V = W_i$.
$\underline{\varphi = \psi_1 \odot \psi_2}$, then we have:

    – if $\psi_1^U, \psi_2^U \leq W_M$ then $\psi_1^U = W_i, \psi_2^U = W_j$ with $i, j \leq M$, so $\varphi^U = W_0$. Then, by induction $\psi_1^V, \psi_2^V \leq W_M$, so $\varphi^V = W_0$.

– if $\psi_1^U \leq W_h$ and $\psi_2^U > W_h$, the formula φ reduces to $\min(\psi_1, \psi_2)$ and $\varphi^U = \psi_1^U = W_i$. By induction $W_i = \psi_1^V \leq W_h$ and $\psi_2^V > W_h$, then $\varphi^V = W_i$. The case $\psi_2^U \leq W_h$ and $\psi_1^U > W_h$ is analogous.

$\varphi = \psi_1 \to \psi_2$, then:

– if $\psi_2^U < \psi_1^U \leq W_M$, then $\varphi^U = W_M \leq W_h$. By induction also $\psi_2^V < \psi_1^V \leq W_M$ so $\varphi^V = W_M \leq W_h$.

– If $\psi_1^U > W_h$ and $\psi_2^U = W_j \leq W_h$ then $\varphi^U = \psi_2^U = W_j$. By induction $\psi_1^V > W_h$ and $\psi_2^V = W_j$ so $\varphi^V = W_j$.   □

We call RDP_n the poset obtained by $\Gamma(\mathcal{K}_n^{RDP})$, using the prefix property stated in the previous Lemma (the operator Γ is defined in Section 2.2). See Figure 3.2 as an example.

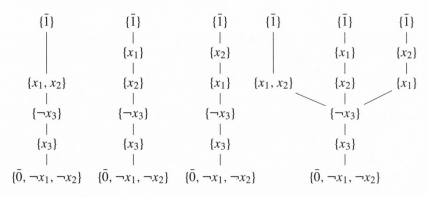

Figure 3.2. Three RDP chains in $\mathcal{K}_3^{RDP}$ and the poset resulting by merging their common prefix.

The poset of sections over RDP_n will be denoted $S(RDP_n)$. The algebraic structure of each chain in $\mathcal{K}_n^{RDP}$ is preserved in the corresponding branch in $\mathcal{B}(RDP_n)$. The algebra $\mathbf{S}(RDP_n) = \langle S(RDP_n), \odot, \wedge, \to, \bot \rangle$, is an RDP algebra where $\bot$ is the least section in $S(RDP_n)$ and the operations $\odot, \wedge, \to$ are defined componentwise. That is, $[p_B]_\mathcal{B} \odot [q_B]_\mathcal{B} = [p_B \odot q_B]_\mathcal{B}$, $[p_B]_\mathcal{B} \wedge [q_B]_\mathcal{B} = [p_B \wedge q_B]_\mathcal{B}$, $[p_B]_\mathcal{B} \to [q_B]_\mathcal{B} = [p_B \to q_B]_\mathcal{B}$, where $\mathcal{B} = \mathcal{B}(S(RDP_n))$. By construction:

Lemma 3.5. *The map*

$$(\varphi^C)_{C \in \mathcal{K}_n^{RDP}} \to [\varphi^B]_{\mathcal{B}(RDP_n)}$$

is a monomorphism from $\prod_{C \in \mathcal{K}_n^{RDP}}$ *to* $S(RDP_n)$. *Hence, by Lemma 2.8 the free algebra* $F_n(RDP)$ *can be embedded in the algebra* $\mathbf{S}(RDP_n)$ *of sections over* RDP_n.

3.1 Normal Forms

In this section we will show that the free n-generated RDP algebra and the algebra of sections $S(RDP_n)$ are isomorphic. We have already shown that between this two algebras there is an embedding (Lemma 3.5). In order to obtain an isomorphism, we need to show that for every section $[p_B]_{\mathcal{B}(RDP_n)}$ in $S(RDP_n)$ it is possible to construct a RDP logic formula that computes $[p_B]_{\mathcal{B}(RDP_n)}$ when evaluated on $S(RDP_n)$. We will build these logical formulas through normal forms (Definition 2.11), hence we need to introduce maxterms for RDP logic.

Let $B = \{W_0 < ... < W_M < ... W_k\}$ be a RDP chain in $\mathcal{B}(RDP)$. For every block W_i with $i > 0$, we fix an element z_i. Moreover, we set $z_0 = \bar{0}$. We define the following formulas:

$$\delta_i := (z_{i+1} \to z_i) \to (\neg(z_{i+1} \to z_i))$$

$$s_i := z_i \to \neg z_i$$

$$C_i := \bigvee_{y \in W_i, y \neq z_i} (z_i \leftrightarrow y) \to \neg(z_i \leftrightarrow y)$$

The behaviour of δ_i, s_i and C_i over B are such that $\delta_i = s_i = C_i = \bar{1}$ if and only if $i \leq M$ otherwise $\delta_i = s_i = C_i = \bar{0}$.

We have seen that each RDP chain C could be represented as a vertical sum $W_0 \oplus C_\downarrow \oplus W_M \oplus C_\uparrow$, where $W_0 \oplus C_\uparrow$ is a Gödel chain. Thanks to this, we can safely apply the terms π and ρ already defined in the Section 2.2.1, on elements of C that belongs to blocks W_i where $i > M$.

For each $p \in\in B$, with $p \in W_l$, if $l > M$ then we set $j = l$ and $h = w$, otherwise we set $j = 0$ and $h = 0$. We define:

$$v_{p_B} := p_B \vee \bigvee_{i=M+1}^{j} \pi_i \vee \bigvee_{i=M+1}^{j-1} \rho_i \vee \bigvee_{i=j+1}^{h} \rho_j^i \vee \bigvee_{i=M+1}^{w} s_i \vee \bigvee_{i=0}^{M} \neg s_i \vee \bigvee_{i=0}^{M-1} \neg \delta_i \vee \bigvee_{i=0}^{M} C_i.$$

Theorem 3.6. *For any* $B = \{W_0 < ... < W_w < \bar{1}\} \in \mathcal{B}(RDP_n)$, v_{p_B} *is a syntactical maxterm for* RDP_n.

Proof. By direct inspection $v_{p_B}^B = p_B$. Moreover, in order to show that v_{p_B} is a syntactical maxterm (see Definition 2.9), we have to prove that for any branch $B' \neq B$ in $\mathcal{B}(RDP_n)$, we have $v_{p_B}^{B'} = \bar{1}$. Let $B' = \{W_0 < ... < W_{k-1} < V_k < ... < V_v\}$ with $V_k \neq W_k$, where $\bar{1} \in V_v$. Hence, $\{W_0 < ... < W_{k-1}\}$ is the longest common prefix of B and B'. We have to distinguish two cases:

- B and B' share a common prefix of length $k - 1 \geq M$.
  B' differs from B on elements belonging to blocks W_l where $l > k - 1 \geq M$. Since $W_0 \oplus C_\uparrow$ and $W_0' \oplus C_\uparrow'$ (the Gödel subchain of C') are Gödel chains, there is an index $M < t \leq w$ such that $\pi_t^{B'} = V_v$ or $\rho_t^{B'} = V_v$ (see Section 2.2.1). Hence, $v_{pB}^{B'} = V_v = \bar{1}^{B'}$.
- B and B' share a common prefix of length $0 \leq k < M$.
  We now assume that $z_k^{B'} < z_{k+1}^{B'} < ... < z_w^{B'}$. Otherwise, there will be an index t, $k \leq t \leq M$ such that $\neg \delta_t^{B'} = V_v$, or an index $M < t \leq w$ such that $\rho_t^{B'} = V_v$. We know $B \neq B'$, so:

  - either there exists a index $k \leq t \leq w$ such that $z_t^B \leq W_M$ and $z_t^{B'} > W_M$, hence $\neg s_t^{B'} = V_v$. Viceversa, $z_t^B > W_M$ and $z_t^{B'} \leq W_M$, hence $s_t^{B'} = V_v$;
  - or there exists an index $k \leq t \leq w$ such that $x \in W_t^B$ but $x \notin W_t^{B'}$, so $C_t^{B'} = V_v$.

We have shown that the maxterm v_{pB}, when evaluated on branches $B' \neq B$, is always equal to the top element. This settles the claim. □

Hence, we can define normal forms for RDP logic. Given $NF(\varphi)$ defined as in Definition 2.11, by Corollary 2.10 and Theorem 2.12 we have:

$$\varphi \equiv_{RDP} NF(\varphi).$$

Theorem 3.7. *The free algebra $F_n(RDP)$ is isomorphic to the algebra of sections $\mathbf{S}(RDP_n)$.*

Proof. By Lemma 2.8 and Lemma 3.5, $F_n(RDP)$ can be embedded in $\mathbf{S}(RDP_n)$. We have to show that for any section $[p_B]_{\mathcal{B}(RDP_n)}$ there is a term φ in n variables such that:

$$[\varphi^B]_{\mathcal{B}(RDP_n)} = [p_B]_{\mathcal{B}(RDP_n)}.$$

By Theorem 3.6 $v_{pB}^{\mathbf{S}(RDP_n)} = [v_{pB}^{B'}]_{B' \in \mathcal{B}(RDP_n)}$ is the semantical maxterm taking value p over B. Then, for any section $[p_B]_{\mathcal{B}(RDP_n)}$ over RDP_n, by Theorem 2.5

$$\left(\bigwedge_{B \in \mathcal{B}(RDP_n)} v_{pB} \right)^{\mathbf{S}(RDP_n)} = \left[\bigwedge_{B \in \mathcal{B}(RDP_n)} v_{pB}^{B'} \right]_{B' \in \mathcal{B}(RDP_n)} = [p_B]_{\mathcal{B}(RDP_n)}. \quad \Box$$

3.2   Combinatorial description

In the previous section we have shown that the algebra $\mathbf{S}(RDP_n)$ is isomorphic to the free n-generated RDP algebra $F_n(RDP)$. Now, we describe the type of RDP_n, showing that the poset is nice. Moreover, from this characterization we obtain a recurrence formula which gives the cardinality of $F_n(RDP)$. As a corollary, we prove that each algebra $F_n(RDP)$ can always be factorized as a direct product, where one of the two factors is $F_n(G)$. Finally, we will show that Gödel logic can be embedded in RDP logic.

Theorem 3.8. RDP_n is nice *and its type is:*

$$\bigsqcup_{u=0}^{n} \binom{n}{u}(H_u)_\perp$$

$$\sqcup$$

$$\bigsqcup_{u=0}^{n}\bigsqcup_{m=1}^{n-u} \binom{n}{u}\binom{n-u}{m}(\mathbf{1}\oplus(H_u)_\perp)$$

$$\sqcup$$

$$\bigsqcup_{u=0}^{n}\bigsqcup_{m=1}^{n-u}\bigsqcup_{d=1}^{n-u-m} \binom{n}{u}\binom{n-u}{m}\binom{n-u-m}{d}\bigsqcup_{c=1}^{d}\left(c!\begin{Bmatrix}d\\c\end{Bmatrix}\mathbf{c}_\perp\oplus(H_u)_\perp\right)$$

$$(3.2)$$

where H_i as in Theorem 2.17.

Proof. RDP chains have the form $C = W_0 \oplus C_\downarrow \oplus W_M \oplus C_\uparrow$. For every chain C in $\mathcal{K}_n^{RDP}$, we define:

- $u = |\{x_i|$ such that $x_i > \neg x_i\}| = |\bigcup_{W\in C_\uparrow} W|$,
- $m = |\{x_i|$ such that $x_i = \neg x_i\}| = |W_M|$,
- $d = |\{x_i|$ such that $x_i \in W_j$, with $0 < j < M\}| = |\bigcup_{W\in C_\downarrow} W|$.

It follows that $n - u - m - d$ is the number of variables in W_0.

Let C be a chain in $\mathcal{K}_n^{RDP}$. We define the following three subsets of $\mathcal{K}_n^{RDP}$:

C belongs to $\mathcal{U}_n$ if and only if $C = W_0 \oplus C_\uparrow$, then $m = d = 0$; C belongs to $\mathcal{M}_n$ if and only if $C = W_0 \oplus W_M \oplus C_\uparrow$, then $d = 0$; and C belongs to $\mathcal{D}_n$, if and only if $d > 0$. Note that $\mathcal{K}_n^{RDP} = \mathcal{U}_n \sqcup \mathcal{M}_n \sqcup \mathcal{D}_n$.

$RDP_n = \Gamma(\mathcal{K}_n^{RDP})$, hence the poset is formed by all chains in $\mathcal{K}_n^{RDP}$ merged at common prefix, by Lemma 3.4. It follows that $RDP_n = \Gamma(\mathcal{K}_n^{RDP}) = \Gamma(\mathcal{U}_n \sqcup \mathcal{M}_n \sqcup \mathcal{D}_n)$. Take three chains $C \in \mathcal{U}_n$, $C' \in \mathcal{M}_n$ and

$C'' \in \mathcal{D}_n$. We have that $C_\downarrow = C'_\downarrow = \emptyset$ and $C''_\downarrow \neq \emptyset$. Moreover, in C the block W_M is empty. Then, the chains C, C' and C'' do not share a common prefix. We conclude that, $RDP_n = \Gamma(\mathcal{K}_n^{RDP}) = \Gamma(\mathcal{U}_n) \sqcup \Gamma(\mathcal{M}_n) \sqcup \Gamma(\mathcal{D}_n)$. Hence, the horizontal sum of these three components gives the type of RDP_n.

Given a RDP chain C, we have seen that the subchain $W_0 \oplus C_\uparrow$ is a Gödel chain. Then, for every choice of u, we are able to describe the type of $\Gamma(\mathcal{U}_n)$ using Theorem 2.17:

$$o(\Gamma(\mathcal{U}_n)) = \bigsqcup_{u=0}^{n} \binom{n}{u} (H_u)_\perp.$$

Each branch C in $\Gamma(\mathcal{M}_n)$ has the form $C = W_0 \oplus W_M \oplus C_\uparrow$. Let C and C' be two branches in $\Gamma(\mathcal{M}_n)$. If C and C' have a common prefix P, then $P = W_0 \oplus W_M \oplus P'$, where P' is a subset of $C_\uparrow$. Given P, the subposet A_P of $\Gamma(\mathcal{M}_n)$, such that each branch of A_P has a common prefix P, has type $\mathbf{1} \oplus (H_u)_\perp$. Hence, for all choices of A_P, the type of $\Gamma(\mathcal{M}_n)$ is given by:

$$\bigsqcup_{u=0}^{n} \bigsqcup_{m=1}^{n-u} \binom{n}{u} \binom{n-u}{m} (\mathbf{1} \oplus (H_u)_\perp).$$

If $d > 0$, it means that one can place d variables in $c = \{1, 2, ..., d\}$ blocks. This means that, for each choice of d, there are chains C in $\mathcal{D}_n$, such that $|C_\downarrow| = c$. The number of such chains correspond to ordered partitions of d variables in c blocks. This number is given by the *Stirling Number of the Second Kind*, in symbols $\left\{ {d \atop c} \right\}$ [2], multiplied by $c!$, the number of ordering of c elements. The type of any $C_\downarrow$ is given by its cardinality, in symbols $o(C_\downarrow) = \mathbf{c}$.

Let $C = W_0 \oplus C_\downarrow \oplus W_M \oplus C_\uparrow$ be a branch in $\Gamma(\mathcal{D}_n)$. Given $P = W_0 \oplus C_\downarrow \oplus W_M$, consider the subposet A_P of $\Gamma(\mathcal{D}_n)$, such that each branch of A_P has a common prefix P. Then, the type of A_P is $\mathbf{c}_\perp \oplus (H_u)_\perp$. So, given u, m and d, the type of the subposet of $\Gamma(\mathcal{D}_n)$ composed by all A_P, for all choices of P, is:

$$\left(\bigsqcup_{i=1}^{d} c! \left\{ {d \atop c} \right\} \mathbf{c}_\perp \oplus (H_u)_\perp \right).$$

In conclusion, for every choice of u, m and d, the type of $\Gamma(\mathcal{D}_n)$ is expressed in the third line of (3.2). $\qquad\qquad\Box$

---

[2] The Stirling Number of the Second Kind $\left\{ {d \atop c} \right\}$ counts the number of partitions into c blocks of a set of n elements ([7]).

Example 3.9. With $n = 3$ variables, let $u = 2, m = 0$ and $d = 0$, we have $\binom{3}{2}\binom{1}{0}\binom{1}{0} = 3$ structures of type $(H_2)_\perp$.

With $u = 2, m = 0$ and $d = 1$ there are $\binom{3}{2}\binom{1}{0}\binom{1}{1} \cdot (1!\{^1_1\}) = 3$ structures of type $(1_\perp \oplus (H_2)_\perp)$ (see Figure 3.3). See Figure 3.4 for the type of RDP_2.

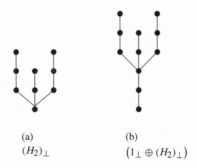

(a)   (b)
$(H_2)_\perp$ $(1_\perp \oplus (H_2)_\perp)$

Figure 3.3. Hasse diagrams of structures in the Example 3.9.

Figure 3.4. The type of the poset RDP_2.

Remark 3.10. Using Iverson's Brackets [3] we can write the type of RDP_n in a more compact form:

$$\bigsqcup_{u=0}^{n} \bigsqcup_{m=0}^{n-u} \bigsqcup_{d=0}^{n-u-m} \binom{n}{u}\binom{n-u}{m}\binom{n-u-m}{d} \cdot$$

$$\left(\left(\bigsqcup_{c=1}^{d} \left(c! \left\{ {d \atop c} \right\} \mathbf{c}_\perp \oplus (H_u)_\perp \right) \sqcup [d = 0]\, ([m > 0]\mathbf{1} \oplus (H_u)_\perp) \right) \right).$$

---

[3] We denote with *Iverson's Brackets* the following condition:

$$[P] = \begin{cases} 1 & \text{if P is true} \\ 0 & \text{otherwise} \end{cases}$$

where P is a proposition.

From the combinatorial description of RDP_n, the cardinality of $F_n(RDP)$ easily follows:

Corollary 3.11.

$$|F_n(RDP)| = \prod_{u=0}^{n} (|S(H_u)| + 1)^{\binom{n}{u}} \cdot \prod_{u=0}^{n} \prod_{m=1}^{n-u} (2 + |S(H_u)|)^{\binom{n}{u}\binom{n-u}{m}}.$$

$$\prod_{u=0}^{n} \prod_{m=1}^{n-u} \prod_{d=1}^{n-u-m} \left(\prod_{c=1}^{d} (c + 2 + |S(H_u)|)^{c!\left\{{d \atop c}\right\}} \right)^{\binom{n}{u}\binom{n-u}{m}\binom{n-u-m}{d}}.$$

Corollary 3.12. $o(RDP_n) = o(G_n) \sqcup o(R_n)$, where:

$$o(R_n) = \bigsqcup_{u=0}^{n} \bigsqcup_{m=0}^{n-u} \bigsqcup_{d=0}^{n-u-m} \binom{n}{u}\binom{n-u}{m}\binom{n-u-m}{d}.$$

$$\left(\bigsqcup_{c=1}^{d} \left(c!\left\{{d \atop c}\right\} \mathbf{c}_\perp \oplus (H_u)_\perp \right) \sqcup [d=0][m>0](\mathbf{1} \oplus (H_u)_\perp) \right).$$

Proof. From (3.2), we immediately sees that, when $d = m = 0$ the type of RDP_n reduces at:

$$\bigsqcup_{u=0}^{n} \binom{n}{u}(H_u)_\perp. \tag{3.3}$$

We can write (3.3) as:

$$\left(\bigsqcup_{u=0}^{n-1} \binom{n}{u}(H_u)_\perp \right) \sqcup (H_n)_\perp = H_n \sqcup (H_n)_\perp = o(G_n),$$

the last equality is given by Theorem 2.17.   □

As a direct consequence, it follows:

Corollary 3.13. $|F_n(RDP)| = |S(RDP_n)| = |S(R_n)| \cdot |S(G_n)| = |S(R_n)| \cdot |F_n(G)|$.

Now, we show that this algebraic relation between Gödel algebras and RDP algebras is reflected also in their logical counterparts, allowing to embed Gödel logic in RDP logic.

Theorem 3.14. *For any formula φ in the language of RDP logic:*

$$G \models \varphi \text{ if and only if } RDP \models \widehat{\varphi},$$

where the formula $\widehat{\varphi}$ is inductively defined as:

- *if $\varphi = x$ then $\widehat{\varphi} = x \odot x$;*
- *if $\varphi = \psi_1 * \psi_2$ then $\widehat{\varphi} = \widehat{\psi_1} * \widehat{\psi_2}$, where $* = \{\odot, \rightarrow, \wedge\}$.*

Proof. Let $\varphi(x_1, ..., x_n)$ be a formula in the language of RDP logic. We show that for every chain $C = W_0 \oplus C_\downarrow \oplus W_M \oplus C_\uparrow$ in $\mathcal{B}(RDP_n)$ if $\varphi^C \leq \neg\varphi^C$ then $\widehat{\varphi}^C = \bar{0}^C$ otherwise $\widehat{\varphi}^C = \varphi^C$ (note that the latter is always the case if $C \in \mathcal{B}(G_n)$).

By induction on the structure of φ. Let $\varphi = x_i$, if we evaluate $\widehat{\varphi}$ on C we have two cases; x_i belongs to some block in $\{W_0 \oplus C_\downarrow \oplus W_M\}$, then $x_i \odot x_i = \bar{0}$, otherwise $x_i \in\in C_\uparrow$ then $x_i \odot x_i = x_i$. Hence, $\widehat{\varphi}^C = \varphi^{W_\perp \oplus C_\uparrow}$, where $W_\perp \oplus C_\uparrow$ is the Gödel chain such that $W_\perp = \{x_i \mid x_i \in W_0$ or $x_i \in C_\downarrow$ or $x_i \in W_M\}$.

Let $\varphi = \psi_1 * \psi_2$ with $* \in \{\odot, \rightarrow, \wedge\}$. We assume that the proposition is valid for ψ_1 and ψ_2. If $\varphi = \psi_1 \rightarrow \psi_2$ we have to distinguish two cases:

- let $\psi_1^C \leq \psi_2^C$. Then $\varphi^C = \bar{1}^C$.
  Suppose $\psi_1^C \leq \neg\psi_1^C$ and $\psi_2^C \leq \neg\psi_2^C$, then by induction hypothesis $\widehat{\psi_1}^C = \bar{0}^C$ and $\widehat{\psi_2}^C = \bar{0}^C$. Hence, $\widehat{\varphi}^C = \widehat{\psi_1}^C \rightarrow \widehat{\psi_2}^C = \bar{1}^C$. Let $\psi_1^C \leq \neg\psi_1^C$ and $\psi_2^C > \neg\psi_2^C$, by induction hypothesis we have $\widehat{\psi_1}^C = \bar{0}^C$ and $\widehat{\psi_2}^C = \psi_2^C$. Hence, $\widehat{\varphi} = \bar{1}^C$. Suppose $\psi_1^C > \neg\psi_1^C$, by (3.1) it can only be $\psi_2^C > \neg\psi_2^C$. Then by induction hypothesis $\widehat{\psi_1}^C = \psi_1^C$ and $\widehat{\psi_2}^C = \psi_2^C$. Hence, $\widehat{\varphi}^C = \psi_1^C \rightarrow \psi_2^C = \bar{1}^C$.
- Let $\psi_1^C > \psi_2^C$. Then $\varphi^C = \psi_2^C$.
  Suppose $\psi_1^C \leq \neg\psi_1^C$, by (3.1) it can only be the case $\psi_2^C \leq \neg\psi_2^C$. Then by induction hypothesis $\widehat{\psi_1}^C = \bar{0}^C$ and $\widehat{\psi_2}^C = \bar{0}^C$. Hence, $\widehat{\varphi}^C = \bar{0}^C = \widehat{\psi_2}^C$. Let $\psi_1^C > \neg\psi_1^C$ and $\psi_2^C \leq \neg\psi_2^C$, by induction hypothesis $\widehat{\psi_1}^C = \psi_1^C$ and $\widehat{\psi_2}^C = \bar{0}^C$. Hence, $\widehat{\varphi}^C = \bar{0}^C = \widehat{\psi_2}^C$. Finally, let $\psi_1^C > \neg\psi_1^C$ and $\psi_2^C > \neg\psi_2^C$. Then, by induction $\widehat{\psi_1}^C = \psi_1^C$ and $\widehat{\psi_2}^C = \psi_2^C$. Hence, $\widehat{\varphi}^C = \widehat{\psi_2}^C$.

For the cases $* = \odot$ and $* = \wedge$, we suppose without loss of generality, that $\psi_1^C \leq \psi_2^C$. If $\varphi = \psi_1 \wedge \psi_2$ then $\varphi^C = \psi_1^C$. By induction hypothesis if $\psi_1^C \leq \neg\psi_1^C$ then $\widehat{\varphi}^C = \widehat{\psi_1}^C = \bar{0}^C$ otherwise $\widehat{\varphi}^C = \psi_1^C$. If $\varphi = \psi_1 \odot \psi_2$ then by induction hypothesis if $\psi_1^C \leq \neg\psi_1^C$ then $\widehat{\psi_1}^C = \bar{0}^C$ and so $\widehat{\varphi}^C = \bar{0}^C$, otherwise $\widehat{\psi_1}^C = \psi_1^C$ and then $\widehat{\varphi}^C = \psi_1^C$. $\square$

3.3 Conclusions

Finally, in the following table we report the cardinalities of the free n-generated algebras and their underlying posets, for the two logics studied in this paper, Gödel logic and RDP logic.

	$|F_n(RDP)|$	$|RDP_n|$	$|F_n(G)|$	$|G_n|$
$n = 1$	72	12	6	5
$n = 2$	94556160000	74	342	17
$n = 3$	$\sim 4,06 \cdot 10^{71}$	510	137186159382	77

RDP logic is not among the most investigated many-valued t-norm logics, however the study of RDP logic is a useful case towards an analysis of WNM logic. In this light, our future research will deal with a categorical analysis of RDP logic and some of its logical properties. Collecting this kind of knowledge about extensions of WNM will afford a good starting point to accomplish the same agenda for WNM itself.

In [10], definitions of probability measures where defined for algebraic systems of continuous functions over [0, 1] and recently these definitions have been extended over non-classical events described by Gödel ([3]) and Nilpotent Minimum logics ([2]). At the moment there is no clear understanding of the meaning of a non-classical event. Concrete representations of different logics could be useful tools towards this goal.

ACKNOWLEDGEMENTS. The author would like to thank Professor Stefano Aguzzoli for his useful suggestions about the paper.

References

[1] S. AGUZZOLI and B. GERLA, *Normal Forms and Free Algebras for Some Extensions of MTL*, Fuzzy Set. Syst. **159** (2008), 1131–1152.

[2] S. AGUZZOLI and B. GERLA, *Probability Measures in the logic of Nilpotent Minimum*, Studia Logica, to appear.

[3] S. AGUZZOLI, B. GERLA and V. MARRA, *Defuzzifying formulas in Gödel logic through finitely additive measures*, In: "Proceedings of FuzzIEEE '08", Hong Kong (2008), 1886–1893.

[4] S. BURRIS and H. P. SANKAPPANVAR, "A course in universal algebra", Springer-Verlag, 1981.

[5] B. DAVEY and H. PRIESTLEY, "Introduction to Lattices and Order", Cambridge University Press, 1991.

[6] F. ESTEVA and L. GODO, *Monoidal t-norm based logic: towards a logic for left-continuous t-norms*, Fuzzy Set. Syst., **124** (2001), 271–288.

[7] R. L. GRAHAM, D. E. KNUTH and O. PATASHNIK, "Concrete Mathematics", Addison-Wesley, 1989.

[8] P. HÁJEK, "Metamathematics of Fuzzy Logic", Kluwer, 1998.

[9] A. HORN, *Free L-algebras*, J. Symbolic Logic **34** (1969), 475–480.

[10] J. KÜHR and D. MUNDICI, *De Finetti theorem and Borel states in [0, 1]-valued algebraic logic*, International J. of Approximate Reasoning **46** (2007), 605–616.

[11] S. JENEI, *A note on the ordinal sum theorem and its consequence for the construction of triangular norms*, Fuzzy Set. Syst. **126** (2002), 199–205.

[12] S. JENEI and F. MONTAGNA, *A proof of standard completeness for Esteva and Godo's logic MTL*, Studia Logica **70** (2002), 183–192.

[13] C. NOGUERA, F. ESTEVA and J. GISPERT, *On varieties generated by Weak Nilpotent Minimum t-norms*, In: Proceedings of Fourth EUSFLAT (2005), 866–871.

[14] S. WANG, *A fuzzy logic for the revised drastic product t-norm*, Soft computing **11** (2007), 585–590.

Uncertainty, indeterminacy and fuzziness: A probabilistic approach

1 Introduction

Uncertainty and ignorance are fundamental and unavoidable features of daily life. In practical reasoning, whether it is aimed at drawing inferences or making decisions, we need to give appropriate weight both to our uncertainty, about facts or events or consequences of our actions, and to the indeterminacy that arises from our ignorance about this matters. Thus elaborating a mathematical framework that allows us to represent and treat uncertainty and ignorance is a question of primary importance.

To measure uncertainty we need probability while to measure indeterminacy we need *imprecise probabilities*. Following de Finetti, we adopt a behavioral interpretation both of probability and imprecise probability. However, under the specific conditions of ignorance that will be discussed in this paper, de Finetti's key assumption ceases to be compelling. As we shall see, it is hardly possible to find, for any given event, a betting rate that we regard as *fair*, *i.e.* such that we are willing to accept a bet *both on and against* the event at that rate. This is bound to happen whenever we have little information on which to base our assessment so that our beliefs about the events of interest are indeterminate. In situations of this sort it seems more prudent and overall more 'rational' to give those events a *probability interval*. Whenever this happens, we speak about imprecise probabilities, that is an upper probability strictly greater than lower probability.

The idea is perhaps best illustrated by means of the following toy examples.

Example 1.1. Suppose that a coin is tossed once. There are two possible outcomes, Heads and Tails. If the coin is known to be fair, we assign probability $1/2$ to each of these outcomes. However, suppose that a coin has an unknown bias. If we represent our ignorance using a single probability measure, then perhaps the best thing to do is to say that Heads and Tails are equally likely, and so each gets probability $1/2$, just as in the case of a fair coin.

However, there seems to be a significant qualitative difference between a fair coin and a coin of unknown bias. Is there some way that this difference can be captured? Precise probabilities do not properly model a state of complete ignorance, meaning a total absence of relevant information. The only reasonable choice seems to assign the interval [0, 1] (a vacuous probability) to each outcome in order to maximize the degree of indeterminacy. This means that the adequate model for complete ignorance concerning the space of events, which is really a rather trivial state of uncertainty, are vacuous probabilities.

Example 1.2. Consider now a bag containing 100 marbles, of which 30 are green while the others are white or red in an unknown proportion. It is perfectly sensible to assign 0.3 to the event *green marble* and 0.7 to the event *white or red marble* but how can we assign a precise probability to the events *white* and *red*?

Through a behavioral interpretation in terms of bets, we notice that also in a situation of partial information precise probabilities seem not to be an adequate model. Indeed, if we must assign a precise probability to each event, the best thing to do is perhaps to say that *white* and *red* are equally likely and have a probability 0.35. In principle, betting 1 on *white* is more convenient than betting 1 on *green*, but if we have to choose among these two bets, it seems reasonable to bet on the event *green* which has a sure probability of 0.3.

However we can use imprecise probability to model this situation in a very natural way, namely by assigning the interval [0, 0.7] to *white* and to *red*.

Note that, for the sake of simplicity, we have only considered examples featuring crisp events, *i.e.* events that will occur (true) or will not occur (false). However, many events of interest are "fuzzy": keeping in mind that uncertainty, ignorance and fuzziness are fundamental features of daily life, we aim at constructing a mathematical theory of imprecise probabilities for fuzzy events, based on a behavioral interpretation and principles of coherence.

In the first section we present the subjective approach to probability proposed by de Finetti and Ramsey, which we take as our starting point. Then we briefly recall two extensions to de Finetti's theorem about coherent probability assignments. The former, recalled in Section 3, is Mundici's result [Mub], which generalizes de Finetti's theorem to many-valued events by introducing a many-valued analogue of a probability: the notion of *states* on an MV-algebra. The latter generalization, introduced in Section 5, is due to Walley who provides an interpretation of subjective probability involving *imprecise probabilities*. In Section 4 we generalize de Finetti's theorem about coherent probability assignments to many-valued events about which an agent can give indeterminate probabilities. In particular we define imprecise probabilities (upper and lower probabilities) axiomatically, we give them an interpretation in terms of bets and finally, we propose a de Finetti-style coherence criterion. Indeed, by using a variant of the Hahn-Banach theorem, we characterize an upper (lower) probability as the maximum (minimum) of a nonempty, weak*-closed set of states. More precisely, we prove the following theorem.

Theorem 1.3. *There are mutually inverse bijections between the set of all pairs consisting of an upper probability and its dual lower probability on a 2-divisible MV-algebra* **A**, *and the family of all non-empty weak*-closed and convex sets of states on* **A**.

Then we observe that if the truth value of an event ϕ is very hard to predict, the bookmaker may want to protect himself from the risk of loosing large sums, by adopting the following strategy: he refuses negative bets and he chooses a betting odds α for betting on ϕ and a betting odds β for betting against ϕ, *i.e.* on $\neg\phi$, where β is greater than $1 - \alpha$. For these games (*non-reversible betting games*), we define the notion of *book* and evaluate the extent to which the classical coherence criteria are still valid.

Then, we prove the following main result. Let us say that a *bad* (respectively a *good*) *bet* is a bet for which there is an alternative system of bets ensuring to the bettor a strictly better (respectively a strictly worse) payoff, independently of the truth values of the events involved.

Theorem 1.4. *There is no bad (respectively good) bet based on a book* Δ *in a non-reversible betting game iff* Δ *can be extended to an upper (respectively lower) probability.*

There is neither a bad bet nor a good bet based on Δ *iff* Δ *can be extended to a state.*

2 De Finetti's coherence criterion

Beliefs come in varying degrees: I am more confident that this coin will land Heads when tossed than I am that Milan will win the Champions League this year, and I am even more confident that $2 + 2 = 4$. But if it seems quite natural to quantify our subjective opinion on a given event by using a number, it is important to explain what this number is. In other words, we need to give an adequate answer to the question "What is the probability of an event?"

The most satisfying answer seems to be the one proposed by de Finetti and Ramsey, which is based on the following observation: it does seem reasonable that one's degrees of belief should be mirrored by one's willingness to bet on the relevant events at suitably high odds.

In order to state this formally, let us consider the following bet on the event ϕ. Given a number $\alpha \in [0, 1]$, a bettor bets a real number λ and pays $\lambda\alpha$ to the bookmaker now. If ϕ turns out to be true, then he will get back λ from the bookmaker, and if ϕ turns out to be false, then he will get nothing from her. If the bettor thinks that the bookmaker's betting odds α for the event ϕ is too much, he can reverse the terms of the bet just betting a negative amount λ of money on ϕ.

We suppose that the bettor bets seriously, *i.e.* only when he believes it financially profitable and not for the sake of gambling as such. If he is willing to accept odds α, we suppose him willing to accept the more favorable odds α', where $\alpha' < \alpha$. This means that there exists a least upper bound to the odds which the bettor is prepared to accept the bet, and a similar argument shows that there is a greatest lower bound to the value for which the bettor is prepared to sell the bet reversing his role with the bookmaker.

De Finetti assumes that

Definition 2.1. For each event ϕ of interest, there are betting odds that you regard as fair, in the sense you are indifferent between buying and selling at that price, that is between betting on or against ϕ.This set of fair betting odds represents your *subjective probability* for the event ϕ.

From this viewpoint there seems to be no restrictions on the choice of one's subjective probability: subjective probability is not an absolute value but a number assigned by an individual representing his degrees of belief. Yet de Finetti's well known Dutch Book argument shows that this is not the case: if an agent wants to avoid the perspective of a sure loss he needs to asses his degrees of belief according to the laws of probability.

In order to state this formally, let us give some definitions:

Definition 2.2. A *book in a reversible betting game* is a finite set

$$\Gamma = \{(\phi_1, \alpha_1) \ldots (\phi_n, \alpha_n)\}$$

where for $i = 1, \ldots, n$ ϕ_i is an event and α_i is the betting odds proposed by the bookmaker for a bet on ϕ_i of the form described above.

If for $i = 1, \ldots, n$, the bettor bets λ_i on ϕ_i, and the truth value of ϕ_i is $v(\phi_i)$ his expected gain will be $\sum_{i=1}^{n} \lambda_i (v(\phi_i) - \alpha_i)$.

Definition 2.3. Let Γ be a fixed book in a reversible betting game. A *winning strategy* for the bettor based on Γ consists of a system of bets on a finite subset of events in Γ which ensures to the bettor a strictly positive payoff independently of the relevant outcomes. A winning strategy for the bettor is often called a *Dutch Book* for Γ.

Notice that it would clearly be irrational to accept a book which leads to sure loss. This motivate the following

Criterion 2.1. A book Γ in a reversible betting game is said to be *rational* or *coherent* if it does not permit a Dutch Book.

The following theorem due to de Finetti shows that an agent's degrees of belief are coherent iff they satisfy the axioms of probability (see [19] for a neat proof).

Theorem 2.4. *A book Γ in a reversible betting game is coherent if and only if there is a probability distribution P such that if $(\phi, \alpha) \in \Gamma$, then $P(\phi) = \alpha$.*

3   A Coherence criterion for many-valued events

In this section we recall Mundici's theorem [16], which generalizes de Finetti's result to many-valued events (in the case of reversible games).

First of all we recall that many-valued events are represented by formulas of Łukasiewicz logic and that MV-algebras constitute a variety of universal algebras which is the equivalent algebraic semantic of the Łukasiewicz logic in the sense of Block-Pigozzi (see [1]). Thus formulas of Łukasiewicz logic (and so many-valued events) can be regarded as elements of an MV-algebra $\mathbf{A}$.

Notice that in this context a *valuation* is an homomorphism from the MV-algebra of the formulas $\mathbf{A}$ into the standard MV-algebra $\mathbf{I} = [0, 1]$ (see [4]).

In order to prove an analogue of de Finetti's theorem for Łukasiewicz logic, Mundici [15] introduces a many-valued analogue of a probability measure, that is the notion of state on an MV-algebra.

Definition 3.1. A *state* on an MV-algebra $\mathbf{A}$ is a map $s : \mathbf{A} \to \mathbf{I}$ such that for all $x, y \in \mathbf{A}$:

(i) $s(1) = 1$.
(ii) If $x \odot y = 0$ then $s(x \oplus y) = s(x) + s(y)$.

It has been argued in detail in [15] why states play the same role for MV-algebras as probabilities for Boolean algebras. Indeed, while only finite additivity appears in the formal definition of a state (see Definition 3.1), the following theorem shows that states on an MV-algebra $\mathbf{A}$ are in one to one correspondence with those regular Borel measures on the Hausdorff compact space $X_{\mathbf{A}}$ of valuations on $\mathbf{A}$ which are σ-additive.

Theorem 3.2 (Panti and Kroupa, [12,17]).   *Let* **A** *an MV-algebra. There is a canonical bijective correspondence between the set of states on* **A** *and the set of Borel regular probability measures on the Hausdorff compact space* $X_{\mathbf{A}}$ *of valuations on* **A**.

Let us give now an interpretation of states (*i.e.* probabilities on many-valued events), in terms of betting behavior.

Given a finite book $\Gamma = \{(\phi_1, \alpha_1), \ldots, (\phi_n, \alpha_n)\}$, the rules of the game are as before, with the only difference that $\phi_1, \ldots, \phi_n$ are now events represented by formulas of Łukasiewicz logic and can take values in the real interval $[0, 1]$. Once again, for every fuzzy event ϕ_i there are betting odds that you regard as fair, in the sense you are indifferent between buying and selling at that price. Those represent your subjective probability for the fuzzy event (*i.e.* your subjective state).

The definition of winning strategy for the bettor are as before. Moreover the Dutch Book criterion carries over with the obvious adaptations concerning the valuations. So a Dutch Book, is again a system of bets $\lambda_1, \ldots, \lambda_n$ on $\phi_1, \ldots, \phi_n$ such that the bettor's payoff $\sum_{i=1}^{n} \lambda_i(v(\phi_i) - \alpha_i)$ is strictly positive independently of the valuation v. Indeed, in [16], Mundici proved the following theorem:

Theorem 3.3. *Let* $\Gamma = \{(\phi_1, \alpha_1), \ldots, (\phi_n, \alpha_n)\}$ *be a finite book over events represented by formulas of Łukasiewicz logic. The following are equivalent:*

(i) *There is a state s on the Lindenbaum algebra of Łukasiewicz logic such that, for $i = 1, \ldots, n, s(\phi_i) = \alpha_i$.*

(ii) Γ *is coherent in the sense explained above.*

Thus coherence is a necessary and sufficient condition to ensure that the betting odds set by the bookmaker are extendible to a state on the MV-algebra of events.

4 Imprecise probabilities and uncertain bets on many-valued events

In this section de Finetti's theorem about coherent probability assignments is extended to many-valued events under a condition of ignorance. Firstly we axiomatically define imprecise probabilities (upper and lower probabilities), then we give those a behavioral interpretation in terms of bets and lastly, we propose a de Finetti-style coherence criterion: a necessary and sufficient condition to ensure that the betting odds proposed by the bookmaker can be extended to an upper probability on the algebra of events.

4.1 Preliminaries

For all concepts of Universal Algebra we refer to [3]. For concepts of many-valued logic, we refer to [10], and for MV-algebras in particular, we refer to [4].

First of all we recall that the Lindenbaum-Tarski algebra of the Łukasiewicz logic is a semisimple MV-algebra **A**, whence it has a functional representation since it is isomorphic to a separating MV-algebra of continuous functions from the Hausdorff compact space $X_\mathbf{A}$ of all valuations on **A** to **I** containing the constant function **1** (see [4] 3.6.8). Thus, from now on, fuzzy events may be regarded as continuous functions from the Hausdorff compact space $X_\mathbf{A}$ in [0,1].

Definition 4.1. An MV-algebra **A** is *2-divisible*, if for every $a \in \mathbf{A}$ there is a $b \in \mathbf{A}$ such that $2b = a$ and $b \odot b = 0$. If such a b exists, it is uniquely determined by a and 2 and will also be denoted by $b = \frac{a}{2}$.

For technical reasons that will be clear later, we work in 2-divisible MV-algebras. This is justified because every MV-algebra is embeddable in a 2-divisible one and all dyadic numbers $q = \frac{a}{2^n}$ in a 2-divisible MV-algebra may be regarded as events that have a priori the same value under every possible valuation, *i.e.* as constant events.

Following [8], we define the notions of upper and lower probability on a 2-divisible MV-algebras as follows:

Definition 4.2. An *upper probability* over a 2-divisible MV-algebra **A** is a map $u : \mathbf{A} \to [0, 1]$ such that for all $x, y \in \mathbf{A}$ and for all dyadic rational numbers $q = \frac{m}{2^n}$ in the unit interval:

(Ax1) is order preserving
(Ax2) $u(1) = 1$.
(Ax3) $u(qx) = qu(x)$.
(Ax4) If $x \odot y = 0$ then $u(x \oplus y) \leq u(x) + u(y)$.
(Ax5) If $q \odot x = 0$ then $u(q \oplus x) = q + u(x)$.

A *lower probability* over a 2-divisible MV-algebra **A** is a map $\ell : \mathbf{A} \to [0, 1]$ with the same properties of an upper probability except (Ax4) that becomes:

(Ax4)′ If $x \odot y = 0$ then $\ell(x \oplus y) \geq \ell(x) + \ell(y)$.

Given a 2-divisible MV-algebra **A** and a lower probability ℓ on it, it is straightforward to prove that there is an upper probability u on **A** such that $\ell(x) = 1 - u(\neg x)$ for every $x \in \mathbf{A}$. Upper probabilities are determined by lower probabilities, and vice versa. From now on we will focus on upper probabilities.

Remark 4.3. A map s is a state on **A** if and only if s is at the same time a lower and an upper probability.

Let us briefly recall now the standard terminology for functionals $u\colon C(X) \to \mathbb{R}$, where X is a given Hausdorff compact space.

Definition 4.4. A functional $u\colon C(X) \to \mathbb{R}$ is:

(1) *order preserving* if $f \le g$ implies $u(f) \le u(g)$;
(2) *normalized* if $u(\mathbf{1}) = 1$;
(3) *subadditive (respectively superadditive)* if $u(f + g) \le u(f) + u(g)$ (respectively $\ge$);
(4) *homogeneous* if $u(rf) = ru(f)$ for all $f \in C(X)$ and all $r \ge 0$;
(5) *sublinear (respectively superlinear)* if u is homogeneous and subadditive (respectively superadditive);
(6) *linear* if u is sublinear and superlinear.

Then we add that a functional $u\colon C(X) \to \mathbb{R}$ is:

(7) *weakly linear* if $u(f + r \cdot \mathbf{1}) = u(f) + r$ for all $r \in \mathbb{R}$ and for all $f \in C(X)$.

Theorem 4.5. *Let* **A** *an MV-algebra*

(i) *There is a canonical bijection between states on* **A** *and order preserving, normalized, homogeneous and linear functionals on* $C(X_{\mathbf{A}})$.

If **A** *is a 2-divisible MV-algebra.*

(ii) *There is a canonical bijection between upper (respectively lower) probabilities on* **A** *and order preserving, normalized, homogeneous, weakly linear and sublinear (respectively superlinear) functionals on* $C(X_{\mathbf{A}})$.

Without entering into detalis, for which we refer to [8], we notice that the hypothesis of 2-divisibitily is fundamental.

Indeed, combining the Stone-Weierstraß Theorem for linear subspaces of $C(X)$ that are also sublattices (see, *e.g.*, [21, Chapter V, 8.1]) with [23, Proposition 1.4] we have the following version of the Stone-Weierstraß Theorem for MV-algebras:

Theorem 4.6. *Let* X *be a compact Hausdorff space. Every 2-divisible MV-subalgebra of the MV-algebra* $C(X, \mathbf{I})$ *separating the points of* X *is dense in* $C(X, \mathbf{I})$ *for the topology of uniform convergence.*

From this result we conclude:

Corollary 4.7. *For every semisimple MV-algebra* **A**, *the 2-divisible hull of its functional representation is dense in* $C(X_\mathbf{A}, \mathbf{I})$ *for the topology of uniform convergence.*

This corollary is one of the key tools for proving Theorem 4.5 (ii), thus only if we work in a 2-divisible MV-algebra **A**, we can reduce the study of probabilities and imprecise probabilities over **A** to the study of their counterparts in the dual of $C(X_\mathbf{A})$, where $X_\mathbf{A}$ is an Hausdorff compact space.

4.2 Tools from functional analysis

We refer to [2] and [20] for all concepts of functional analysis.

Let X be an Hausdorff compact space and $\mathcal{M}(X)$ the dual vector space of $C(X)$, that is the set of all linear functionals $\mu : C(X) \to \mathbb{R}$ which are continuous with respect to the supremum norm. We endow $\mathcal{M}(X)$ with the *weak* topology*, which is the weakest topology such that the evaluation maps $\mu \mapsto \mu(f)$ are continuous for all $f \in C(X)$. This topology can also be seen as the topology of pointwise convergence (μ_i converges to μ iff $\mu_i(f)$ converges to $\mu(f)$ for all $f \in C(X)$) induced by the product topology on $\mathbb{R}^{C(X)}$.

Let $\mathcal{P}(X)$ be the set of all order preserving, normalized and linear functionals on $C(X)$.

Lemma 4.8. *For every weak*-continuous linear functional* $F : \mathcal{M}(X) \to \mathbb{R}$ *there is a uniquely determined* $f \in C(X)$ *such that* $F(\mu) = \mu(f)$ *for all* $\mu \in \mathcal{M}(X)$.

Combining this Lemma (see [21, Chapter IV, 1.2]) with the Hahn-Banach theorem and the Separation Theorem we have the following theorems (for details see [8]):

Theorem 4.9.

(i) *For every weak*-closed subset K of $\mathcal{P}(X)$, the functional $K^+ : C(X) \to \mathbb{R}$ defined by $K^+(f) = \max_{\mu \in K} \mu(f)$ is well defined. Moreover it is an order preserving, normalized, sublinear and weakly linear functional.*

(ii) *Conversely, for every order preserving, normalized, sublinear and weakly linear functional $u : C(X) \to \mathbb{R}$, $u_\leq = \{\mu \in \mathcal{P}(X) : \mu \leq u\}$ is a weak*-closed convex subset of $\mathcal{P}(X)$.*

The maps $K \mapsto K^+$ and $u \mapsto u_\leq$ are mutually inverse bijections.

Theorem 4.10.

(i) *For every weak*-closed subset K of $\mathcal{P}(X)$ the functional defined by $K^-(f) = \min_{\mu \in K} \mu(f)$ is well defined. Moreover it is an order preserving, normalized, superlinear and weakly linear functional.*

(ii) *Conversely, for every order preserving, normalized, superlinear and weakly linear functional $u: C(X) \to \mathbb{R}$, $u_{\geq} = \{\mu \in \mathcal{P}(X) : \mu(f) \geq u(f)$ for all $f \in C(X)\}$ is a non-empty weak* closed convex subset of $\mathcal{P}(X)$.*

The maps $K \mapsto K^-$ and $u \mapsto u_{\geq}$ are mutually inverse bijections.

Thanks to the Theorem 4.5 we can adapt the results of this section to imprecise probabilities characterizing imprecise probabilities in terms of sets of states. In particular an upper probability is the maximum of a non-empty convex set of states, closed with respect to the weak* topology and vice versa; a lower probability is the minimum of a non-empty convex set of states, closed with respect to the weak* topology and vice versa.

Lastly we notice that the essential technical tool for proving an analogue of de Finetti's theorem for fuzzy events under condition of ignorance is the following theorem that characterizes those real-valued functions u defined on some subset $D \subseteq C(X)$ that can be extended to superlinear or sublinear functionals.

For $f \in C(X)$, let

$$\text{Max}(f) = \sup_{x \in X} f(x) = \max_{x \in X} f(x),$$

$$\text{Min}(f) = \inf_{x \in X} f(x) = \min_{x \in X} f(x).$$

Theorem 4.11. *Let X be a compact Hausdorff space. Consider a function $u: D \to \mathbb{R}$ defined on a nonempty subset D of $C(X)$.*

(i) *There is a normalized, order preserving, weakly linear and sublinear functional $\widetilde{u}: C(X) \to \mathbb{R}$ extending u if and only if the following condition $(\bar{U})$ is satisfied: for all $f_0, f_1, \ldots, f_n \in D$ and all $r_1, \ldots, r_n \in \mathbb{R}_+$*

$$\text{Min}\left(\sum_{i=1}^{n} r_i f_i - f_0\right) \leq \sum_{i=1}^{n} r_i u(f_i) - u(f_0).$$

(ii) *There is a normalized, order preserving, weakly linear and super-linear functional $\underline{u}\colon C(X) \to \mathbb{R}$ extending u if and only if the following condition $(\underline{U})$ is satisfied: for all $f_0, f_1, \ldots, f_n \in D$ and all $r_1, \ldots, r_n \in \mathbb{R}_+$,*

$$\sum_{i=1}^{n} r_i u(f_i) - u(f_0) \leq \operatorname{Max}\left(\sum_{i=1}^{n} r_i f_i - f_0\right).$$

(iii) *There is a positive, normalized, linear functional $\widetilde{u}\colon C(X) \to \mathbb{R}$ extending u if and only if the following condition (L) is satisfied: for all $f_0, f_1, \ldots, f_n \in D$ and all $r_0, r_1, \ldots, r_n \in \mathbb{R}$,*

$$\operatorname{Min}\left(\sum_{i=0}^{n} r_i f_i\right) \leq \sum_{i=0}^{n} r_i u(f_i) \leq \operatorname{Max}\left(\sum_{i=0}^{n} r_i f_i\right).$$

Proof. See [8]. □

4.3   An interpretation of imprecise probabilities in terms of bets: the bookmaker's viewpoint

In this subsection we try to find a connection between the abstract results of the preceding sections and the interpretation in terms of bets of imprecise probabilities. To begin with, let us consider the following example which justifies the use of imprecise probabilities and of non-reversible games:

Example 4.12. Let us suppose that we know that the bingo drawing is biased. If we don't know how it is biased, the event "A high number will be drawn out" is a fuzzy event characterized by a condition of indeterminacy due to our ignorance about the bias.

Let us adopt now the viewpoint of a real bookmaker: he surely will not be so honest as to propose betting odds that he regards as fair, in the sense he is indifferent between buying and selling at that price the bet on an event. In particular we consider a case in which the bookmaker sets the betting odds for an event ϕ under condition of indeterminacy, as the one of the example. Having no idea about the probability of ϕ, the bookmaker may not be willing to quantify with a single number his subjective opinion about the event as the risk of substantial loss could just be too high. He might argue that under theese circumstances the best thing to do is to choose conditions so disadvantageous for the bettor as to lead him to not bet. This is equivalent to use imprecise probabilities, *i.e.*

to associate to ϕ an interval of probability, choosing a betting odds β for betting on ϕ and a betting odds $\alpha \leq \beta$ for betting against ϕ. The bettor has two options:

(1) he pays $\lambda\beta$ (with $\lambda > 0$) and he will get back $\lambda v(\phi)$;
(2) he receives $\lambda\alpha$ and he will pay $\lambda v(\phi)$,

where $v(\phi)$ is the truth value of ϕ. Games such that the bookmaker cannot switch roles with the bettor are referred to as *non-reversible betting games*.

Remark 4.13. In a state of complete ignorance, meaning a total absence of relevant information about the outcomes of an event, the bookmaker can choose as interval of probability [0, 1].

Let us adopt the viewpoint of the bookmaker, and suppose that he proposes betting odds seriously, *i.e.* only when he believes it can be financially profitable. If he is willing to propose odds β, we suppose he is also willing to propose the more favorable odds β', where $\beta < \beta'$, since he receives more. In the same manner, if he is willing to propose odds α, we suppose him willing to propose the more favorable odds α', where $\alpha' < \alpha$, since he pays a smaller sum.

Now we are able to give a subjective interpretation to imprecise probabilities, based on a behavioral interpretation in terms of bets. In particular we give the following definitions:

Definition 4.14. The *upper probability* of an event ϕ is the infimum betting odds β which a rational non-reversible bookmaker would propose for betting on ϕ at rate β.

The *lower probability* of ϕ can be interpreted as the supremum betting odds α which a rational and non-reversible bookmaker would propose for betting against ϕ at rate α.

Definition 4.15. A *book in a non-reversible betting game* is a finite set of the form $\Delta = \{(\phi_1, [\alpha_1, \beta_1]), \ldots, (\phi_n, [\alpha_n, \beta_n])\}$, where for $0 \leq \alpha_i \leq \beta_i \leq 1$.

Let us notice that if $\alpha_i = \beta_i$ for all $i = 1, \ldots, n$ we are in the case of reversible betting game: indeed, the bettor is supposed willing to bet at the same odds whether the stakes λ are positive or negative, that is, whether he is willing to buy the bet betting on ϕ or to sell the bet betting against ϕ.

Remark 4.16. If the bettor bets a stake λ against ϕ when the proposed betting odds is α, his payoff will be $\Pi_{\text{bettor}} = \lambda(\alpha - v(\phi))$.

On the other hand, if he bets the same stake λ on $\neg\phi$ when the betting odds proposed is $1 - \alpha$, he pays $\lambda(1 - \alpha)$ and he will get back $\lambda v(\phi)$. Thus his payoff will be:

$$\Pi_{\text{bettor}} = -\lambda(1-\alpha)+\lambda v(\neg\phi) = -\lambda(1-\alpha)+\lambda(1-v(\phi)) = \lambda(\alpha-v(\phi)).$$

This means that betting against ϕ at a betting odds α is equivalent (in the sense that gives the same payoff to the bettor) to betting on $\neg\phi$ at a betting odds $1 - \alpha$.

This remark has two important consequences. The first one is that given a book Δ in a non-reversible game, every element $(\phi, [\alpha, \beta]) \in \Delta$ can be replaced by the pairs $(\phi, \beta), (\neg\phi, 1 - \alpha)$. Then Δ is a set of pairs (ϕ, α) which may be considered to be a partial function, since for each formula ϕ there is at most one real number α such that $(\phi, \alpha) \in \Delta$. Thus we may regard Δ as a function u_Δ from a subset D of the set of events into $[0, 1]$. Let us recall that every event ϕ may be regarded as a continuous function from a compact Hausdorff space X_A into the unital real interval $\mathbf{I}$, where X_A is the set of all valuations into $\mathbf{I} = [0, 1]_{MV}$. Hence we may suppose $D \subseteq C(X_A)$ and $u_\Delta \in \mathcal{M}(X)$ and this will be an important tool for using the abstract results of the previous section for proving a de Finetti-style theorem for imprecise probabilities and non-reversible betting games.

The second consequence is that we can modify the rules for betting by requiring the bettor to bet always on the chosen event. Thus

(1) If a bettor wants to bet on ϕ, he pays $\lambda\beta$ with $\lambda > 0$ and he will receive $\lambda v(\phi)$.
(2) If he wants to bet against ϕ, he bets on $\neg\phi$; hence he pays $\lambda(1 - \alpha)$ with $\lambda > 0$ and he will receive $\lambda v(\neg\phi)$

where v is a valuation in $[0, 1]$.

4.4 An adequate coherence criterion

Now we would like to know if there is an adequate coherence criterion in the non-reversible case, where "adequate" means necessary and sufficient to ensure that the betting odds proposed by the bookmaker can be extended to an upper probability on the algebra of events. Let us give the following definitions:

Definition 4.17. Let Δ be a book in a non-reversible betting game.

(i) A *winning strategy* (a *loosing strategy*, respectively) for the bettor based on Δ consists of a finite subset $\Gamma = \{(\phi_1, \alpha_1), ..., (\phi_n, \alpha_n)\}$ of

Δ and of a system of bets $\lambda_1, \ldots, \lambda_n$ on $\phi_1, \ldots, \phi_n$, respectively, such that for every valuation v the corresponding payoff $\sum_{i=1}^{n} \lambda_i (v(\phi_i) - \alpha_i)$ of the bettor is strictly positive (strictly negative, respectively).

(ii) A *bad bet* (a *good bet*, respectively) consists of an element (ϕ, α) of Δ and of a bet δ on ϕ such that there is a finite subset $\Gamma = \{(\phi_1, \alpha_1), \ldots, (\phi_n, \alpha_n)\}$ of Δ and a system $\lambda_1, \ldots, \lambda_n$ of bets on $\phi_1, \ldots, \phi_n$ which ensures to the bettor a better payoff (a worse payoff, respectively) independently of the valuation, that is,

$$\sum_{i=1}^{n} \lambda_i (v(\phi_i) - \alpha_i) > \delta (v(\phi) - \alpha) \quad \text{for every valuation } v$$

(in the definition of good bet, $>$ must be replaced by $<$).

We recall that in the reversible case a criterion of rationality is the non existence of a Dutch Book, *i.e.* of a winning strategy for the bettor. Besides that, the following result is straightforward:

Theorem 4.18. *Let Γ be a book in a reversible game. The following are equivalent:*

(1) *There is no winning strategy for the bettor.*
(2) *There is no loosing strategy for the bettor.*
(3) *There is no bad bet for the bettor.*
(4) *There is no good bet for the bettor.*

Proof. (1) $\Leftrightarrow$ (2). If $\Gamma = \{(\phi_1, \alpha_1), \ldots, (\phi_n, \alpha_n)\}$ and $\lambda_1, \ldots, \lambda_n$ constitute a winning strategy for the bettor, then $\Gamma \{(\phi_1, \alpha_1), \ldots, (\phi_n, \alpha_n)\}$ and $-\lambda_1, \ldots, -\lambda_n$ constitute a loosing strategy for him, and vice-versa.

(2) $\Leftrightarrow$ (3). If betting λ_i on ϕ_i, for $i = 1, \ldots, n$, is a loosing strategy for the bettor, then betting λ_1 on ϕ_1 is a bad bet for him: a better alternative is betting $-\lambda_i$ on ϕ_i, for $i = 2, \ldots, n$. Conversely, if betting λ on ϕ is a bad bet and betting λ_i on ϕ_i, for $i = 1, \ldots, n$ constitutes a better alternative, then betting λ on ϕ and $-\lambda_i$ on ϕ_i, $i = 1, \ldots, n$, is a loosing strategy.

(3) $\Leftrightarrow$ (4). Betting λ on ϕ is a bad bet iff betting $-\lambda$ on ϕ is a good bet. $\qquad \square$

Let us consider the following examples and see what happens in a non reversible game.

Example 4.19. Let

$$\Gamma = \left\{ \left(\phi, \frac{1}{3}\right), (\neg\phi, 1), \left(\psi, \frac{1}{3}\right), (\neg\psi, 1), (\phi \vee \psi, 1), (\neg(\phi \vee \psi), 1) \right\}.$$

The bettor has no winning strategy for this book because if both ϕ and ψ are false, then he cannot win anything.

Since it is entirely reasonable to expect that both bookmakers and bettors aim at maximising their returns while minimizing their losses, the book Γ does not look rational. Indeed, the bookmaker might make it more attractive for the bettor by reducing the betting odds on $\phi \vee \psi$ to $\frac{2}{3}$, without any loss of money if the bettor plays his best strategy.

Thus the non-existence of a Dutch Book is a necessary, but not a sufficient condition for rationality. Moreover, the non-existence of a loosing strategy, as well as the non-existence of a good bet are too strong conditions. One should not expect them to hold.

Indeed, let us consider the following example:

Example 4.20. Let Δ a book containing the items (ϕ, α), $(\neg\phi, \beta)$ with $\alpha + \beta > 1$.
The bettor has a loosing strategy, namely, betting 1 on both ϕ and $\neg\phi$, and he has also a good bet, namely, betting nothing. But this book cannot be considered irrational: it can be obtained from an element $(\phi, [0.3, 0.6])$ that is obviously acceptable if we use imprecise probabilities.

In the next theorem, we examine the criterion of the non-existence of bad bets, and we relate it to upper probabilities. In particular we say that

Criterion 4.1. A book Δ in a non-reversible betting game is said to be *rational* or *coherent* if there is no bad bet for the bettor.

and this is an adequate rationality criterion. Indeed, the following theorem holds:

Theorem 4.21. *Let Δ be a book in a non-reversible game over a divisible MV-algebra $\mathbf{A}$ of events. Then the following are equivalent:*

(1) *There is no bad bet for the bettor based on Δ.*
(2) *There is an upper probability u over the MV-algebra $\mathbf{A}$ of events such that, if $(\phi, \alpha) \in \Delta$, then $u(\phi) = \alpha$.*

Proof. See Appendix. □

Dually, one can prove the following result:

Theorem 4.22. *Let Δ be a book in a non-reversible game over a divisible MV-algebra $\mathbf{A}$ of events. Then the following are equivalent:*

(1) *There is no good bet for the bettor based on Δ.*
(2) *There is a lower probability ℓ over the MV-algebra $\mathbf{A}$ of events such that, if $(\phi, \alpha) \in \Delta$, then $\ell(\phi) = \alpha$.*

One may wonder why the non-existence of a good bet, which, as we said above, is not a rationality criterion, corresponds to a significant property, that is, extendability to a lower probability. However, if betting $\lambda \geq 0$ on ϕ is a good bet in a game in which only positive bets are allowed, then betting $-\lambda$ on ϕ is a bad bet in a game in which only *negative* bets are allowed. Hence, accordance with a lower probability corresponds to a reasonable rationality criterion.

Lastly, we relate the non-existence of either a good bet or a bad bet to states.

Theorem 4.23. *Let Δ be a book in a non-reversible game over a divisible MV-algebra $\mathbf{A}$ of events. Then the following are equivalent:*

(1) *There is no good bet and no bad bet based on Δ.*
(2) *There is a state s on $\mathbf{A}$ such that, if $(\phi, \alpha) \in \Delta$, then $s(\phi) = \alpha$.*

To conclude, we make some remarks on this last theorem. Recall that a map s on the 2-divisible MV-algebra $\mathbf{A}$ is a state if and only if is at the same time an upper and a lower probability. Thus if the betting odds proposed by the bookmaker can be extended to a state on the MV-algebra of events, for each event ϕ, the lower probability of ϕ is equal to its upper probability. This means that the bookmaker is able to quantify with a single number his subjective opinions concerning the occurrence of events: the events considered are fuzzy events not under condition of indeterminacy, that have been already studied by Mundici, as noted in the Introduction. The main difference from Mundici's result is that Theorem 4.23 proposes an adequate rationality criterion in the case of non-reversible betting game. This seems to be closer to representing the behavior of a real bookmaker, that surely will not agree to reverse his role with the bettor.

5 Walley's imprecise probabilities

In this section we recall the basic theory of coherent upper and lower probabilities introduced by Walley in [24] and [25] and we compare it with our theory of imprecise probabilities presented in the previous section.

5.1 Lower and upper previsions

The usual approach in constructing a theory of probability is to start with axioms for probability and then define expectation, with the mathematical properties of linear prevision, as a derived concept. Following de Finetti, Walley regards prevision as more fundamental than probability. The main reason for doing so is that coherent lower probabilities may have many different extensions to coherent lower previsions.

For this reason, we start with upper and lower previsions and we define upper and lower probabilities only as the special cases in which the range of random quantities is $\{0, 1\}$.

Let us introduce the key notion on which Walley's theory is based:

Definition 5.1. Let Ω a set of possible states which may represent the outcomes of experiments or any conceivable facts about the world.

A *gamble* is a bounded real-valued function defined on domain Ω satisfying the following conditions. Let $\mathcal{L}(\Omega)$ the set of all gambles on Ω:

(i) For all $\lambda \in \mathbb{R}$ and $X \in \mathcal{L}(\Omega)$, λX is the gamble defined by $(\lambda X)(\omega) = \lambda X(\omega)$.
(ii) For all $X, Y \in \mathcal{L}(\Omega)$, $X + Y$ is the gamble $(X + Y)(\omega) = X(\omega) + Y(\omega)$.

A gamble should be interpreted as a reward in units of utility whose value depends on the uncertain state ω: if you accept the gamble X, then at some later time the true state ω will be determined and you will receive the reward $X(\omega)$.

In saying that a gamble X is *desirable* to the bettor, we mean that he has a disposition to accept X whenever it is offered him. When X is not desirable, the bettor does not necessarily have a disposition to reject it, he may be undecided about whether to accept or reject. The following conditions, which appear in [24], are proposed as rationality constraints on the set of desirable gambles to the bettor. They require, in effect, that the scale in which the rewards $X(\omega)$ are measured behaves like a linear utility scale.

(*D0*) If $\sup X = \sup\{X(\omega)|\omega \in \Omega\} < 0$ then X is not desirable.
(*D1*) If $\inf X = \inf\{X(\omega)|\omega \in \Omega\} > 0$ then X is desirable.
(*D2*) If X is desirable and λ is a positive real number then λX is desirable.
(*D3*) If X and Y are each desirable then $X + Y$ is desirable.

We can now introduce lower previsions and their corresponding upper previsions (see [25]).

Definition 5.2. A *lower previsions* $\underline{P}$ is a real-valued function defined on the set $\mathcal{L}(\Omega)$ of all gambles.

Recall that the gambles $X \in \mathcal{G}$ should be regarded as uncertain rewards. The lower prevision $\underline{P}$ then models bettor's attitudes (behavioral dispositions) concerning the gambles in $\mathcal{G}$. Specifically, Walley's behavioral interpretation of the lower prevision $\underline{P}$ is that for each gamble X the bettor is currently willing to pay any price less than $\underline{P}(X)$ in return for X. So the lower prevision $\underline{P}(X)$ can be regarded as a supremum buying price μ for which it is asserted that the gamble $X - \mu$ is desirable to the bettor.

Definition 5.3. Suppose that $\underline{P}$ is a lower prevision. Its *conjugate upper prevision* $\overline{P}$ is defined on the same domain $\mathcal{L}(\Omega)$ by $\overline{P}(X) = -\underline{P}(-X)$.

Since $\underline{P}(X)$ is interpreted as supremum price μ for which $X - \mu$ belongs to the class $\mathcal{D}$ of desirable gambles, we have

$$\overline{P}(X) = -\sup\{\mu : -X - \mu \in \mathcal{D}\}$$
$$= \inf\{-\mu : -X - \mu \in \mathcal{D}\}$$
$$= \inf\{\alpha : \alpha - X \in \mathcal{D}\}.$$

Thus $\overline{P}(X)$ is an infimum price α for which $\alpha - X$ is desirable, *i.e.* such that the bettor is willing to sell X in return for price α. Just as the lower prevision $\underline{P}(X)$ is a supremum buying price for X, the upper prevision $\overline{P}(X)$ is an infimum selling price for X. This model does not say anything about whether the bettor will buy or sell X if the price lies between $\underline{P}(X)$ and $\overline{P}(X)$; either course of action may be reasonable.

Remark 5.4. Upper previsions are determined by lower previsions, and vice versa. The theory can therefore be developed in terms of either upper or lower previsions. We will continue to emphasize lower previsions.

Let us give the following definition:

Definition 5.5. Any lower prevision such that $\underline{P}(X) = \overline{P}(X)$ for all gambles X is called a *linear prevision*. In this case we will write $P(X)$ instead of $\underline{P}(X)$.

Following the behavioral interpretation of lower and upper previsions, the requirement of precision $\underline{P}(X) = \overline{P}(X)$ means that the bettor is willing to buy X for any price less than $P(X)$, to sell X for any price greater than $P(X)$ and is indifferent between buying and selling if the price is exactly $P(X)$. Thus de Finetti's original idea can be presented in terms of linear previsions, since $P(X)$ is the fair price for X.

As we have already seen in Section 2, de Finetti's theory is based on a notion of coherence which, in the case of linear previsions, is equivalent to the criterion of **avoiding sure loss**. We now define "coherence" and "avoiding sure loss" in the case in which $\underline{P}(X) < \overline{P}(X)$.

5.2 Avoiding sure loss and Coherence

Definition 5.6. Let $\underline{P}$ a lower prevision. Say that a lower prevision avoids sure loss if $\sup[\sum_{j=1}^{n}(X_j - \underline{P}(X_j))] \geq 0$ whenever $n \geq 1$ and $X_1, \ldots, X_n$ are gambles.

$D0$ and $D3$ are sufficient conditions for avoiding sure loss. Since the lower prevision $\underline{P}(X)$ is the supremum buying price μ for which it is asserted that the gamble $X - \mu$ is desirable to the bettor, each gamble $X - (\underline{P}(X) - \delta) = (X - \underline{P}(X)) + \delta$ is desirable to the bettor for every $\delta > 0$. Thanks to $D3$ the finite sum $\sum_{j=1}^{n}((X_j - \underline{P}(X_j)) + \delta)$ is desirable. Axiom $D0$ then implies that $\sup \sum_{j=1}^{n}((X_j - \underline{P}(X_j)) + \delta) \geq 0$. If δ tends to 0 we have the claim.

When rewards are expressed in a linear utility scale, we regard $D3$ and $D0$ as necessary conditions of rationality, thus avoiding sure loss is a normative criterion of rationality.

We now compare Walley's definition of coherence (see [24]) with the criterion of avoiding sure loss.

Definition 5.7. We say that a lower prevision $\underline{P}$ is *coherent* if it satisfies the three axioms:

($P1$) $\underline{P}(X) \geq \inf X$ when $X \in \mathcal{G}$;
($P2$) $\underline{P}(\lambda X) = \lambda \underline{P}(X)$ when $X \in \mathcal{G}$ and $\lambda > 0$;
($P3$) $\underline{P}(X + Y) \geq \underline{P}(X) + \underline{P}(Y)$ when $X, Y \in \mathcal{G}$.

The three axioms can be justified through the behavioral interpretation of lower previsions and the earlier axioms for desirability $D1 - D3$. Indeed, coherence axioms $P1, P2, P3$ correspond to desirability axioms $D1, D2, D3$ respectively.

Like avoiding sure loss, coherence is a normative criterion of rationality, albeit a much stronger one. The following example taken from [24] illustrates the difference between avoiding sure loss and coherence, let us consider the following situation.

Example 5.8. Let X_j denote the gamble which pays 1 unit if a die shows the number j, and otherwise pays nothing. Suppose you assess the lower previsions $\underline{P}(X_1) = \underline{P}(X_2) = \underline{P}(X_1 + X_2) = \frac{1}{4}$. These assessments are incoherent because they violate axiom $P3$. However they avoid sure loss as defined above, they cannot be combined to produce a sure loss.

Finally we observe that linear previsions allow us to characterize coherent lower previsions. Indeed, in many problems the simplest way to establish coherence of a given lower prevision is to show that it is a lower envelope of a closed convex set $\mathcal{C}$ of linear previsions, *i.e.* $\underline{P}(X) = \min\{P(X) : P \in \mathcal{C}\}$, and invoke the following theorem (see [24]):

Theorem 5.9. *$\underline{P}$ is a coherent lower prevision on $\mathcal{G}$ if and only if it is the lower envelope of a closed convex set of linear previsions defined on $\mathcal{G}$.*

5.3 An interpretation of imprecise probabilities in terms of bets: the bettor's viewpoint

Again following de Finetti, upper and lower probabilities are defined as special cases of upper and lower previsions.

Let us introduce some notation. We use capital letters $A, B, C, \ldots$ to denote subsets of Ω, called events. If A is an event, we use the same symbol A to denote its indicator function on domain Ω, defined by $A(\omega) = 1$ if $\omega \in A$ and $A(\omega) = 0$ is $\omega \in A^c$, where A^c denotes the set-theoretic complement of A.

Let us remark that an indicator function is just a $\{0, 1\}$ real valued gamble, thus we identify each event with a gamble.

Definition 5.10. When a lower prevision $\underline{P}$ is defined on the set of all $0 - 1$ valued gambles, we call $\underline{P}$ a *lower probability*, and call $\underline{P}(A)$ the lower probability of event A.

We notice that it is convenient to modify the earlier definition of the conjugate upper prevision $\overline{P}$, now called *upper conjugate probability*, so that $\overline{P}$ is defined on $\mathcal{A}^c = \{A^c : A \in \mathcal{A}\} = \{1 - A : A \in \mathcal{A}\}$ rather than on $-\mathcal{A} = \{-A : A \in \mathcal{A}\}$. Define $\overline{P}$ on $\mathcal{A}^c$ by $\overline{P}(A) = 1 - \underline{P}(A^c)$.

In the following proposition we list the basic properties of upper and lower probabilities.

Proposition 5.11. *Suppose $\mathcal{A}$ is the class of all events, $\underline{P}$ is a lower probability on domain $\mathcal{A}$ and $\overline{P}$ is the conjugate upper probability, i.e. $\overline{P}(A) = 1 - \underline{P}(A^c)$ The following properties hold:*

(1) $0 \le \underline{P}(A) \le \overline{P}(A) \le 1$.

(2) $\underline{P}(\emptyset) = \overline{P}(\emptyset) = 0$ *and* $\underline{P}(\Omega) = \overline{P}(\Omega) = 1$.

(3) *If $A \supseteq B$ then* $\underline{P}(A) \ge \underline{P}(B)$ *and* $\overline{P}(A) \ge \overline{P}(B)$.

(4) *If $A \cap B = \emptyset$ then*

$$\underline{P}(A) + \underline{P}(B) \le \underline{P}(A \cup B) \le \underline{P}(A) + \overline{P}(B) \le \overline{P}(A \cup B)$$
$$\le \overline{P}(A) + \overline{P}(B).$$

(5) $\underline{P}(A \cup B) + \underline{P}(A \cap B) \ge \underline{P}(A) + \underline{P}(B)$.

In [25] Walley proves that properties $1-4$ are necessary but not sufficient for coherence and that property 5 is instead sufficient but not necessary for coherence.

As in the case of previsions, it can be given a behavioral interpretation of lower and upper probabilities. The gamble A pays 1 unit if A occurs and nothing otherwise; $\underline{P}(A)$ is interpreted as a supremum buying price for A, and by duality, $\overline{P}(A)$ is interpreted as an infimum selling price for the gamble A.

6 Conclusions and further work

In those cases in which the lower probability is strictly smaller than the upper probability, De Finetti's key assumption of fairness is no longer valid. Thus the roles of bettor and bookmaker cannot be interchanged, and it is important to specify from which viewpoint we interpret imprecise probabilities.

We notice that Walley's behavioral interpretation of imprecise probabilities is given from the bettor's viewpoint, whereas our interpretation, presented in the previous section, is given from the bookmaker's viewpoint.

Our results do not carry over if we change the point of view of the bookmaker with that of the bettor. In the framework of Section 4 a *lower probability* of an event A is interpreted as the supremum betting odds which a rational and non-reversible bookmaker would propose for betting against A and by duality the *upper probability* of A is the infimum betting odds which a rational and non-reversible bookmaker would propose for betting on A. On the other hand Walley's lower probability $\underline{P}(A)$ of an event A is interpreted as the infimum betting rate which a bookmaker would propose for betting on A. Indeed, if a bettor bets on A at rate α, he pays α to the bookmaker, thus if the bookmaker is willing to propose odds α, we suppose that he is willing to propose the more favorable odds α' with $\alpha < \alpha'$. Dually Walley's upper probability $\overline{P}(A)$ of an event A is interpreted as the supremum betting rate which a rational bookmaker would propose for betting against A. Indeed, if a bettor bets against an event A at rate β, the bookmaker pays him β; thus if the bookmaker is willing to propose odds β, we suppose that he is willing to propose the more favorable odds β' with $\beta' < \beta$.

Moreover, we point out that in our framework, we define a notion of coherence in terms of betting behavior and then we prove that coherence is a necessary and sufficient condition to ensure that betting odds obey axioms of upper and lower probabilities. In Walley's approach coherence is axiomatically defined but no equivalent interpretation in terms of betting behavior is given: he only proves that coherence is not equivalent to avoiding sure loss.

These considerations open up to two problems that we intend to investigate in further works:

(i) Is there any interesting duality between our notion of upper and lower probabilities and Walley's?

(ii) Can we give Walley's notion of coherence a behavioral interpretation equivalent to the axiomatic one?

7 Appendix

Following [8] we prove the Theorem 4.21:

Proof. We use the fact that an MV-subalgebra $\mathbf{A}$ of events has a representation $\phi \mapsto \widehat{\phi}$ onto a MV-subalgebra $\widehat{\mathbf{A}}$ of the MV-algebra $C(X_\mathbf{A}, \mathbf{I})$ of all continuous functions $f \colon X_\mathbf{A} \to \mathbf{I}$, where $X_\mathbf{A}$ is the space of valuations of $\mathbf{A}$ and $\widehat{\phi}(v) = v(\phi)$ (see [4] 3.6.8). Every upper probability u on $\mathbf{A}$ has a unique representation by an upper probability $\widehat{u}$ on $\widehat{\mathbf{A}}$ and $\widehat{u}$ has a unique extension to an order preserving, normalized, homogeneous, weakly linear and sublinear functional $\widetilde{u} \colon C(X_\mathbf{A}) \to \mathbb{R}$. Moreover, every state on $\mathbf{A}$ has a unique extension to a linear, normalized and order preserving functional on $C(X_\mathbf{A})$.

We now derive the claim from Theorem 4.11, (1). We translate the existence of a bad bet (a good bet respectively) into the functional language: There is a bad bet for the book Δ if there are $(\phi_1, \alpha_1), \dots, (\phi_n, \alpha_n), (\psi, \alpha) \in \Delta$ and $\delta, \lambda_1, \dots, \lambda_n, \delta \geq 0$ such that

$$\sum_{i=1}^n \lambda_i \widehat{\phi}_i(v) - \delta \widehat{\psi}(v) > \sum_{i=1}^n \lambda_i \alpha_i - \delta \alpha \quad \text{for every } v \in X_\mathbf{A}$$

equivalently

$$\text{Min} \left(\sum_{i=1}^n \lambda_i \widehat{\phi}_i - \delta \widehat{\psi} \right) > \sum_{i=1}^n \lambda_i \alpha_i - \delta \alpha$$

where $\text{Min}(f) = \min_{v \in X_\mathbf{A}} f(v)$ for $f \in C(X_\mathbf{A})$. Thus, there is no bad bet for the book Δ if, for all $(\phi_1, \alpha_1) \dots, (\phi_n, \alpha_n), (\psi, \alpha) \in \Delta$ and $\lambda_1, \dots, \lambda_n, \delta \geq 0$,

$$\text{Min} \left(\sum_{i=1}^n \lambda_i \widehat{\phi}_i - \delta \widehat{\psi} \right) \leq \sum_{i=1}^n \lambda_i \alpha_i - \delta \alpha.$$

We now prove that (1) implies (2): Let u be an upper probability on $\mathbf{A}$ and suppose that the book is of the form $\Delta = \{(\phi, u(\phi)) : \phi \in D\}$. Let $\widetilde{u}$ be the unique order preserving, weakly linear, sublinear functional on $C(X_\mathbf{A})$ extending u. By Theorem 4.11 condition $(\widetilde{U})$ holds: for all $\phi_1, \dots, \phi_n, \psi \in D$ and $\lambda_1, \dots, \lambda_n, \delta \geq 0$,

$$\text{Min} \left(\sum_{i=1}^n \lambda_i \widehat{\phi}_i - \delta \widehat{\psi} \right) \leq \sum_{i=1}^n \lambda_i u(\widehat{\phi}_i) - \delta u(\widehat{\psi})$$

which tells us that there is no bad bet for the book Δ.

We now turn to the proof that (2) implies (1). Let Δ be a book for which there is no bad bet. Then Δ is a set of pairs (ϕ, α) which may be considered to be a partial function, *i.e.* for each formula ϕ there is at most one real number α such that $(\phi, \alpha) \in \Delta$. Thus we may regard Δ as a function u from a subset D of the set of events into $[0, 1]$. We recall that every event ϕ may be regarded as a continuous function from a compact Hausdorff space $X_{\mathbf{A}}$ into the unit interval $\mathbf{I}$, where $X_{\mathbf{A}}$ is the set of all valuations into $\mathbf{I} = [0, 1]_{MV}$, and for every valuation v, $\phi(v) = v(\phi)$. Hence, denoting the set of all continuous functions from $X_{\mathbf{A}}$ into $\mathbb{R}$ by $C(X_{\mathbf{A}})$, we may suppose $D \subseteq C(X_{\mathbf{A}})$. Moreover, it makes sense to speak of linear combinations of events, even if they are no longer events, but elements of $C(X_{\mathbf{A}})$. We notice that, as Δ admits no bad bet, the function $u \colon D \to \mathbb{R}$ satisfies condition $(\tilde{U})$: For $\phi_1, \ldots, \phi_n, \psi \in D$ and $\lambda_1, \ldots, \lambda_n, \delta \in \mathbb{R}_+$,

$$\mathrm{Min}\left(\sum_{i=1}^{n} \lambda_i \widehat{\phi_i} - \delta \widehat{\psi}\right) \le \sum_{i=1}^{n} \lambda_i \widehat{u}(\widehat{\phi_i}) - \delta \widehat{u}(\widehat{\psi}).$$

By Theorem 4.11, there is an order preserving, weakly linear, sublinear functional $\tilde{u} \colon C(X) \to \mathbb{R}$ extending $\widehat{u}$. The restriction of $\tilde{u}$ to $\widehat{\mathbf{A}}$ yields the desired upper probability extending the given $u \colon D \to [0, 1]$. $\qquad \square$

References

[1] W. BLOK and D. PIGOZZI, *Algebraizable logics*, Mem. Amer. Math. Soc. **396**(77) (1989).

[2] H. BREZIS, "Analyse Fonctionnelle : Théorie et Applications", Masson Editeur, 1983.

[3] S. BURRIS and H.P. SANKAPPANAVAR, "A Course in Universal Algebra", Springer Verlag, New York, 1981.

[4] R. CIGNOLI, I. D'OTTAVIANO and D. MUNDICI, "Algebraic Foundations of Many-valued Reasoning", Kluwer, Dordrecht, 2000.

[5] G. COLETTI and R. SCOZZAFAVA, "Probabilistic Logic in a Coherent Setting", Kluwer, Dordrecht, 2002.

[6] B. DE FINETTI, "Theory of Probability", Vol. I, John Wiley and Sons, Chichester, 1974.

[7] R. FAGIN and J. Y. HALPERN, *Uncertainty, belief and probability*, Computational Intelligence **7**(3) (1991), 160–173.

[8] M. FEDEL, K. KEIMEL, F. MONTAGNA, W. ROTH, *Imprecise probabilities, bets and functional analytic methods in Lukasiewicz logic*, submitted.

[9] M. FEDEL, "Un Approccio Probabilistico alla Rappresentazione dell'Incertezza: Probabilità Imprecise, Scommesse e Funzionali

Normalizzati su Spazi di Riesz", Master Thesis, available at
http://www.crm.sns.it/lori/

[10] P. HÁJEK, "Metamathematics of Fuzzy Logic", Kluwer, Dordrecht, 1998.

[11] J. Y. HALPERN, "Reasoning about Uncertainty", MIT Press, 2003.

[12] T. KROUPA, *Every state on a semisimple MV algebra is integral*, Fuzzy Sets and Systems **157**(20) (2006), 2771–2787.

[13] J. KÜHR and D. MUNDICI, *De Finetti theorem and Borel states in [0,1]-valued algebraic logic*, International Journal of Approximate Reasoning **46**(3) (2007), 605–616.

[14] I. LEVI, *Imprecision and indeterminacy in probability judgment*, Philosophy of Science, **52**(3) (1985), 390–409.

[15] D. MUNDICI, *Averaging the truth value in Lukasiewicz logic*, Studia Logica **55**(1) (1995), 113–127.

[16] D. MUNDICI, *Bookmaking over infinite-valued events*, International Journal of Approximate Reasoning **46** (2006), 223–240.

[17] G. PANTI, *Invariant measures in free MV algebras*, Communications in Algebra (to appear), available at Arxiv preprint math. LO/0508445, 2005.

[18] J.B. PARIS, "The Uncertain Reasoner's Companion - A Mathematical Perspective", Cambridge University Press, 1994.

[19] J.B. PARIS, *A note o the Dutch Book method*, In: "Proceedings of the 2nd International Symposium on Imprecise Probabilities and their Applications", 2001.

[20] W. RUDIN, "Functional Analysis", 2nd Edition, McGraw-Hill Publ., 1991.

[21] H. H. SCHAEFER, "Topological vector spaces", 3rd Printing, Springer Verlag, New York Heidelberg Berlin, 1971.

[22] C. SMITH, *Consistency in Statistical Inference and Decision*, Journal of the Royal Statistical Society, Series B (Methodological), **23**(1) (1961), 1–37.

[23] I. YAACOM and A. USVYATSOV, *Continuous first order logic and local stability*, Transactions of the American Mathematical Society, to appear.

[24] P. WALLEY, "Statistical Reasoning with Imprecise Probabilities", Monographs on Statistics and Applied Probability, Vol. 42, Chapman and Hall, London, 1991.

[25] P. WALLEY, *Measures of uncertainty in expert systems*, Artificial Intelligence **83** (1996), 1–58.

4

RATIONALITY

Tractable depth-bounded logics and the problem of logical omniscience

Marcello D'Agostino

1 Introduction

Theories of rationality, in their broad sense, are notoriously marred by highly unrealistic assumptions about the reasoning power of agents. This applies also to several logical theories that are supposed to play a significant role in the development of models of rationality. For example, any such model is bound to make a pervasive use of notions such as "information", "knowledge" and "belief" which prompt for conceptual clarification. Any attempt at such a clarification must involve an analysis of the logical rules that govern their use, and these rules must interact in a meaningful way with those that govern the relation of logical consequence in a usual propositional or first-order language. However, if the relation of logical consequence is taken to be that of classical logic, the interaction turns out to be highly problematic. According to the standard logic of knowledge (epistemic logic) and belief (doxastic logic), as well as to the more recent attempts to axiomatize the "logic of being informed" (information logic),[1] if an agent i knows, or believes, or is informed that a sentence A is true, and B is a logical consequence of A, then i is supposed to know, or believe, or be informed also that B is true. This is often described as paradoxical and labelled as "the problem of logical omniscience". Let $\Box_i$ express any of the propositional attitudes at issue, referred to the agent i. Then, the "logical omniscience" assumption can be expressed by saying that, for any finite set Γ of sentences,

$$\text{if } \Box_i A \text{ for all } A \in \Gamma \text{ and } \Gamma \vdash B, \text{ then } \Box_i B, \qquad (1.1)$$

where $\vdash$ stands for the relation of logical consequence. Observe that, letting $\Gamma = \emptyset$, it immediately follows from (1.1) that any rational agent i is

---

[1] For a survey on epistemic and doxastic logic see [20, 24]. For information logic, or "the logic of being informed", see [17, 29].

supposed to be aware of the truth of all classical tautologies, that is, of all the sentences of a standard logical language that are "consequences of the empty set of assumptions". In most axiomatic systems of epistemic, doxastic and information logic assumption (1.1) emerges from the combined effect of the "distribution axiom", namely,

$$\Box_i (A \rightarrow B) \rightarrow (\Box_i A \rightarrow \Box_i B) \qquad \textbf{(K)}$$

and the "necessitation rule":

$$\text{if } \vdash A, \text{ then } \vdash \Box_i A. \qquad \textbf{(N)}$$

On the other hand, the paradoxical flavour of (1.1) seems an inescapable consequence of the standard Kripke-style semantical characterization of the logics under consideration. The latter is carried out in terms of structures of the form $(S, \tau, R_1, \ldots, R_n)$, where S is a set of possible worlds, τ is a function that associates with each possible world s an assignment $\tau(s)$ of one of the two truth values (0 and 1) to *each* atomic sentence of the language, and each R_i is the "accessibility" relation for the agent i. Intuitively, if s is the actual world and $sR_i t$, then i would regard t as "possible". Then, a forcing relation $\models$ is introduced to define truth for the complex sentences of the language, starting from the initial assignment to the atomic sentences. The forcing relation incorporates the usual semantics of classical propositional logic and defines the truth of $\Box_i A$ as "A is true in all the worlds that i regards as possible". In this framework, given that the notion of truth in a possible world is an extension to the modal language of the classical truth-conditional semantics for the standard logical operators, (1.1) appears to be both compelling and, at the same time, counter-intuitive.

Now, under this reading of the consequence relation $\vdash$ as based on classical logic, (1.1) may perhaps be satisfied by an "idealized reasoner", in some sense to be made more precise,[2] but it is out of the question that it is not satisfied, and is not likely to ever be satisfiable, in practice. Even restricting ourselves to the domain of propositional logic, the theory of computational complexity tells us that the decision problem for Boolean logic is co-NP-complete [3], that is, among the hardest problems in co-NP. Although not a proved theorem, it is a widely accepted conjecture that there exists no decision procedure for such problems that

---

[2] It should be noted that the appeal to an "idealized reasoner" has usually the effect of sweeping under the rug a good deal of interesting questions, including how idealized such a reasoner should be. Idealization may well be a matter of degree.

runs in polynomial (*i.e.*, feasible) time. This means that any real agent, even if equipped with an up-to-date computer running a decision procedure for Boolean logic, will never be able to feasibly recognize that certain Boolean sentences logically follow from sentences that she regards as true. So, the clash between (1.1) and the classical notion of logical consequence, which arises in any real application context, may only be solved either by waiving the assumption stated in (1.1), or by waiving the consequence relation of classical logic in favour of a weaker one with respect to which it may be safely assumed that the modality $\Box_i$ is closed under logical consequence for any practical reasoner.

Both options have been discussed in the literature.[3] Observe that, according to the latter, the problem of logical omniscience does not lie in assumption (1.1) in itself, but rather in the standard (classical) characterization of logical consequence for a propositional language that is built in the possible-world semantics originally put forward by Jaakko Hintikka as a foundational framework for the investigation of epistemic and doxastic logic. Frisch [18] and Levesque [23] were among the first authors to explore this route and argue for a notion of "limited inference" based on "a less idealized view of logic, one that takes very seriously the idea that certain computational tasks are relatively easy, and others more difficult" [23, page 355]. A more recent (and related) proposal can be found in [13], where the authors suggest to replace classical logic with a non-standard one, deeply rooted in relevance logic and called NPL (for "Nonstandard Propositional Logic"), to mitigate the problem of logical omniscience. The mitigation consists mainly in the existence of a polynomial time decision procedure for the CNF fragment of the proposed logical system [13, Theorem 7.4]. However, the decision problem for the unrestricted language of NPL is still co-NP-complete (Theorem 6.4). Moreover, NPL shares with relevance logic and with Levesque's notion of limited inference the invalidity of disjunctive syllogism (from $A \vee B$ and $\neg A$ one cannot infer B) which sounds disturbing to most classical ears. Finally, the NPL-based approach does not allow, in a natural way, for the possibility of defining *degrees* of logical omniscience, that may apply to *increasingly idealized* reasoning agents, in terms of correspond-

---

[3] See [24, Section 4], [20, Section 4] and [22] for a survey and proper references. See also: [26] for an interesting third view that draws on the tradition of subjective probability, and [1] for an approach based on proof size. A general semantic framework in which several different approaches can be usefully expressed is that based on "awareness structures", which draws on the distinction between "explicit" and "implicit" knowledge, to the effect that an agent may implicitly know that a sentence is a logical consequence of a set of assumptions, without being *aware* of it. See [31, 32] for an insightful discussion of this framework and proper references to the literature.

ingly stronger consequence relations. On the other hand, the possibility of characterizing in a uniform way such a hierarchy of approximations to the "perfect reasoner" (which may well be a classical one) would certainly allow for all the flexibility needed by a suitable model of practical rationality.[4]

In this paper we set out to make a contribution towards the solution of the logical omniscience problem that is much in the spirit of [13]. We too maintain that the problem can be properly solved by restricting the classical notion of logical consequence rather than by waiving assumption (1.1). We suggest, however, that an interesting alternative solution could be based on *Depth-Bounded Boolean Logics*, a novel incremental approach to the characterization of classical propositional logic that construes it as the limit of an infinite sequence of weaker *tractable* logics.[5] Agents committed to these logics can be seen as *approximations* to the idealized reasoning agent of standard epistemic, doxastic and information logic. The full decision problem for each of the approximating logics is solvable in polynomial time — although its complexity grows as we proceed along the approximation sequence — with *no restriction* to any particular syntactic fragment. Moreover, the meaning of the logical operators is *the same for all logics* and is explained in purely informational terms — that is, in terms of informational interpretations of "true" and "false" — in such a way that the most basic inference principles of classical propositional logic, including disjunctive syllogism, are preserved throughout the sequence.

Although this paper does not contain any new technical result on Depth-Bounded Boolean Logics (being parasitical on [7] and [6] in this respect), it sets out to shed new light on their potential applications, try-

---

[4] The idea of approximating full classical propositional logic via a converging sequence of tractable subsystems has received considerable attention in the field of automated deduction (see, for instance, [4,8,9,14,15,30]). See also [2] for an early paper that proposes to use approximated Boolean reasoning to deal with the problem of logical omniscience. In this work we are interested in tractable approximations that can themselves be described as *consequence relations* in some of the currently accepted senses (mainly, in Tarski's sense), equipped with a uniform, and possibly intuitive, account of the meaning of the logical operators. A brief discussion of the main differences between our approach and the cited ones can be found in [7] and [6]. Here we observe only that some of the systems proposed in the literature are not consequence relations, in that they do not satisfy unrestricted transitivity, while others (like the notion of "limited inference" of [23] and the NPL system of [13]) do not satisfy disjunctive syllogism, namely, one of the most basic inference principles of classical logic that is immediately recognized as sound, *pace* relevance logicians, by any far from idealized agent, including Chrysippus' dog (see [16]).

[5] This approach was introduced in [7] and is still under development. The denomination "Depth-Bounded Logics" itself was not used in the original paper and has been put forward in [6], which contains a more detailed semantical account and a network-based proof-theoretic characterization that adjoins the "natural deduction" system presented in [7].

ing them on an open problem that is widely debated in the logical literature. When technical results are mentioned, the proofs are omitted and the reader is referred to the source papers for more details. In Section 2 we make a diagnosis of the problem of logical omniscience and list some desiderata that a characterization of logical consequence should meet to be a liable candidate for replacing classical logic in the attempt to solve the problem. Then, in Section 3, we outline an "informational interpretation" of the logical operators that is more in tune than their classical interpretation with the propositional attitudes investigated in epistemic, doxastic and information logic. In Section 4 we present the sequence of Depth-Bounded Boolean Logics from the semantical viewpoint, and in Section 5 we illustrate a variant of the proof-theoretic presentation introduced in [7] in the style of "natural deduction". Finally, in the concluding section, we outline future developments of the ideas presented in this paper.

2   A diagnosis and a wish list

What goes wrong in the interplay between classical logic and the propositional attitudes investigated in epistemic, doxastic or information logic? A possible clue points at the well-known philosophical fact that the inner working of classical logic cannot be fully explicated by referring only to an agent's information state. More precisely, the classical meaning of the logical operators makes an essential reference to recognition-transcendent notions of truth and falsity as properties of sentences that apply to them quite independently of the information available to us. This is the concession that classical logic makes to metaphysical realism and is expressed by the principle of bivalence: every sentence is determinately either true or false, no matter whether we are able or not to recognize its truth or falsity on the basis of the available evidence. However, it seems plausible that a consequence relation that is likely to turn assumption (1.1) into a compelling property should be explicable in purely informational terms. Rather than being truth-preserving with respect to *possible worlds*, requiring that the conclusion be true in all possible worlds in which all the premises are true, it should be truth-preserving with respect to *information states*, requiring that the conclusion be true in all the information states in which all the premises are true. Now, if a notion of information state is suitably defined — to the effect that an agent can be *realistically* assumed to be totally in control of his own information state — it will be a truism that the propositional attitude of "being informed" be preserved under logical consequence, once this latter notion has been taken in the informational sense that we have just outlined. Moreover, it will be quite plausible that an agent should also know (or believe) all the logical consequences, still in the informational sense, of the sentences that he knows (or believes). For, the problem with (1.1) is that the agent might not be *aware* that the truth of a sentence A logically follows from the truth of the sentences in Γ that he knows or believes as true: were he aware of this, the propositional attitudes would certainly apply also to A. But, if the consequence relation $\vdash$ and the notion of information state are defined in such a way that

D1 $\Gamma \vdash A$ means that A is true in all information states in which all the sentences in Γ are true,

D2 any agent can be assumed to be totally in control of his own information state, namely, to have feasible access to the information contained in it,

then assumption (1.1) becomes unassailable even when $\Box_i$ is interpreted in terms of knowledge or belief. Observe also that, taken together, D1 and D2 imply that the notion of logical consequence we are looking for should be feasible.

Hence, a good reason for distributing the modalities $\Box_i$ over logical consequence is that the latter — defined in accordance with D1, let us call this *the informational sense of logical consequence* — does not add anything *new* to an agent's information state, it does not provide any genuinely new information. This is possible only because a sentence's being a logical consequence of a set of assumptions means that its truth can be established, given the truth of the assumptions, by virtue only of the meaning of the logical operators occurring in the sentences, and so it depends solely on the conventions governing our use of language. For example, an agent may establish by external means that $A \vee B$ is true and A is false, which will force her information state to include the truth of B solely by virtue of the accepted meaning of $\vee$. Once such meaning has been properly explicated, one should literally be able to "see" that the truth of the conclusion is contained in the truth of the premises. Accordingly, Wittgenstein dreamt of a logically perfect language in which "we can in fact recognize the formal properties of propositions by mere inspection of the propositions themselves" (Tractatus 6.122) and "every tautology itself shows that it is a tautology" (6.127(b)). However, if we take for granted the classical meaning of the logical operators — as defined by the usual semantic clauses, in terms of recognition-transcendent notions of truth and falsity satisfying the principle of bivalence — the explication process required to reveal the soundness of an arbitrary inference turns out to be be unfeasible, and the theory of NP-completeness strongly suggests that this situation cannot be improved upon: there cannot be any perfect language in Wittgenstein's sense.

At this point, one may conjecture that intuitionistic logic could be apt to provide a suitable solution to our problem. First, the intuitionistic meaning of the logical operators may indeed be explained (with some difficulty) in terms of a notion of truth as provability or verifiability that is not recognition-transcendent.[6] Moreover, the well-known Kripke semantics defines logical consequence as truth-preserving over states that Kripke himself intuitively described as "points in time (or 'evidential situations'), at which we may have various pieces of information".[7] However, replacing classical with intuitionistic logic would be no solution, because the decision problem for the propositional fragment of the latter is PSPACE-complete (that is, among the hardest problems in PSPACE)[8] and, again, it is highly plausible that there is no polynomial time decision procedure for solving such problems. Prima facie, this may sound like a refutation of our initial diagnosis. Here we have a notion of logical consequence that is indeed construed in terms of information states; and yet, by adopting it, the logical omniscience problem would even be aggravated, since PSPACE-completeness is usually regarded as stronger evidence of intractability than co-NP-completeness. But a closer analysis reveals that the notion of truth at an "information state" that is embodied in Kripke's semantics has features that make it clash with our desideratum D2.

The first problematic feature is that the truth of some complex sentences at an information state s cannot be established without "visiting" information states that are essentially richer than s. For example, in order to recognize that a conditional $A \rightarrow B$ is true at a state s in which A is not true, a reasoning agent must ideally transfer from s to a "virtual" state s^* in which the antecedent A is true and any other sentence has the same value as in s; that is, the agent reasons *as if* his state were s^*, observes that in s^* the consequent B must be true as well, and concludes that $A \rightarrow B$ must be true in his real information state s.

This use of "virtual information" is part of our common reasoning practice and is not too problematic as long as the structure of the sentence whose truth is being evaluated keeps simple. However, when rec-

---

[6] These issues, and all the subtleties that they involve, have been thoroughly discussed in the logical literature, especially in the writings of Michael Dummett; the reader is referred to [12] for an overall picture. On the informational view of intuitionistic logic and its relations with epistemic logic see [37].

[7] See [21, page 100].

[8] See [33].

ognizing the sentence as true requires weaving in and out of a complex recursive pattern of virtual information states, the situation may soon get out of control, as shown by the fact that the decision problem for the pure $\{\rightarrow\}$-fragment of intuitionistic logic is also PSPACE complete [33, 34]. The necessity of venturing out of one's actual information state in order to recognize the truth of certain sentences is what makes such inference steps "non-analytic" in a sense very close to Kant's original sense:[9] we essentially need to go *beyond the data*, using "virtual information", *i.e.*, simulating situations in which we hold information that, in fact, we do not hold. Although all virtual information is eventually removed, to the effect that the conclusion depends only on the information initially available, it remains true that such inference steps could not be performed at all without (temporarily) trespassing on richer information states.

The second problematic feature is the treatment of disjunction. In Kripke semantics a disjunction $A \vee B$ is true at an information state s if and only if either A is true at s or B is true at s. This reflects the intuitionistic notion of truth as (conclusive) verification, more precisely, the idea that the truth of a sentence coincides with the existence of a canonical proof for it, that is, a proof obtained "by the most direct means". In a natural deduction system this is a proof whose last step is the application of an introduction rule.[10] Indeed, in intuitionistic terms, we have a canonical proof of $A \vee B$ if and only if we have either a canonical proof of A or a canonical proof of B. However this does not seem to be a compelling feature of our understanding of $\vee$ in relation to a more ordinary notion of "information state", in which the truth of a sentence may be licensed by some weaker kind of epistemic condition. It is not difficult to come up with intuitive examples in which we hold enough information to assert a disjunction as true, but we do not hold enough information to assert either of the two disjuncts as true. Suppose we put two bills of 50 and 100 euros in two separate envelopes and then we shuffle the envelopes so as to loose track of which contains which. If we pick up one of them, we certainly hold the information that it contains either a 50-euro bill or a 100-euro bill, but we do not hold the information that it contains a 50-euro bill, nor do we hold the information that it contains a 100-euro bill.[11]

---

[9] See [7] on this point.

[10] See [12, Chapter 11] and [28] for a thorough discussion.

[11] This example is particularly tricky in that we could claim that we have, in some sense, arrived at the disjunction in a canonical way, except that the information has decayed during the process of shuffling the envelopes.

Beth's semantics for intuitionistic logic[12] seems to offer a more natural account of the truth of disjunctive statements based on an information state: a disjunction $A \vee B$ is true at the actual state provided we have the means of recognizing that necessarily one or the other disjunct will eventually become true at some future information state. However, at least in non-mathematical reasoning, we may well be in a position to assert a disjunction even if we have no means of recognizing that we will eventually reach an information state that will enable us to establish the truth of one of the two disjuncts.[13] If we do have such means, this should be regarded as additional information, not as information that is incorporated in the original assertion of the disjunction. So, it seems inappropriate to assume, as part of the ordinary meaning of $\vee$, that the assertion of $A \vee B$ can be licensed only if we know in advance that some future information state will infallibly put us in a position to assert one of the two disjuncts. As Michael Dummett puts it:

> I may be entitled to assert "A or B" because I was reliably so in-
> formed by someone in a position to know, but if he did not choose
> to tell me which alternative held good, I could not apply an or-
> introduction rule to arrive at that conclusion. [...] Hardy may sim-
> ply not have been able to hear whether Nelson said "Kismet hardy"
> or "Kiss me Hardy", though he heard him say one or the other:
> once we have the concept of disjunction, our perceptions them-
> selves may assume an irremediably disjunctive form [12, pages
> 266–267]. [...]
>
> Unlike mathematical information, empirical information de-
> cays at two stages: in the process of acquisition, and in the course
> of retention and transmission. An attendant directing theatre-goers
> to different entrances according to the colours of their tickets might
> even register that a ticket was yellow or green, without registering
> which it was, if holders of tickets of either colours were to use the
> same entrance; even our observations are incomplete, in the sense
> that we do not and cannot take in every detail of what is in our
> sensory fields. That information decays yet further in memory and
> in the process of being communicated is evident. In mathematics,

---

[12] For an exposition, see [10].

[13] In classical terms, the situation could be easily described as follows: according to our information we are able to assert that at least one of P and Q is true, but we are not able, and we may never be able, to assert that either of the two alternatives is true. However, even this description is inadequate for our purposes, since the sentence "at least one of P and Q is true" can be understood only with reference to the classical notion of truth as independent of our information state.

any effective procedure remains eternally available to be executed; in the world of our experience, the opportunity for inspection and verification is fleeting [12, pages 277–278].

One way of going around this difficulty consists in postulating that, whenever the assertion of a disjunctive statement can be made at all, one or the other disjunct *could have* been asserted by a properly informed agent. Again, such a way out of the difficulty requires what we have called "virtual information", going *beyond* the information that is actually held by the agent that is making the assertion.[14]

The rôle played by virtual information is apparent in the so-called "discharge rules" of natural deduction, a proof-theoretic presentation of logical consequence that is very close to the intuitionistic explanations of the logical operators.[15] For example, in the $\lor$-elimination rule,

$$\frac{\Gamma \vdash A \lor B \quad \Gamma, A \vdash C \quad \Gamma, B \vdash C}{\Gamma \vdash C,}$$

each of the discharged assumptions A and B represents a piece of information that needs not be included in all information states that verify the sentences in Γ (when it is, the rule application is indeed redundant).

The above considerations strongly suggest that the notion of information state underlying both Kripke's and Beth's semantics, and the way the meaning of the logical operators is explicated thereby, is inadequate for our purposes, since it essentially requires a reference to "virtual information" that is not actually contained in the agent's information state with respect to which this meaning is being explicated. As a consequence, the contents of an information state are somehow overstated, in such a way that the desideratum D2 above cannot be guaranteed. So, in the light of the analysis carried out in this section, we can add to D1 and D2 a further desideratum for a putative consequence relation that is likely to solve the problem of logical omniscience. Given that the truth or falsity of a sentence in an information state is established either by external means or by virtue of the meaning of the logical operators, we may request that:

D3 the meaning of the logical operators should be fixed by appropriate conventions expressed exclusively in terms of an agent's information state, so that the agent in question can be assumed to be totally

---

[14] On this point see [12, especially Chapter 12] and [11, especially Chapter 14].

[15] See [19, 27]; see also [36] for an excellent exposition

in control of this meaning; in particular, any reference to virtual information should be avoided.

Because of D3, we expect the logical operators that may emerge from this effort to bear a *weaker* meaning than the one they usually bear in classical semantics: we may call this *the informational meaning of the logical operators*. The underlying guess is that the inference steps that cannot be justified by virtue of such weaker informational meaning, *i.e.*, without appealing to what we have called "virtual information", are exactly those that essentially increase the computational complexity of deductive inference. In other words, a meaning-theory satisfying D3 is likely to yield a notion of information state satisfying D2 and so, via D1, a feasible notion of logical consequence that may dissolve the paradox that lurks in the usual reading of (1.1).

This does not imply that one cannot define tractable logics in which the use of a certain amount of virtual information is tolerated. It implies, however, that if virtual information is essentially used *to define the meaning of the logical operators*, so that its unbounded use *has* to be tolerated if one wants to fully exploit this meaning in drawing logical conclusions, then the resulting logic is likely to be intractable. On the other hand, a weaker definition of this meaning, in accordance with D3, would pave the way for gradually re-introducing virtual information, by imposing an upper bound on its recursive use. Then, we may look for some purely *structural principle*, expressing such bounded manipulation of virtual information, which can allow us to define a sequence of tractable logics, depending on the chosen upper bound, that converge to classical propositional logic and can legitimately be said to share *the same logical operators*. In this way the idealized reasoning agent of epistemic, doxastic and information logic could be arbitrarily approximated by realistic agents of increasing deductive power, but *all speaking the same language*. These considerations lead to our final desideratum:

D4 classical propositional logic should be characterized as the infinite union $\bigcup_{k \in \mathbb{N}} \{\vdash_k\}$ of approximating logics $\vdash_k$, such that (i) $\vdash_k \subset \vdash_{k+1}$, for every $k \in \mathbb{N}$, (ii) the meaning of the logical operators should be the same for all the approximating logics $\vdash_k$, and (iii) each logic $\vdash_k$ should be defined by an upper bound on the recursive use of some purely structural principle expressing the manipulation of virtual information.

It is apparent that a key role, in this road map, is played by D3. So, we now turn to the main question it raises: is there such a thing as "the informational meaning of the logical operators" and how can it be properly defined?

3   The informational meaning of the logical operators

Whatever the nature of the information concerned may be, an information state should provide a *partial* valuation v of the sentences of a standard propositional language into $\{0, 1\}$, where "0" stands for "false" and "1" stands for "true", describing the effect of the information held by an agent on the assertion or rejection of sentences. Intuitively, $v(A) = 1$ means that the agent's information state licenses the assertion of A as true, while $v(A) = 0$ means that it licenses the rejection of A as false. The sentences for which v is undefined are those that, on the basis of the information available, can neither be asserted as true, nor rejected as false. We shall write "$v(A) = \perp$" for "A is undefined in v".[16] Within an information state, we can distinguish a set of basic sentences whose truth-value is established by external means, the truth-values of the other sentences being established, starting from the basic ones, by virtue of the very meaning of the logical operators. The latter is usually fixed by defining, within the set of all possible valuations, those which are *admissible*. In classical semantics this is done by specifying the following set of if-and-only-if conditions:

C1 $v(\neg A) = 1$ if and only if $v(A) = 0$;

C2 $v(A \wedge B) = 1$ if and only if $v(A) = 1$ and $v(B) = 1$;

C3 $v(A \vee B) = 1$ if and only if $v(A) = 1$ or $v(B) = 1$;

C4 $v(A \rightarrow B) = 1$ if and only if $v(A) = 0$ or $v(B) = 1$;

C5 $v(\neg A) = 0$ if and only if $v(A) = 1$;

C6 $v(A \wedge B) = 0$ if and only if $v(A) = 0$ or $v(B) = 0$;

C7 $v(A \vee B) = 0$ if and only if $v(A) = 0$ and $v(B) = 0$;

C8 $v(A \rightarrow B) = 0$ if and only if $v(A) = 1$ and $v(B) = 0$.

A valuation satisfying the above conditions is said to be *saturated*. More specifically, we say that a valuation is *upward saturated*, if it satisfies the "if-part" of the above conditions, and *downward saturated* if it satisfies the "only-if" parts. A *Boolean valuation* is a saturated valuation that satisfies the additional condition of being *total*, *i.e.* defined for all sentences.[17]

---

[16] Although $\perp$ is often used as a logical constant standing for the "absurd", here we use it in the sense of "undefined", as customary in domain theory.

[17] Observe that, for total valuations, conditions C5–C8 are redundant, in that they can be derived from conditions C1–C4. On the other hand, every valuation satisfying C1–C8 that is total over the *atomic* sentences of the language, is total over the whole language and, therefore, is a Boolean valuation.

Now, not only cannot it be assumed that valuations representing information states be total, but the discussion in the previous section shows that some of the only-if conditions should also be dropped. In particular, the only-if parts of C3, C4 and C6 cannot be justified. That $A \vee B$ is assigned the value 1 by an agent's information state, does not guarantee that either A is assigned the value 1 or B is assigned the value 1, and so the only-if part of C3 fails. Dually, if $A \wedge B$ is assigned the value 0, this does not imply that either A is assigned the value 0 or B is assigned the value 0, and so the only-if part of C6 fails too.[18] Since we are dealing with classical (Boolean) conditional, the failure of the only-if part of C4 is a side-result of the definition of $A \to B$ as $\neg A \vee B$. Thus, if the meaning of $\vee$ and $\wedge$ (and, derivatively, of $\to$) has to be understood, in accordance with D3, by exclusive reference to an agent's *actual* information state (*i.e.*, without allowing for any use of virtual information) there is no way in which one can express this meaning by means of if-and-only-if clauses like the ones usually employed in classical semantics. Moreover, the standard way in which this meaning is fixed in the so-called proof-theoretic semantics, by specifying suitable introduction and elimination rules — the former playing a prominent rôle in this task and the latter being, in some sense, derivative[19] — is also inadequate for our purposes because some of the rules that are taken as part of the meaning of $\vee$ and $\to$ would make essential use of virtual information, as the usual natural deduction rules for $\to$-introduction and $\vee$-elimination.[20]

The problem is then: how do we fix what we have called "the informational meaning of the logical operators"? What kind of conventions can be distilled from linguistic practice in order to determine this weaker kind of meaning, other than the standard if-and-only-if conditions on admissible valuations or the standard Gentzen-style intelim rules? If we maintain that to grasp the meaning of a logical operator consists in acquiring some kind of linguistic information, it seems plausible that the latter can be unfolded by determining which possibilities are *ruled out* if one wants to use that operator consistently. From this point of view, a solution to our problem may consist in taking the informational meaning of a logi-

---

[18] Observe that the intuitionistic meaning of $\vee$, via the notion of canonical verification, satisfies the only-if part of C3 (under the intuitionistic interpretation of truth and falsity) and so essentially agrees with its classical meaning. By way of contrast, the intuitionistic meaning of $\wedge$ does not satisfy the only-if part of C6.

[19] Proof-theoretic semantics dates back to [19]. See [28] for a discussion and proper references. See also [25] for interesting remarks on this topic.

[20] See also [7] on this point.

cal operator to be fixed by a set of *negative constraints* on the valuations describing an agent's information state, specifying which ones should be ruled out as *inadmissible*. For example, a valuation that assigns the value 1 to $A \vee B$ and, at the same time, the value 0 to both A and B, should be excluded as inadmissible, revealing a mismatch between the valuation and the accepted (informational) meaning of $\vee$. An agent cannot, at the same time, hold the information that $A \vee B$ is true and the information that A and B are both false, without immediately realizing that this information is inconsistent. On the other hand, a valuation that assigns 1 to $A \vee B$, while being undefined for both A and B, is perfectly admissible and corresponds to a legitimate use of the word "or". Similarly, an agent cannot, at the same time, consistently hold the information that $A \wedge B$ is false and the information that A and B are both true, while a valuation that assigns 0 to $A \wedge B$, while being undefined for both A and B, is admissible and complies with the ordinary use of $\wedge$. A set of negative constraints that formally agree with the classical truth-tables, and can therefore be taken as distilling their informational content, is shown in Figure 3.1, where A and B stand for sentences of arbitrary complexity and each line, in the table for a given operator, represents a forbidden assignment, so that any valuation containing such an assignment is inadmissible.

$\neg A$	A		$A \vee B$	A	B		$A \wedge B$	A	B		$A \to B$	A	B
1	1		1	0	0		1	0	1		1	1	0
0	0		0	1	1		1	0	0		0	1	1
			0	1	0		1	0	$\perp$		0	0	1
			0	1	$\perp$		1	1	0		0	$\perp$	1
			0	0	1		1	$\perp$	0		0	0	0
			0	$\perp$	1		0	1	1		0	0	$\perp$

Figure 3.1. Constraints on admissible partial valuations. Each line represents a *forbidden* local configuration of values.

Although these constraints may be said to reflect the classical meaning of the logical operators, to the extent that this meaning can be expressed in terms of an agent's actual information state, they do not immediately justify *any* logical inference. In the next section, however, we shall observe the gradual emergence of inference rules once these constraints are combined with *purely structural principles* that can be naturally associated with the "depth" of the inferential process involved.

4   The semantics of Depth-Bounded Boolean Logics

We start by observing how some basic inferences can be recognized as valid, on the grounds of the meaning-constraints, by virtue of a minimal structural principle that we call "the single candidate principle", or SCP for short, after the well-known strategy for solving simple sudoku puzzles:

Infer that A is true (false) if the other option is immediately *ruled out by some of the accepted constraints that define the meaning of the logical operators.*

In what follows, we shall use the lower case letters p, q, r, etc. as variables for atomic sentences and continue using the capital letters A, B, C, etc. as variables for arbitrary sentences. As before, we shall use capital Greek letters Γ, Δ, Λ, etc. as variables for sets of sentences. Consider a valuation v such that $v(p \vee q) = 1$ and $v(p) = 0$, while $v(q) = \bot$. We can legitimately say that the value of q in this valuation is implicitly determined by the values of $p \vee q$ and p and by our understanding of the meaning of $\vee$ based on the constraints specified in Figure 3.1. For, there is *no admissible refinement* of v, that is, a refinement compatible with the meaning constraints, such that $v(q) = 0$: given the actual values of the other sentences, the assignment of 0 to q would be *immediately*[21] recognized as inadmissible by any agent that understands $\vee$ via the specified meaning constraints. In other words, the value 1 is *deterministically dictated* by v, since it is the only defined value that q can possibly take. The value 1 can therefore be assigned to q by exclusion of the other defined value.

By contrast, consider a typical example of "reasoning by cases": if our information state, described by a partial valuation v, is such that $v(p \vee q) = 1$, $v(p \rightarrow r) = 1$ and $v(q \rightarrow r) = 1$, then the piece of information that r is true is also, in some sense, implicitly contained in the information currently available, but we cannot specify this sense without introducing virtual information concerning p or q. We can reason as follows:

1. p must be either objectively true or objectively false (although this cannot be established on the basis of the current information);
2. assuming that we were informed about the objective truth-value of p

   (a) if we were informed that p is true, then the constraints on the meaning of $\rightarrow$ would rule out the possibility that r is false and so, by the SCP, r should be assigned the value 1;
   (b) if we were informed that p is false, then

---

[21] This means that the immediate refinement of v obtained by assigning 0 to q is inadmissible.

    i. the constraints on the meaning of ∨ would rule out the possibility that q is false and so, by SCP, q should be assigned the value 1;

    ii. if q were assigned the value 1, by the meaning constraints on → and SCP, r should be assigned the value 1.

3. Hence, r must be assigned the value 1 whatever the objective truth-value of p may be; therefore, the information that r is true is "implicitly contained" in our initial information state.

It is apparent that this sense of "implicitly contained" is essentially different from the sense in which the conclusion of disjunctive syllogism is implicitly contained in any information state that verifies the premises, because it requires the introduction of virtual information in steps 2(a) and 2(b). These steps cannot be internally justified on the basis of the agent's actual information state, but involve simulating the possession of definite information about the objective truth-value of p, by enumerating the two possible outcomes of the process of acquiring such information, *neither of which* is deterministically dictated by v. The inference displays, intuitively, a deeper reasoning process than the one displayed by disjunctive syllogism, and we relate this depth to the necessity of manipulating virtual information concerning p. An even deeper inference process would be displayed if steps 2(a) and 2(b) themselves contained, in a recursive fashion, further use of virtual information in order to obtain the common conclusion. In the remains of this section, we shall elaborate on these intuitive remarks in a more systematic way, leading to a classification of inferences according to their logical depth, starting from the most basic ones that do not require any use of virtual information.

Let $\mathcal{L}$ be a standard language for propositional logic and let $\mathcal{A}$ be the set of all admissible partial valuations of $\mathcal{L}$, namely, all the partial functions $\mathcal{L} \to \{0, 1\}$ that do not violate the negative constraints in Figure 3.1. The set $\mathcal{A}$ is partially ordered by the usual approximation relation $\sqsubseteq$ defined as follows: $v \sqsubseteq w$ (read "w is a *refinement* of v" or "v is an *approximation* of w") if and only if w agrees with v on all the formulas for which v is defined. Being a partial function, a partial valuation v is a set of pairs of the form $\langle A, i \rangle$, where A is a sentence of the given language and i is equal to 0 or 1, subject to the restriction that, for no $A \in \mathcal{L}$, $\langle A, 1 \rangle$ and $\langle A, 0 \rangle$ are both in v. Each pair in v can be thought of as a "piece of information", and the partial valuation itself as an attempt to put together such pieces of information in a way that is compatible with the intended meaning of the logical operators. The partial ordering $\sqsubseteq$ is a meet-semilattice with a bottom element equal to $\emptyset$, the valuation which is undefined for all formulas of the language. It fails to be a lattice because the join of two admissible valuations may be inadmissible.

Let $\mathcal{L}^*$ be the *evaluated language* based on $\mathcal{L}$, *i.e.* the set of all ordered pairs $\langle A, i \rangle$ such that A is a formula of $\mathcal{L}$ and $i \in \{0, 1\}$. Let $\Vdash_0$ be a relation $\mathcal{A} \times \mathcal{L}^*$ satisfying the following condition:

$$v \Vdash_0 \langle A, i \rangle \text{ if and only if } v \cup \{\langle A, |i - 1|\rangle\} \notin \mathcal{A}. \qquad \text{(SCP)}$$

Clearly, (SCP) expresses the structural property that we have called "single candidate principle". Notice that, by definition, $v \Vdash_0 \langle A, i \rangle$ for all $\langle A, i \rangle \in v$.

The image of $\Vdash_0$ under a partial valuation v represents all the information that can be recognized as "implicitly contained" in v without any need to introduce virtual information or, as we also say, *at depth* 0. This information may be seen to stem immediately from the (informational) meaning of the logical operators via a basic structural principle such as SCP. Now, we define a 0-depth information state as an admissible partial valuation that is closed under $\Vdash_0$:

Definition 4.1. A partial valuation v is a 0-*depth information state* if and only if $v \in \mathcal{A}$ and, for all $A \in \mathcal{L}$ and $i \in \{0, 1\}$,

$$\text{if } v \Vdash_0 \langle A, i \rangle, \text{ then } \langle A, i \rangle \in v.$$

Remark 4.2. It may be the case that v is an admissible valuation, but v cannot be embedded into any 0-depth information state. For example, suppose that v is such that $v(A \vee B) = 1$, $v(A) = 0$, $v(B \wedge C) = 0$, $v(C) = 1$ and $v(B) = \bot$; then v is admissible (it does not violate any of the meaning constraints), but $v \Vdash_0 \langle B, 1 \rangle$ and $v \cup \langle B, 1 \rangle$ is not in $\mathcal{A}$, because it violates one of the meaning constraints for $\wedge$. So, there is no 0-depth information state v' such that $v \sqsubseteq v'$.

We can now define the consequence relation $\vdash_0$ as truth-preserving over 0-depth information states.

Definition 4.3. For every finite set Γ of formulas and every formula A, $\Gamma \vdash_0 A$ if and only if $v(A) = 1$ for all 0-depth information states v such that $v(B) = 1$ for all $B \in \Gamma$.

The reader can check that $\vdash_0$ is a consequence relation in Tarski's sense, that is, it satisfies the following conditions for all formulas A, B and all finite sets Γ of formulas:

$$A \vdash_0 A \qquad \qquad \text{(Reflexivity)}$$

$$\text{if } \Gamma \vdash_0 A, \text{ then } \Gamma, B \vdash_0 A \qquad \qquad \text{(Monotonicity)}$$

$$\text{if } \Gamma \vdash_0 A, \text{ and } \Gamma, A \vdash_0 B, \text{ then } \Gamma \vdash_0 B. \qquad \text{(Transitivity)}$$

Moreover, $\vdash_0$ is also *substitution-invariant*, that is, it satisfies:

$$\text{if } \Gamma \vdash_0 A, \text{ then } \sigma(\Gamma) \vdash_0 \sigma(A), \qquad \text{(SubInv)}$$

for every uniform substitution σ.

The logic $\vdash_0$ is the basic element in our hierarchy of depth-bounded logics approximating full Boolean logic, namely, the one that allows for no use of virtual information and therefore contains all the logical inferences that can be validated by virtue only of the informational meaning of the logical operators, as fixed by the negative meaning constraints, and of the *purely structural* principle SCP. We call it *the Boolean Logic of depth 0*. As a result of Theorems 5.1 and 5.3 below, *this logic is tractable*. We may regard it as a minimum requirement on a reasoning agent that she is able to recognize 0-depth logical consequences.[22]

Say that a set Γ of formulas is 0-*depth inconsistent* if there is no 0-depth information state that verifies all the formulas in Γ. (Since $\vdash_0$ is a proper subsystem of classical logic, a set of formulas may be classically inconsistent and 0-depth consistent at the same time.) Observe that the logic $\vdash_0$ validates, by definition, the controversial *ex-falso quodlibet principle*: if Γ is a 0-depth inconsistent set of formulas, then $\Gamma \vdash_0 A$ for every formula A. In this logic, however, the principle in question is not nearly as dangerous as it is in full classical logic because the 0-depth inconsistency of Γ can be feasibly detected (see Theorem 5.3 below). On the other hand, if Γ is classically inconsistent, but 0-depth consistent, the ex-falso principle does not apply to Γ, because there is some 0-depth information state that verifies all the formulas in Γ.

It is also worth noticing that $\vdash_0$, like Belnap's four-valued logic and the NPL system of [13], *has no tautologies*. This is not surprising, however, since a tautology is a sentence that is a "logical consequence of the empty set of assumptions" and so, in order to establish its truth in any information state, we must make essential use of virtual information, to the effect that the information state itself cannot be of depth 0. On the other hand, $\vdash_0$ validates a good deal of classical inference schemes, including *modus ponens* $(A \rightarrow B, A \vdash_0 B)$, *modus tollens* $(A \rightarrow B, \neg B \vdash_0 \neg A)$, *disjunctive syllogism* $(A \vee B, \neg A \vdash_0 B)$ and its dual $(\neg(A \wedge B), A \vdash_0 \neg B)$. Moreover, the transitivity of $\vdash_0$ and its being based on a systematic view of the meaning of the logical operators, are features that make it a respectable, although minimalist, logical system. This system achieves tractability in a natural way, which is expressed in its semantics, rather than by tampering procedurally with a system for classical logic.

---

[22] One could say that the agent's "awareness" should be closed under 0-depth consequences.

Given two admissible valuations v and v', we say that v' *is a refinement of* v *on* P if (i) $v \sqsubseteq v'$ and (ii) P is defined in v'. Let now $\Vdash_k$, for $k = 1, 2, \ldots$, be a relation $\mathcal{A} \times \mathcal{L}^*$ satisfying the following condition:

$v \Vdash_k \langle A, i \rangle$ if and only if there exists an atomic p such that

$v'(A) = i$ for every information state v' of depth $k - 1$ (PB(k))

that refines v on p.

An information state of depth k can be simultaneously defined as an admissible valuation closed under $\Vdash_k$.

Definition 4.4. A partial valuation v is a k-*depth information state* if and only if $v \in \mathcal{A}$ and, for all $A \in \mathcal{L}$ and $i \in \{0, 1\}$,

$$\text{if } v \Vdash_k \langle A, i \rangle, \text{ then } \langle A, i \rangle \in v.$$

The image of $\Vdash_k$ under a partial valuation v represents all the information that can be recognized as "implicitly contained" in v by means of the meaning constraints augmented with a *purely structural* principle, PB(k), that allows for bounded use of virtual information. The parameter k represents the maximal number of nested introductions of atomic virtual information that are allowed at each step of the process. From a classical viewpoint, PB(k) allows for expansions of the current information state by means of at most k nested applications of the classical "Principle of Bivalence".

Remark 4.5. Given an admissible valuation v, it may be the case that, for some atomic p and some k, there is no information state of depth $k - 1$ that refines v on p. Under these circumstances PB(k) implies that $v \Vdash_k \langle A, i \rangle$, whatever A and i may be. Then, if a refinement of v were closed under $\Vdash_k$, it could not be admissible and so v cannot be embedded in an information state of depth k. Moreover, if $v \cup \{\langle p, i \rangle\}$ cannot be embedded in an information state of depth $k - 1$, while $v \cup \{p, |i - 1|\rangle\}$ can, then (PB(k)) implies that $v \Vdash_k \langle A, i \rangle$ whenever $v \cup \{\langle p, |i-1|\rangle\} \Vdash_{k-1} \langle A, i \rangle$.

The consequence relation $\vdash_k$ can then be defined as truth-preserving over information states of depth k.

Definition 4.6. For every finite set Γ of formulas and every formula A, $\Gamma \vdash_k A$ if and only if $v(A) = 1$ for all k-depth information states v such that $v(B) = 1$ for all $B \in \Gamma$.

For example, the argument given above for the inference expressing the principle of "reasoning by cases" shows the validity of the sequent

$p \vee q, p \to r, q \to r \vdash_1 r$. We call $\vdash_k$ *the Boolean Logic of depth k* and it follows from Definitions 4.3 and 4.6 that $\vdash_k \subset \vdash_{k+1}$ for every $k \in \mathbb{N}$.

Each $\vdash_k$ is a consequence relation in Tarski's sense. Moreover, in each $\vdash_k$ the meaning of the logical operators is fixed by the same negative constraints, namely, those of Section 3. It is not difficult to show that:

Theorem 4.7. *Let $\vdash_C$ be the consequence relation of classical propositional logic. Then:*

$$\vdash_C = \bigcup_{k \in \mathbb{N}} \vdash_k .$$

A set Γ of sentence is *k-depth inconsistent* if there is no k-depth information state that verifies all the formulas in Γ. As for k-depth tautologies, unlike $\vdash_0$, every k-depth logic with $k > 0$ has its related set of tautologies. In particular, if A is a formula containing k atomic sentences, A is a classical tautology if and only if A is true in all information states of depth k, that is, if and only if $\vdash_k A$. (This is a straightforward consequence of the fact that every k-depth information state can simulate the complete truth table of any formula with k atomic sentences.)

Again, each Boolean Logic of depth k is tractable (this follows from Theorems 5.5 and 5.6 in the next section), although the complexity of the decision procedure essentially grows with k. Hence, each $\vdash_k$ can be regarded as a feasible approximation to the unrealistic deductive power of classical propositional logic.[23]

5 Natural deduction for Depth-Bounded Boolean Logics

Let $\mathcal{L}_s$ be *the signed language* based on $\mathcal{L}$, namely, the set of all expressions of the form $T\,A$ and $F\,A$, for $A \in \mathcal{L}$. The elements of $\mathcal{L}_s$ are called *signed formulas*. The intuitive interpretation of signed formulas is the usual one: $T\,A$ means "A is true" and $F\,A$ means "A is false". We shall use the lower case Greek letters, φ, ψ, χ, etc. as variables for arbitrary signed formulas and the capital letters X, Y, Z, etc. as variables for sets of signed formulas. We construe a deduction of the signed formula φ from the assumptions in X as a *sequence of signed formulas* starting with signed formulas in X and ending with φ, such that every intermediate element instantiates the conclusion of some schematic inference rule whose premises are instantiated by previous elements of the sequence. In the

---

[23] The same remarks concerning the *ex-falso quodlibet* principle made above for $\vdash_0$ apply to $\vdash_k$ as well: if $\vdash_k$ is regarded as realistic, then an agent can detect the k-depth inconsistency of any finite set Γ of sentences and act upon it. On the other hand, if the inconsistency can be detected only at depth $m > k$, the *ex-falso quodlibet* principle does not apply to Γ at depth k.

context of $\vdash_0$ we do not need any device for discharging hypothesis, since any use of virtual information is banned. So, our inference rules will be of the simplest type, namely, principles licensing the assertion of a signed sentence of a certain form given the prior assertion of a finite number of other sentences of related forms.[24] Such simple inference rules will therefore be represented as follows:

$$S_1 \ A_1$$
$$\vdots$$
$$\frac{S_{n-1} \ A_{n-1}}{S_n \ A_n}$$

where each S_i (with $i = 1, \ldots, n$) is either "T" or "F".

Little reflection shows that, by the combined action of the meaning constraints of Section 3 and of the structural principle that we have named "Single Candidate Principle", the introduction and elimination rules of Figures 5.1 and 5.2 are all sound for $\vdash_0$. By the same means, the "mingle" elimination rules of Figure 5.3 are also shown to be sound for $\vdash_0$ and cannot be derived from the other rules. We shall use the expression "intelim rules" to refer to all these inference rules collectively.

$$\frac{F\ A}{T\ A \to B}\ T \to \mathcal{I} \qquad \frac{T\ B}{T\ A \to B}\ T \to \mathcal{I}2 \qquad \frac{\begin{array}{c} T\ A \\ F\ B \end{array}}{F\ A \to B}\ F \to \mathcal{I}$$

$$\frac{T\ A}{T\ A \vee B}\ T \vee \mathcal{I}1 \qquad \frac{T\ B}{T\ A \vee B}\ T \vee \mathcal{I}2 \qquad \frac{\begin{array}{c} F\ A \\ F\ B \end{array}}{F\ A \vee B}\ F \vee \mathcal{I}$$

$$\frac{\begin{array}{c} T\ A \\ T\ B \end{array}}{T\ A \wedge B}\ T \wedge \mathcal{I} \qquad \frac{F\ A}{F\ A \wedge B}\ F \wedge \mathcal{I}1 \qquad \frac{F\ B}{F\ A \wedge B}\ F \wedge \mathcal{I}2$$

$$\frac{T\ A}{F\ \neg A}\ F \neg \mathcal{I} \qquad \frac{F\ A}{T\ \neg A}\ T \neg \mathcal{I}$$

Figure 5.1. Introduction rules for signed sentences.

An *intelim sequence* based on a set X of signed formulas is a sequence $\varphi_1, \ldots, \varphi_n$ of signed formulas such that each element φ_i of the sequence either (i) is an element of X, or (ii) results from preceding elements of the sequence by an application of one of the intelim rules in Figures 5.1, 5.2 and 5.3. An intelim sequence is *closed* if it contains both $T\ A$ and $F\ A$ for some formula A, otherwise it is *open*. It can be easily shown that

---

[24] Clearly, the assertion of a signed formula $F\ A$ is tantamount to the rejection of A.

$$\frac{\begin{array}{c}T\ A \to B\\T\ A\end{array}}{T\ B}\ T \to \mathcal{E}1 \qquad \frac{\begin{array}{c}T\ A \to B\\F\ B\end{array}}{F\ A}\ T \to \mathcal{E}2 \qquad \frac{F\ A \to B}{T\ A}\ F \to \mathcal{E}1 \qquad \frac{F\ A \to B}{F\ B}\ F \to \mathcal{E}2$$

$$\frac{\begin{array}{c}T\ A \lor B\\F\ A\end{array}}{T\ B}\ T \lor \mathcal{E}1 \qquad \frac{\begin{array}{c}T\ A \lor B\\F\ B\end{array}}{T\ A}\ T \lor \mathcal{E}2 \qquad \frac{F\ A \lor B}{F\ A}\ F \lor \mathcal{E}1 \qquad \frac{F\ A \lor B}{F\ B}\ F \lor \mathcal{E}2$$

$$\frac{T\ A \land B}{T\ A}\ T \land \mathcal{E}1 \qquad \frac{T\ A \land B}{T\ B}\ T \land \mathcal{E}2 \qquad \frac{\begin{array}{c}F\ A \land B\\T\ A\end{array}}{F\ B}\ F \land \mathcal{E}1 \qquad \frac{\begin{array}{c}F\ A \land B\\T\ B\end{array}}{F\ A}\ F \land \mathcal{E}2$$

$$\frac{F\ \neg A}{T\ A}\ F \neg \mathcal{E} \qquad \frac{T\ \neg A}{F\ A}\ T \neg \mathcal{E}$$

Figure 5.2. Elimination rules for signed sentences.

$$\frac{T\ A \lor A}{T\ A}\ T \lor \mathcal{E}3 \qquad \frac{F\ A \land A}{F\ A}\ F \land \mathcal{E}3$$

Figure 5.3. "Mingle" rules for $\lor$ and $\land$.

every closed intelim sequence can be extended to an *atomically closed* one, *i.e.* one that contains both $T\ p$ and $F\ p$ for some *atomic* formula p. An *intelim deduction* of a *signed* formula φ from the set of *signed* formulas X is an intelim sequence based on X ending with φ. An *intelim deduction* of an *unsigned* formula A from the set of *unsigned* formulas Γ is an intelim deduction of $T\ A$ from $\{T\ B \mid B \in \Gamma\}$. We say that a *signed* formula φ is *intelim-deducible* from a set of *signed* formulas X, if there is an intelim deduction of φ from X. We also say that an *unsigned* formula A is *intelim-deducible* from the set Γ of *unsigned* formulas if $T\ A$ is intelim-deducible from $\{T\ B \mid B \in \Gamma\}$. Figure 5.4 contains an example of an intelim deduction that proves the sequent:

$$\neg u \lor s, u, s \to r \lor \neg u, \neg(q \land r), p \to q, \neg p \to t, t \land r \lor z \to v \vdash_0 v.$$

Finally, an *intelim refutation* of a set of formulas Γ is a closed intelim sequence based on $\{T\ B \mid B \in \Gamma\}$. When there is an intelim refutation of Γ, we say that Γ is *intelim-inconsistent*.

The notion of intelim deduction is adequate for the logic $\vdash_0$ semantically presented in the previous section and allows for a particularly strong normalization procedure. These properties are stated in Theorems 5.1 and 5.2 whose proofs can be adapted from [7] and [6].

Theorem 5.1. *For every finite set Γ of formulas and every formula A:*

1. $\Gamma \vdash_0 A$ *if and only if A is intelim-deducible from Γ;*
2. Γ *is 0-depth inconsistent if and only if Γ is intelim-inconsistent.*

□

1	$T \neg u \lor s$	
2	$T u$	
3	$T s \to r \lor \neg u$	
4	$T \neg(q \land r)$	
5	$T p \to q$	
6	$T \neg p \to t$	
7	$T t \land r \lor z \to v$	
8	$F \neg u$	$F \neg \mathcal{I}(2)$
9	$T s$	$T \lor \mathcal{E}1(1, 8)$
10	$T r \lor \neg u$	$T \to \mathcal{E}1(3, 9)$
11	$T r$	$T \lor \mathcal{E}2(10, 8)$
12	$F q \land r$	$T \neg \mathcal{E}(4)$
13	$F q$	$F \land \mathcal{E}2(12, 11)$
14	$F p$	$T \to \mathcal{E}2(5, 13)$
15	$T \neg p$	$T \neg \mathcal{I}(14)$
16	$T t$	$T \to \mathcal{E}1(6, 15)$
17	$T t \land r$	$T \land \mathcal{I}(11, 16)$
18	$T t \land r \lor z$	$T \lor \mathcal{I}1(17)$
19	$T v$	$T \to \mathcal{E}1(7, 18)$

Figure 5.4. An example of intelim deduction.

Say that a signed formula ψ occurring in an intelim deduction π of A from Γ is *redundant* if $\psi \neq T A$ and ψ is not used as a premise of any application of an intelim rule in π. Call *non-redundant reduction* of π the intelim deduction of A from Γ obtained from π by removing the redundant signed formulas. Then, we say that an intelim deduction of A from Γ is *regular* if its non-redundant reduction is open, and *irregular* otherwise. In other words, irregular deductions of A from Γ are deductions in which information which is explicitly inconsistent, and recognized as such, has been used to obtain a given conclusion.

Finally, say that an intelim deduction π of A from Γ enjoys the *subformula property* if every signed formula occurring in π has the form $T B$ or $F B$, where B is a subformula of A or of some formula in Γ.

The following theorem states a basic normalization property for intelim deductions (the *length* $|\pi|$ of an intelim sequence π is defined as the total number of symbols occurring in π):

Theorem 5.2. *Let Γ be a finite set of formulas. Then:*

1. *every regular intelim deduction π of A from Γ can be transformed into an intelim deduction π' of A from Γ such that (i) π' enjoys the subformula property, and (ii) $|\pi'| \leq |\pi|$;*

2. *every intelim refutation π of Γ can be transformed into an intelim refutation π' of Γ such that (i) π' enjoys the subformula property, and (ii) $|\pi'| \leq |\pi|$.*

Theorem 5.2 suggests that irregular intelim deductions may not be normalizable. And this is indeed the case, as shown by the intelim deduction of q from $\{p, \neg p\}$ shown in Figure 5.5, which cannot be normalized. In

1   $T\ p$
2   $T\ \neg p$
3   $F\ p$ $T\neg \mathcal{E}(2)$
4   $T\ p \vee q$ $T \vee \mathcal{I}1(1)$
5   $T\ q$ $T \vee \mathcal{E}1(4, 3)$

Figure 5.5. An intelim deduction of q from $\{p, \neg p\}$.

some sense, however, normalization fails exactly when it ought to, that is, when we have already obtained a closed intelim sequence and try to use two signed formulas that explicitly contradict each other in order to obtain a certain "conclusion" from them. But, to quote Michael Dummett again, "once a contradiction has been discovered, no one is going to go *through* it: to exploit it to show that the train leaves at 11:52 or that the next Pope will be a woman".[25] On the other hand, intelim refutations are always normalizable.[26] Notice that Theorem 5.2 marks a clear distinction from normalization theorems that can be proved for full classical (or intuitionistic) logic, where normal proofs may be longer, and sometimes exponentially longer, than non-normal ones (the same holds true for cut-free proofs versus cut-based proofs in the sequent calculus).

Theorem 5.2 paves the way for efficient decision procedures. One of them is presented in [7], and improved on in [6], where it is used to show the following:

Theorem 5.3. *Intelim-deducibility and intelim-refutability are tractable problems. Whether a formula A is intelim-deducible from a finite set Γ of formulas and whether a finite set Γ of formulas is intelim-refutable are both questions that can be decided in time $\mathcal{O}(n^2)$.*

We now turn to the general proof-theoretic presentation of the logics $\vdash_k$ with $k > 0$.

---

[25] [12, page 209].

[26] If we add a new structural rule corresponding to the *ex-falso quodlibet* principle — to the effect that an arbitrary signed formula can be deduced from two contradictory ones — and modify the deducibility relation accordingly, then the subformula property holds in general.

Given a signed formula SA, with S equal to T or F, let us denote by $\overline{S}A$ its *conjugate*, that is, FA if $S = T$ and TA if $S = F$. An *intelim sequence of depth k*, with $k > 0$ based on a set X of signed formulas is a sequence of signed formulas $\varphi_1, \ldots, \varphi_n$ such that for each element φ_i, with $i \leq n$, one of the following conditions is satisfied:

1. $\varphi_i \in X$,
2. φ_i is intelim$(k-1)$-deducible from $\varphi_1, \ldots, \varphi_{i-1}, Tp$ and $\varphi_1, \ldots, \varphi_{i-1}, Fp$, for some *atomic* formula p,
3. φ_i is intelim$(k-1)$-deducible from $\varphi_1, \ldots, \varphi_{i-1}, S\,p$, for some *atomic* formula p such that $\varphi_1, \ldots, \varphi_{i-1}, \overline{S}\,p$ is intelim$(k-1)$-inconsistent.[27]

Remark 5.4. Observe that, since intelim deducibility (at any depth) is clearly monotonic, clause 2 in the above definition covers the case in which φ_i is intelim$(k-1)$ deducible from $\varphi_1, \ldots, \varphi_{i-1}$ alone, with no need for the virtual information concerning the atomic formula p. As a result any intelim(k) sequence, with $k \in \mathbb{N}$, is also an intelim$(k+1)$ sequence.

An *intelim(k) deduction* of a *signed* formula φ from the set of *signed* formulas X is an intelim(k) sequence based on X ending with φ. An *intelim(k) deduction* of an *unsigned* formula A from the set of *unsigned* formulas Γ is an intelim(k) deduction of $T A$ from $\{T B \mid B \in \Gamma\}$. We say that a *signed* formula φ is *intelim(k)-deducible* from a set of *signed* formulas X, if there is an intelim(k) deduction of φ from X. We also say that the *unsigned* formula A is *intelim(k)-deducible* from the set Γ of *unsigned* formulas if $T A$ is intelim(k)-deducible from $\{T B \mid B \in \Gamma\}$.

An intelim(k) sequence $\varphi_1, \ldots, \varphi_n$ is *closed* if, for some atomic p, there are closed intelim$(k-1)$ sequences for both $\varphi_1, \ldots, \varphi_n, T\,p$ and $\varphi_1, \ldots, \varphi_n, F\,p$. (The notion of a closed intelim(0) sequence is the same as that of a closed intelim sequence.) Finally, an *intelim(k) refutation* of a set of formulas Γ is a closed intelim(k) sequence based on $\{T B \mid B \in \Gamma\}$. When there is an intelim(k) refutation of Γ, we say that Γ is *intelim(k) inconsistent*.

An intelim(k) sequence can be conveniently represented using boxes for the auxiliary sequences of depth greater than 0 that may be needed to establish that a signed formula φ_i can be appended to the sequence by virtue of clauses 2–3 of the above definition. To represent such a step, two parallel boxes may be opened (when necessary, see Remark 5.4) that contain, respectively, the auxiliary $(k-1)$-depth sequences based on

---

[27] See Remark 4.5 above.

1 $T\ p \vee q$ Assumption

2 $T\ p \to r \vee s$ Assumption

3 $T\ q \to r \vee s$ Assumption

4 $T\ r \to t$ Assumption

5 $T\ s \to u$ Assumption

6 $T\ p$	
7 $T\ r \vee s$	$T \to \mathcal{E}1(2,6)$

$F\ p$	
$T\ q$	$T \vee \mathcal{E}2(1,6)$
$T\ r \vee s$	$T \to \mathcal{E}1(3,7)$

9 $T\ r \vee s$

10 $T\ r$	
11 $T\ t$	$T \to \mathcal{E}1(10,4)$
12 $T\ t \vee u$	$T \vee \mathcal{I}1(11)$

$F\ r$	
$T\ s$	$T \vee \mathcal{E}1(9,10)$
$T\ u$	$T \to \mathcal{E}1(5,11)$
$T\ t \vee u$	$T \vee \mathcal{I}2(12)$

14 $T\ t \vee u$

Figure 5.6. An intelim deduction of depth 1.

$\varphi_1, \ldots, \varphi_{i-1}, T\ p$ and $\varphi_1, \ldots, \varphi_{i-1}, F\ p$, for some atomic p. The usual scoping rules for boxes are employed here: each formula occurring in a box can be used, as premiss of a rule application, in every box contained in it and *cannot* be used in any other box. The whole deduction should be regarded as being contained in a root box that contains all the others, although we shall not usually draw the borders of this most external box. Figure 5.6 and Figure 5.7 show two examples of intelim deductions of depth 1 and 2 respectively.

Theorem 5.5. *For every finite set Γ of formulas and every formula A:*

1. $\Gamma \vdash_k A$ *if and only if A is intelim(k)-deducible from Γ;*
2. Γ *is k-depth inconsistent if and only if Γ is intelim(k)-inconsistent.*

A normalization theorem analogous to Theorem 5.2 can be shown for intelim(k) deductions. The restriction of virtual information to *atomic* formulas, in the definition of intelim(k) sequences, as well as in the se-

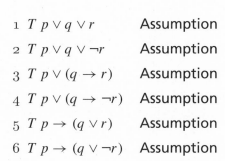

1 $T\ p \vee q \vee r$ Assumption

2 $T\ p \vee q \vee \neg r$ Assumption

3 $T\ p \vee (q \to r)$ Assumption

4 $T\ p \vee (q \to \neg r)$ Assumption

5 $T\ p \to (q \vee r)$ Assumption

6 $T\ p \to (q \vee \neg r)$ Assumption

7 $T\ p \to (q \to r)$ Assumption

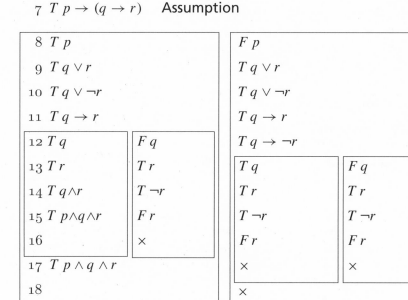

8 $T\ p$		$F\ p$	
9 $T\ q \vee r$		$T\ q \vee r$	
10 $T\ q \vee \neg r$		$T\ q \vee \neg r$	
11 $T\ q \to r$		$T\ q \to r$	
12 $T\ q$	$F\ q$	$T\ q \to \neg r$	
13 $T\ r$	$T\ r$	$T\ q$	$F\ q$
14 $T\ q \wedge r$	$T\ \neg r$	$T\ r$	$T\ r$
15 $T\ p \wedge q \wedge r$	$F\ r$	$T\ \neg r$	$T\ \neg r$
16	×	$F\ r$	$F\ r$
17 $T\ p \wedge q \wedge r$		×	×
18		×	

19 $T\ p \wedge q \wedge r$

Figure 5.7. An intelim deduction of depth 2. The justification of the inference steps is omitted for reasons of space. The symbol "×" indicates a closed intelim sequence.

mantical definition of $\vdash_k$, ensures that the normalization procedure is essentially the same as in the 0-depth case and does not increase the length of proofs, since spurious atomic formulas are easily removed with no loss of deductive power. The price is that, unlike $\vdash_0$, the consequence relations $\vdash_k$, with $k > 0$, are not substitution-invariant. On the other hand, the family $\{\vdash_k\}_{k \in \mathbb{N}}$, as a whole, *is* substitution invariant in the sense that $\Gamma \vdash_k A$ implies $\sigma(\Gamma) \vdash_{k+j} \sigma(A)$ for some j depending on the substitution σ (which is obvious, given Theorem 4.7 above and the fact that classical logic is substitution-invariant). To make each single $\vdash_k$ substitution-invariant, one should modify the definitions so as to allow for the use of virtual information of arbitrary complexity. In this case, normalization can still be achieved, but the length of normal proofs can be exponentially longer than that of non-normal ones. A normalization theorem for an intelim system with unbounded use of virtual information, which is therefore sound and complete for full classical propositional logic, is given in [5].[28] Finally, it can be shown that the relations of intelim(k)-deducibility and intelim(k)-refutability admit of a polynomial time decision procedure.

Theorem 5.6. *Intelim(k)-deducibility and intelim(k)-refutability are tractable problems for every fixed k. Whether a formula A is intelim(k)-deducible or whether a finite set Γ of formulas is intelim(k)-refutable can be decided in time $\mathcal{O}(n^{2k+2})$.*

6   Concluding remarks

We have argued that Depth-Bounded Boolean Logics can be profitably adopted as a basis for constructing epistemic, doxastic and information logics that are not affected by the problem of logical omniscience, but a good deal of work has yet to be done in this direction. First, one should devise an alternative to the standard characterization of the propositional attitudes $\Box_i$ based on possible worlds. Our analysis suggests that, for this task, possible worlds could be replaced by feasible depth-bounded information states, as defined in Section 4, and that the semantics of $\Box_i$ could be fixed via some sort of invariance over information states, depending on the propositional attitude under consideration. For example, knowledge could be characterized as information that is persistent through information states that may change in a non-monotonic way. Moreover, the incremental characterization of classical logical consequence that is typical of Depth-Bounded Boolean Logics may allow for the design of multimodal systems where different (human or artificial) agents may be endowed with unequal deductive power, maybe on the basis of the unequal computing resources available to them, but share the same understanding of the logical operators (displayed in the common 0-depth logic). Another interesting topic of future research is the treatment of inconsistent information. As remarked in Sections 4 and 5, Depth-Bounded Boolean Logics seem likely to provide an interesting way out of the qualms about the *ex-falso quodlibet* principle of classical logic. Furthermore, the failure of

---

[28] In the cited paper the intelim rules are presented for *unsigned* formulas. The difference is largely a matter of taste, but we now believe that the use of signed formulas is conceptually more transparent when dealing with a classical-like negation operator.

substitution-invariance for the logics of depth greater than 0 prompts for an exploration of the possible remedies outlined in Section 5. Last, but not least, the possibility of augmenting Depth-Bounded Boolean Logics with suitable quantifier rules, in a way that preserve their useful properties (first of all tractability), is an important topic of future investigation.

ACKNOWLEDGEMENTS. I wish to thank the organizers and all the participants of the "One-day workshop on the logical foundations of rational interaction" hosted in Pisa by the Centro di Ricerca Matematica Ennio de Giorgi, on 4 November 2009, and in particular Hykel Hosni who pointed me at the potential connections between my work and the problem of logical omniscience.

References

[1] S. ARTEMOV and R. KUZNETS, *Logical omniscience as a computational complexity problem*, In: "TARK '09: Proceedings of the 12th Conference on Theoretical Aspects of Rationality and Knowledge", pages 14–23, New York, NY, USA, 2009. ACM.

[2] M. CADOLI and M. SCHAERF, *Approximate reasoning and non-omniscient agents*, In: "TARK '92: Proceedings of the 4th conference on Theoretical aspects of reasoning about knowledge", pages 169–183, San Francisco, CA, USA, 1992. Morgan Kaufmann Publishers Inc.

[3] S. A. COOK, *The complexity of theorem-proving procedures*, In: "STOC. Conference Record of Third Annual ACM Symposium on Theory of Computing", 1971, Shaker Heights, Ohio, USA, pages 151–158. ACM, 1971.

[4] J. M. CRAWFORD and D. W. ETHERINGTON, *A non-deterministic semantics for tractable inference*, In: "AAAI/IAAI", pages 286–291, 1998.

[5] M. D'AGOSTINO, *Classical natural deduction*, In: Sergei N. Artëmov, Howard Barringer, Artur S. d'Avila Garcez, Luís C. Lamb and John Woods (eds.), "We Will Show Them!" (1), pages 429–468. College Publications, 2005.

[6] M. D'AGOSTINO, M. FINGER and D. M. GABBAY, *Depth-bounded boolean logics*, submitted. Preliminary version available on request.

[7] M. D'AGOSTINO and L. FLORIDI, *The enduring scandal of deduction. Is propositionally logic really uninformative?* Synthese **167** (2009), 271–315.

[8] M. DALAL, *Anytime families of tractable propositional reasoners*, In: "Proceedings of the Fourth International Symposium on AI and Mathematics (AI/MATH-96)", 1996, 42–45.

[9] M. DALAL, *Anytime families of tractable propositional reasoners*, Annals of Mathematics and Artificial Intelligence **22** (1998), 297–318.

[10] M. DUMMETT, "Elements of Intuitionism", Clarendon Press, Oxford, 1977.

[11] M. DUMMETT, "Truth and Other Enigmas", Duckworth, 1978.

[12] M. DUMMETT, "The Logical Basis of Metaphysics" Duckworth, 1991.

[13] R. FAGIN, J. Y. HALPERN, and M. Y. VARDI, *A nonstandard approach to the logical omniscience problem*, Artificial Intelligence **79** (1995), 203–240.

[14] M. FINGER, *Polynomial approximations of full propositional logic via limited bivalence*, In: "9th European Conference on Logics in Artificial Intelligence (JELIA 2004)", Vol. 3229 of "Lecture Notes in Artificial Intelligence", pages 526–538, Springer, 2004.

[15] M. FINGER and D. GABBAY, *Cut and pay*, Journal of Logic, Language and Information **15** (3) (2006), 195–218.

[16] L. FLORIDI, *Scepticism and animal rationality: the fortune of chrysippus' dog in the history of western thought*, Archiv für Geschichte der Philosophie **79** (1) (1997), 27–57.

[17] L. FLORIDI, *The logic of being informed*, Logique et Analyse **49** (196) (2006), 433–460.

[18] A. M. FRISCH, *Inference without chaining*, In: "IJCAI'87: Proceedings of the 10th international joint conference on Artificial intelligence", pages 515–519, San Francisco, CA, USA, 1987. Morgan Kaufmann Publishers Inc.

[19] G. GENTZEN, *Unstersuchungen über das logische Schliessen*, Math. Zeitschrift **39** (1935), 176–210. English translation in [35].

[20] J. Y. HALPERN, *Reasoning about knowledge: a survey*, In: Dov M. Gabbay, C. J. Hogger, and J.A. Robinson (eds.), "Handbook of Logic in Artificial Intelligence and Logic Programming", Vol. 4, pages 1–34. Clarendon Press, Oxford, 1995.

[21] S. A. KRIPKE, *Semantical analysis of intuitionistic logic I*, In: J. Crossley and M. A. E. Dummett (eds.), "Formal Systems and Recursive Functions", pages 92–130. North-Holland Publishing, Amsterdam, 1965.

[22] M. S. KWANG, *Epistemic logic and logical omniscience: a survey*, International Journal of Intelligent Systems **12** (1) (1997), 57–81.

[23] H. J. LEVESQUE, *Logic and the complexity of reasoning*, Journal of Philosophical Logic **17** (4) (1988), 355–389.

[24] J.-J. C. MEYER, *Modal epistemic and doxastic logic*, In: Dov M. Gabbay and Franz Guenthner (eds.), "Handbook of Philosophical Logic", Vol. 10, pages 1–38. Kluwer Academic Publishers, 2nd edition, 2003.

[25] E. MORICONI, *Normalization and meaning theory*, Epistemologia **23** (2000), 281–304.

[26] R. PARIKH, *Sentences, belief and logical omniscience, or what does deduction tell us?*, The Review of Symbolic Logic **1** (4) (2008).

[27] D. PRAWITZ, "Natural Deduction. A Proof-Theoretical Study" Almqvist & Wilksell, Uppsala, 1965.

[28] D. PRAWITZ, *Meaning approached via proofs*, Synthese **148** (3) (2006), 507–524.

[29] G. PRIMIERO, *An epistemic logic for being informed*, Synthese **167** (2) (2009), 363–389.

[30] M. SCHAERF and M. CADOLI, *Tractable reasoning via approximation*, Artificial Intelligence **74** (2) (1995), 249–310.

[31] G. SILLARI, *Quantified logic of awareness and impossible possible worlds*, Review of Symbolic Logic **1** (4) (2008), 1–16.

[32] G. SILLARI, *Models of awareness*, In: G. Bonanno, W. van der Hoek, and M. Wooldridge (eds.), "Logic and the Foundations of Games and Decisions", University of Amsterdam, 2008, 209–240.

[33] R. STATMAN, *Intuitionistic propositional logic is polynomial-space complete*, Theoretical Computer Science **9** (1979), 67–72.

[34] V. SVEJDAR, *On the polynomial-space completeness of intuitionistic propositional logic*, Archive for Mathematical Logic **42** (7) (2003), 711–716.

[35] M. SZABO (ed.), "The Collected Papers of Gerhard Gentzen", North-Holland, Amsterdam, 1969.

[36] N. TENNANT, "Natural Logic", Edinburgh University Press, 1990.

[37] J. VAN BENTHEM, *The information in intuitionistic logic*, Synthese **167** (2) (2009), 251–270.

Rational behaviour at trust nodes

Hykel Hosni and Silvia Milano

1 Introduction and motivation

The importance of trust in rational interaction has been widely recognized in the literature, yet there seems to be no general consensus on its precise meaning. Rather than attempting at a general definition of rational choice involving trust, in this paper we restrict our attention to some specific aspects of strategic interaction based on what we call *trust nodes*.

In many interactive decision problems, rational agents need to form beliefs about the trustworthiness of other agents. The starting point of our investigation is the idea, supported by a vast body of empirical evidence, that beliefs about trustworthiness are rather specific kinds of beliefs. As a consequence, a key step towards characterizing rational choice in problems involving trust consists in isolating the parameters which make *specific* a decision based on trust. Before going into the details of our proposal, however, it is useful to try and make clear the level at which we intend to enter the discussion on trust-related rational behaviour.

What *should* a rational agent do when facing a decision whose outcome depends on the trustworthiness of other agents? From the normative point of view nothing much can be said except the totally obvious recommendation that an agent should trust another just in case this latter is trustworthy. But even if we were to make sense of such a platitude, why should it ever be rational to trust other agents in the first place?

The way much economic literature puts it, at least since since Arrow's seminal work is that

> Virtually every commercial transaction has within itself an element of trust, certainly any transaction conducted over a period of time. [1]

The impressive literature which developed out of this line of research converged rapidly towards establishing a positive correlation between the average level of trust of a community and its aggregate economic performance (see, *e.g.* [1, 11]). At this point one would be tempted to say that, modulo a certain degree of approximation, a reason for trusting others is that by doing so we eventually increase our own welfare. Recent studies on *individual* rather than aggregate economic performance in relation to trust however, show that things are not as straightforward [5], for a number of reasons, some of which will be discussed in a short while.

There is a clear moral here that deserves our full attention. Unlike other aspects of rational behaviour, a reasonably meaningful theory of trust cannot dispense from being informed by empirical evidence. It is only by carefully observing how people behave in certain specific situations that we can hope for an adequate account of what 'the right amount of trust' [5] might turn out to be. Unfortunately, however, the available empirical evidence is seldom methodologically unquestionable, as the authors of [5] and [15] openly admit. The attempt to overcome this difficulty ultimately motivates the perspective of the present note which focusses on the simplest scenario in which trust can be isolated according to some specific parameters which jointly account for what we call a *trust node*. Our contribution is therefore mainly methodological: by looking at a specific class of problems and by analysing some relevant empirical data, we put forward a proposal about what determines the specificity of a trust node with respect to the general problem of rational choice. We show that our proposal is consistent with the available data and we outline some initial intuitions about how trust nodes could be given a rather general game theoretic interpretation.

The paper is organised as follows. In Section 2 we introduce decisions involving trust by considering a real-world example: eBay transactions. As we shall see, there are a number of advantages in restricting our attention to this problem, chief among them the availability of empirical data about how people actually behave when interacting on eBay. In Section 2.2 we define trust nodes in terms of Informal Institutions, Reputation and Cheap Talk. We move on to analyse eBay data in the light of our characterization of trust nodes in Section 2.4 and point out how those field data suffer from important methodological limitations. We take up this question in Section 3 in which we review some controlled experiments on the so-called *trust game*. We then point out how the variations on the trust game provide us with good evidence about the appropriateness of characterizing trust nodes by means of Informal Institutions, Reputation and Cheap Talk. Finally, in Section 4 we draw some conclusions and point to some future lines of research aimed at providing a game-theoretic characterization of rational behaviour at trust nodes.

2 Trust nodes

As a simple example of interaction involving trust, let us consider the decision problem depicted in Figure 2.1 which represents a stylised eBay transaction. At the top node, the buyer decides whether a given item is valuable to her. At the second node she has to make a decision about the trustworthiness of the seller. If the seller is not reliable, the buyer risks loosing her money, so she should not trust him and resolve not to bid. On the other hand if the seller *is* reliable, the buyer faces the crucial node of her decision where she decides whether to bid or not. It is clear that in this simple scenario decisions involving trust play a fundamental role towards determining the final outcome of the transaction.

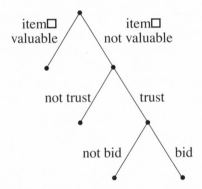

Figure 2.1. The eBay game: A diagrammatic representation of the decisions taken by the buyer.

Despite its simplicity, the eBay game is representative of a wide class of (economic) interactions involving trust and as such it proves to be a methodologically sound tool to investigate rational behaviour involving trust.

First of all, it is a situation for which we have a vast body of empirical data coming from field observations [6, 10, 15]. This meets the recommendation outlined in the previous Section to the effect that any adequate account of rational behaviour involving trust needs grounding on empirical data.

Secondly, eBay provides an operational example of what is usually referred to as an *informal institution*. Intuitively, an informal institution can be thought of as an environment governed by rules whose violation does not result in a predefined penalty mechanism. Note that we are not looking at eBay as an individual acting in the stock market, which is clearly subject to specific regulations and enforceable penalties. Rather, we insist on the point of view of a *user* to whom eBay gives a framework, so to speak, to engage properly in transactions with other users. As we shall see in full detail in the next section informal institutions provide, in this environment, the key structure required to enable trusting behaviour.

A further reason for focussing on the eBay game consists in its generality. In the vast majority of cases eBay transactions take place among strangers from all over the world. As noted in [13] trust *in strangers* is of particular interest in the perspective of providing a relatively general characterization of rational behaviour based on trust. Moreover, since the interaction takes place virtually over the Internet, the resulting data are free of the complications arising in the interpretation of data based on face-to-face transactions, such as the effects of pre-existing personal relations and instinctive reactions to the physical person[1].

Along with such important advantages, the eBay game and the data relating to it, present some methodological limitations. Specifically, as pointed out *e.g.* in [15], field experiments on eBay data suffer from a problem of omitted variables. This is mostly due to the fact that it is difficult to anticipate, at design time, what will account as meaningful information in the unstructured communication which usually takes place between buyers and sellers[2].

In order to overcome those limitations we shall move to the analysis of controlled experiments on the *trust game*. This latter, as shown in Section 3 below, allows us to reproduce the relevant features of the eBay game in a controlled environment in which the problem of omitted variables disappears. The resulting characterization, as we shall argue in the concluding section of this paper, opens to a game-theoretic analysis of rational behaviour at trust nodes.

2.1 A behavioural definition of trust

According to [9] behaviour based on trust can be analysed by referring to two components: preferences and beliefs. The preference component of trust articulates into *risk* and *social preferences*, which jointly determine how an agent responds to a given decision problem in terms of *risk aversion* and *betrayal aversion*. A comprehensive survey of the role of social

---

[1] This is not to say, of course, that those issues are devoid of any interest. It seems to us, however, that their analysis can be postponed until a reasonably stable characterization of the key parameters defining rational behaviour involving trust is available.

[2] In eBay this might take various forms, from private email exchange, to chat rooms to the eBay community forum (see Figure A.1 in the Appendix).

preferences in trust can be found in [9] and [13] which analyse empirical evidence obtained through standard survey questionnaires. Those papers provide compelling evidence, based on biological as well as economical considerations, to the effect that preferences regarding social risk, also referred to as *betrayal aversion*, are qualitatively different from preferences regarding ordinary risk (*risk aversion*).

The authors of [5], on the other hand, give priority to the belief component of trust which they interpret as the belief held by an agent about the trustworthiness of the others. The paper points out that an agent's degree of trust is positively correlated to her own trustworthiness. Interestingly analogous findings are reported in [2] drawing on the analysis of controlled experiments on the dictator game. In particular the authors conclude that there is a tight correlation between the subject's own projected behaviour and her beliefs about the *fairness* of other subjects.

Following the latter line of research this paper concentrates on the *belief component* of trust. Our reason for doing so lies in our attempt to distinguish rational beliefs about the trustworthiness of the others from rational beliefs *tout court*. As a consequence, our primary interest is the formation of trust beliefs as the rational response to certain social inputs. Specifically, then, the starting point of our discussion is the standard behavioural definition of trust given by [9][3]:

> An individual [...] trusts if she voluntarily places resources at the disposal of another party (the trustee) without any legal commitment from the latter. In addition the act of trust is associated with an expectation that the act will pay off in terms of the investor's goals.

A number of things are worth noticing at this point, though we shall come back to them in full details later on. First of all if we imagine an interaction involving trust as an extensive form game, the trustor makes 'the first move' by sharing some resources with the trustee. Secondly, the concept of trust is viewed as an 'all-or-nothing' choice: an agent decides whether to trust another or not. Finally, and most importantly, trust is intrinsically connected to the absence of *pre-determined* penalties. The import of that is enormous, as we shall see. It is enough to anticipate at this point that if the trustee betrays the truster, this latter has no legally enforceable way of imposing a penalty to the former. However this is not to say that a trustee can betray free of any consequences. What exactly those consequences

---

[3] As the author acknowledges this is indeed an adaptation from [8].

might turn out to be is specified by some appropriate *Informal Institution*, the key ingredient of trust nodes, that we are now ready to define.

2.2   Trust nodes defined

Interactions involving trust can be represented both in strategic and extensive forms.[4] As it will be clear in a short while, it is natural, for the purposes of this work, to concentrate on the latter case. In particular we assume that an agent must decide whether to trust or not on the basis of the available information and the expected behaviour of the other party. We refer to a decision of this kind as a *trust node*.

At each trust node we assume that the agent grounds her decision on three sources of information:

Informal Institutions context-based constraints on interaction
Reputation information about relevant past interactions
Cheap talk non-committal, relevant pre-play communication

These parameters characterise a trust node in the sense that if we abstract away from them, we cannot distinguish a decision based on trust from any other decision under uncertainty.

Informal Institutions play a key role in determining the overall beahaviour of a rational agent facing a trust node. First and foremost they determine the rules of interaction among agents. Such rules can take a whole variety of forms, yet what is common among them all is the fact that they determine some form of *social sanction* for betrayers. This is clearly meant to deter cheating behaviour.

To illustrate the key properties of trust nodes, we now move on to consider a concrete example: the *eBay Game*.

2.3   Trust nodes in eBay transactions

We begin with a stylized model of eBay interactions in which all the relevant facts about a single transaction can be represented in a five-node decision tree. It is essential to the model that an agent facing a trust node has access to the information coming from previous eBay transactions.

The initial state is determined by the Informal Instituion: eBay. At this initial node the rules governing the interaction as well as the information relevant to determine the agents' reputation, notably their public profiles,

---

[4] Strategic games involving trust, *e.g.* the prisoners' dilemma and the stug hunt, can be either one-shot or repeated.

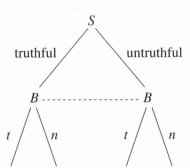

Figure 2.2. Key steps (nodes 2-4 in the description given in the body of the paper) of eBay transactions. S (B, respectively) labels the node at which the seller (buyer, respectively) has a choice, while t (n, respectively) stands for 'the buyer trusts' (the buyer does not trust, respectively) the seller.

are given. This information will subsequently be taken to be common knowledge among buyers and sellers.

At node 2, labelled by S in Figure 2.2 the seller puts an item on sale. This is done by providing a description for the item which aims at making it palatable to the buyer. Cheap talk enters substantially at this stage since there is no independent party guaranteeing the truthfulness of the information provided by the seller.

Node 3 is the trust node. As illustrated in 2.2 the Buyer (B) faces the decision whether to trust the Seller or not without *knowing* if his cheap talk is truthful. This uncertainty is represented by the information state –the dashed line– which joins the two nodes labelled B. The buyer's decision whether to bid or not reduces at this point to the decision whether to trust the seller's cheap talk. If the buyer resolves not to bid, the interaction ends. Otherwise at node 4 the Seller, depending on his choice of strategy, delivers the item or not[5].

Finally, at node 5, the Buyer has an opportunity to express her feedback and by doing so directly affects the Seller's public profile and reputation. This is clearly where the Informal Institution introduces a strong incentive for the Seller to reciprocate the Buyer's initial trust.

This leads us to distinguish, in general, between the group of sellers and that of buyers. Sellers, as a group, have a collective interest in keeping eBay a safe market place, for doing otherwise would result in deterring potential buyers from bidding. To prevent this they aim at having

---

[5] In some cases, the quality of the delivery (*i.e.* whether it is timely, whether the item is well packed, etc.) might also be relevant to the construction of a seller's public profile.

as good a reputation as possible by committing themselves to truthful cheap-talk. Note that this clearly gives individual sellers an opportunity to free-ride on the group's trustworthiness by delivering untruthful cheap talk. The feedback mechanism on which eBay is based aims precisely at discouraging cheap talk of this sort. Hence the relevance of the reputation that sellers acquire in the long run for a buyer facing a trust node.

Whilst as a group buyers can effectively deter sellers from cheating through the feedback mechanism, individuals are always open to being cheated. Therefore by grounding their decision at a trust node on the three informational variables described above, buyers can minimize their expectation of being cheated but can never rule out this very possibility. This is ultimately why eBay is an informal institution and hence, by definition, a place where successful interaction requires trust.

2.4 Analysis of eBay data

That eBay's informal institutional setting incentives the pursuit of good reputation for sellers is supported by strong empirical evidence [6]. This is mainly due to the fact that the eBay marketplace imposes costs to sellers for entering the market, for instance the costs of setting up a site to present their items, and support electronic payments. To build and maintain a good reputation is costly and, as [15] report, it takes a long time. Sellers are only encouraged to pursue a good reputation if they aim at 'keeping up with the business' for a long enough time, as has been pointed out by [6], because if the seller behaves trustworthily, at each single round he gains less than he would if he cheated. On the other hand, if he cheats, he risks loosing future opportunities: data reported in [6] clearly show that negative feedback reduces the bids received by a seller, thus accelerating his exit from eBay. These data also show that not all negative feedback is equal: the first negative tends to have a major impact if compared to subsequent ones. Interestingly enough, as [6, 15] point out, the sellers' reputation does not seem to have a positive correlation with the price buyers are willing to pay for their items. A good reputation just seems to attract more bids, justifying our choice of centering trust nodes on a buyer's bidding decision. Moreover, this reflects our idea of trust as an *all-or-nothing* decision: buyers only place a bid if they think they can *actually* trust the seller.

The fact that cheating behaviour on the part of sellers is not completely ruled out, is nevertheless consistent with the eBay game. This is due to the asymmetry of information between sellers and buyers in a given transaction. The seller has some private information about his future intentions, namely whether he intends to continue his activity on eBay or not. This clearly affects his behaviour, for if the seller is committed to keep his place in the market the informal institution of eBay provides a strong disincentive towards adopting a cheating behaviour [6]. Since this is private information, the buyer can only guess what the seller's behaviour will be by looking at his reputation profile.

The asymmetry of information is but one variable which cannot be adequately observed in field experiments. Other key variables tend to be omitted in field experimental data. Among those listed in [15] the following are particularly relevant: the specific way an item is presented, the visual impact of the seller's profile, private communication through emails or previous history of transaction between known sellers and buyers. Whilst those variables arguably play an important role toward determining behaviour at trust nodes, they can hardly be extracted from field observations: the authors of [15] report an attempt to overcome this difficulty by fine-tuning the parameters of their field experiment, but acknowledge that the problem is not ruled out even in their specific setting. The authors also claim that much literature on eBay sellers' reputation, while generally overlooking these variables, tends to ascribe their effects to reputation, thus failing to isolate the specific implications of the feedback mechanism.

The problems of omitted variables and the asymmetry of information described above highlight two important methodological limitations of field observations which ultimately results in noisy information. One potential remedy to this, which we now turn to consider, consists in using data from controlled experiments on the Trust Game.

3   Controlled experiments on the *Trust Game*

We begin by recalling the *basic Trust Game* which is usually taken to be the simplest and most widely used extensive game form involving trust. Empirical data from controlled experiments on this game are therefore abundant in the literature [3,4,13] . The goal of this section is to compare the evidence coming from these data with the field observation data on eBay discussed above.

What justifies us in using the Trust Game data is the fact that as far as trust based transactions are concerned the basic Trust Game can be viewed as an individual occurrence of a trust node. It is in fact immediate to see that if we ignore Reputation and Cheap talk in the eBay game, what we are left with is just an instance of the basic Trust Game. Conversely, if we add specifically designed Reputation and Cheap talk to the basic Trust Game we end up with a representation of the eBay game.

The basic trust game involves two players: the Investor and the Allocator. At the initial node the Investor is endowed with a sum of money and has to decide whether to invest or not. Investing means passing the endowment to the Allocator. In this transaction, the endowment is multiplied by some (small) factor. This is meant to capture the idea that the investment is successful in terms of utility. The extensive form of the basic trust game is depicted in Figure 3.1.

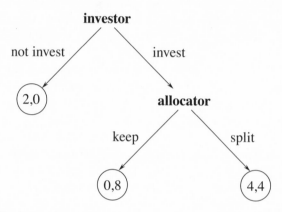

Figure 3.1. Basic trust game.

At the last node, the Allocator has the following choice: either keep or split (evenly) the sum of money. According to usual argument based on backward induction (see *e.g.* [14,16]) the Investor should rationally never choose to invest. Thus, according to this prescription, trusting behaviour is never rationally justified. This is clearly at odds with the empirical evidence drawn from eBay and indeed with the crucial importance that much economic thought, as we remarked in the opening section of this paper, has given to trust.

Two kinds of solutions to this impasse have been proposed in the literature. The former consists in taking a bounded-rationality approach to the problem. Here the focus is on the individual agent who is taken to be incapable, in general, of performing an error-free backward induction. This clearly rules out the possibility of taking the prediction of backward induction as a general prescription for the agent, especially in long interactions. An obvious limitation of this approach lies in its failure to explain the observed correlation between average trust and aggregated economic performance [5].

The way out the impasse which goes back to [12] is more directly related to our analysis of trust. According to it, trusting behaviour can be facilitated by the belief that at least some individuals in the group are acting cooperatively (independently of their reason for doing so)[6]. Consistently with the analysis of [17], Informal Institutions, Reputation and Cheap Talk are the key variables which in the long run lead to the formation of such beliefs. To get an idea of the role of those variables in facilitating trusting behaviour, recall that in the Trust Game, the Investor must decide whether to trust the Allocator or not. The idea is that she does so by considering her belief about the correlation between informal institutions and the expected payoff of the Allocator, an insight which is supported by the experimental findings reported by [4]. What we concluded in the previous section about the role of Informal Institutions in the eBay game carries over to the basic Trust Game.

As shown in Tables A.1 and A.2 in the Appendix similar findings are reported in connection to Reputation[7]. Indeed, if the Allocator can be identified reputation is directly relevant to the Investor when facing her decision. A rationally justified choice of trusting the Allocator therefore depends both on the Investor's individual assessment of the Allocator's reputation and on the effectiveness of the informal institutions in discouraging cheating behaviour.

Informal Institutions arguably determine a framing effect on the Investors' decision at trust nodes. In order to illustrate this point, it is useful to introduce a distinction between the two kinds of trust beliefs. The former pertains the information that an agent has about the Informal Institution and can be usefully thought of in terms of *background information*. This includes in particular beliefs about the social conventions governing the Institution in which the interaction is taking place. For instance, in the eBay game, this includes all the information an agent has about online auctions.

The latter kind of beliefs which can be thought of as *interaction-specific information* relate to the particular agent, say a seller, the buyer is interacting with at a specific trust node. Those beliefs will typically refer to the seller's reputation and the truthfulness of his cheap talk, in short, the information that a buyer is likely to consider to be directly relevant in assessing the trustworthiness of the seller *once the Informal Institution has been fixed*.

What distinguishes background from interaction-specific information is mainly the fact that the former is more entrenched than the latter, that

---

[6] See [7] and the references therein for a discussion centered on Reputation.

[7] It must be noted that although not inconsistent with them, the role of Cheap Talk has not been proved directly relevant by the empirical findings of [4], at least as concerns anonymous and structured cheap talk.

is to say it is held by the agent with a higher degree of confidence and therefore it is less likely to be revised. This is supported by empirical evidence as reported by [5]. Interaction-specific information, on the other hand, can be substantially revised as a consequence of new incoming evidence, like, for instance, a private email about the current transaction.

The interplay of background and interaction-specific information opens an interesting perspective on the formalization of rational reasoning at trust nodes, as we shall point out in the next, concluding, section.

4 Conclusions

We proposed a characterization of trust nodes in terms of Informal Institutions, Reputation and Cheap Talk. The analysis of empirical evidence, both in the form of field observations and in the form of controlled experiments, strongly suggests that despite being all relevant, those three parameters carry different weights in determining rational behaviour at trust nodes. Indeed, a major role in this is played by Informal Institutions, whose primacy can be described in two ways. First of all they essentially constitute the frame in which an agent forms her background beliefs. It is at this level that a buyer, to take again the eBay scenario, forms expectations about a seller's behaviour. In the absence of enforceable contracts (and consequently, of legally enforceable penalties) it is precisely this background information which is used by the buyer to decide whether to trust the seller or not. This brings us to the second fundamental role played by Informal Institutions: they take us away from the riddle of 'backward induction'. In other words, it is mostly owing to appropriate Informal Insitutions that a trustor can exploit interaction-specific information about the trustee's past behaviour and present Cheap Talk. This explains not only how trust in strangers is possible, but also how it can be profitable. Again this is supported by empirical evidence. As noted in [13] the subjects' answers to specifically designed questionnaires provide us with a good measure of their *trust in strangers*. It turns out that a subject's actual behaviour in the Trust Game can be predicted with a high degree of precision just by looking at her answers to the questionnaire. This is but another piece of evidence supporting the primacy of Informal Institutions as the key background framing reasoning at trust nodes.

Informal Institutions are essential also in a scenario in which the subject knows the identity of the trustee. Indeed, in this case, the subject can use interaction-specific information, notably the trustee's reputation, only because its meaning is being fixed by the Informal Institution.

These considerations suggest that Informal Institutions play a fundamental role in facilitating trust. Since they give meaning to those beliefs that agents form about each other's trustworthiness, they capture the essential strategic component of trust. It is therefore promising to focus on Informal Institutions in order to formulate the game-theoretic principles that lead agents to converge on mutual expectations of trust.

Appendix

Tables and Figures

Treatment	Rounds	Cheap Talk?	Observation of previous action?	Number of sessions	Total number of subjects
All	1 – 5	No	No	13	240
C	6 – 15	No	No	3	60
W	6 – 15	Yes	No	4	78
D	6 – 15	No	Yes	3	54
WD	6 – 15	Yes	Yes	3	48

Table A.1. This table is taken from [4] and illustrates the variations on the Trust Game used in the controlled experiment reported in the paper. Four distinct treatments are considered. C corresponds to the *basic trust game*. W is the Cheap Talk treatment. D includes reputation whilst WD is the combination of both.

Sessions	Frequency of invest			
	Rounds 1-5		Rounds 6-15	
C	0.567	(85/150)	0.400	(120/300)
W	0.585	(114/195)	0.405	(158/390)
D	0.519	(70/135)	0.633	(171/270)
WD	0.517	(62/120)	0.712	(171/240)

Table A.2. This table, again from [4] illustrates the experimental results by reporting the frequencies of the trusting behaviour ('invest') in the various treatments defined in Table A.1. Note, as we discussed in the main body of the paper that trusting behaviour occurs more frequently in the reputation treatments.

Figure A.1. A snapshot of the Community homepage on eBay. This is a concrete example of informal institution, the key structure in the assessment of the users' trustworthiness. Note that it allows users to gather information about the *reputation* of other eBay users as well as opportunities to implement *chep talk*.

ACKNOWLEDGEMENTS. We would like to thank the participants of the *Probability, Uncertainty and Rationality* workshop for their remarks on the presentation of this material. Thanks also to Stefano Marmi for his helpful suggestions on the ideas presented here and his comments on an earlier draft of this paper.

References

[1] K. J ARROW, *Gifts and Exchanges*, Philosophy and Public Affairs **1** (1972), no. 4, 343–362.

[2] C. BICCHIERI and E. XIAO, *Do the Right Thing: But Only if Others Do So*, Journal of Behavioral Decision Making **208** (2008), 191–208.

[3] J. BRACHT and N. FELTOVICH, *Efficiency in the trust game: an experimental study of precommitment*, International Journal of Game Theory **37** (2007), no. 1, 39–72.

[4] J. BRACHT and N. FELTOVICH, *Whatever you say, your reputation precedes you: Observation and cheap talk in the trust game*, Journal of Public Economics **93** (2009), no. 9-10, 1036–1044.

[5] J. BUTLER, P. GIULIANO and L. GUISO, *The Right Amount of Trust*, European University Institute Working Papers ECO **33** (2009).

[6] L. CABRAL and A. HORTACSU, *The dynamics of seller reputation evidence from eBay*, Journal of Industrial Economics, (forthcoming).

[7] C. F CAMERER and E. FEHR, *When does 'economic man' dominate social behavior?*, Science **311** (2006), no. 5757, 47–52.

[8] J. COLEMAN, "Foundations of Social Theory", The Belknap Press of Harvard University Press, 1990.

[9] E. FEHR, *On the Economics and Biology of Trust*, IZA DP No. 3895 (2008).

[10] D. HOUSER and J. WOODERS, *Reputation in Auctions: Theory, and Evidence from eBay*, Journal of Economics and Management Strategy **15** (2006), no. 2, 353–369.

[11] S. KNACK and P. KEEFER, *Does Social Capital Have An Economic Payoff*, Quarterly Journal of Economics **112** (1997), no. 4, 1251–1288.

[12] D. KREPS, P. MILGROM, P. ROBERTS and R. WILSON, *Rational Cooperation in the Finitely Repeated Prisoners' Dilemma*, Journal of Economic Theory **27** (1982), 245–252.

[13] M. NAEF and S. MARCH, *Measuring Trust: Experiments and Surveys in Contrast and Combination*, IZA DP (2009), no. 4087.

[14] E. RASMUSEN, "Games and Information", Basil Blackwell, fourth ed., 2005.

[15] P. RESNICK, R. ZECKHAUSER, J. SWANSON and K. LOCKWOOD, *The value of reputation on eBay: A controlled experiment*, Experimental Economics **9** (2006), no. 2, 79–101.

[16] R. SELTEN, *The chain store paradox*, Theory and Decision **9** (1978), 127–159.

[17] B. SKYRMS, *Signals, Evolution and the Explanatory Power of Transient Information*, Philosophy of Science **69** (2002), 407–428.

CRM Series
Publications by the Ennio De Giorgi
Mathematical Research Center Pisa

The Ennio De Giorgi Mathematical Research Center in Pisa, Italy, was established in 2001 and organizes research periods focusing on specific fields of current interest, including pure mathematics as well as applications in the natural and social sciences like physics, biology, finance and economics. The CRM series publishes volumes originating from these research periods, thus advancing particular areas of mathematics and their application to problems in the industrial and technological arena.

Published volumes

1. Matematica, cultura e società 2004 (2005). ISBN 88-7642-158-0
2. Matematica, cultura e società 2005 (2006). ISBN 88-7642-188-2
3. M. GIAQUINTA, D. MUCCI, *Maps into Manifolds and Currents: Area and $W^{1,2}$-, $W^{1/2}$-, BV-Energies*, 2006. ISBN 88-7642-200-5
4. U. ZANNIER (editor), *Diophantine Geometry*. Proceedings, 2005 (2007). ISBN 978-88-7642-206-5
5. G. MÉTIVIER, *Para-Differential Calculus and Applications to the Cauchy Problem for Nonlinear Systems*, 2008. ISBN 978-88-7642-329-1
6. F. GUERRA, N. ROBOTTI, *Ettore Majorana. Aspects of his Scientific and Academic Activity*, 2008. ISBN 978-88-7642-331-4
7. Y. CENSOR, M. JIANG, A. K. LOUISR (editors), *Mathematical Methods in Biomedical Imaging and Intensity-Modulated Radiation Therapy (IMRT)*, 2008. ISBN 978-88-7642-314-7
8. M. ERICSSON, S. MONTANGERO (editors), *Quantum Information and Many Body Quantum systems*. Proceedings, 2007 (2008). ISBN 978-88-7642-307-9
9. M. NOVAGA, G. ORLANDI (editors), *Singularities in Nonlinear Evolution Phenomena and Applications*. Proceedings, 2008 (2009). ISBN 978-88-7642-343-7

Matematica, cultura e società 2006 (2009). ISBN 88-7642-315-4

10. H. HOSNI, F. MONTAGNA (editors), *Probability, Uncertainty and Rationality*, 2010. ISBN 978-88-7642-347-5

Volumes published earlier

Dynamical Systems. Proceedings, 2002 (2003)
   Part I: *Hamiltonian Systems and Celestial Mechanics*.
ISBN 978-88-7642-259-1
   Part II: *Topological, Geometrical and Ergodic Properties of Dynamics*.
ISBN 978-88-7642-260-1
Matematica, cultura e società 2003 (2004). ISBN 88-7642-129-7
Ricordando Franco Conti, 2004. ISBN 88-7642-137-8
N.V. KRYLOV, *Probabilistic Methods of Investigating Interior Smoothness of Harmonic Functions Associated with Degenerate Elliptic Operators*, 2004. ISBN 978-88-7642-261-1
Phase Space Analysis of Partial Differential Equations. Proceedings, vol. I, 2004 (2005). ISBN 978-88-7642-263-1
Phase Space Analysis of Partial Differential Equations. Proceedings, vol. II, 2004 (2005). ISBN 978-88-7642-263-1

Fotocomposizione "CompoMat" Loc. Braccone, 02040 Configni (RI) Italia
Finito di stampare nel mese di febbraio 2010
dalla CSR, Via di Pietralata 157, 00158 Roma